ERYTHROCYTES
Physiology and Pathophysiology

ERYTHROCYTES
Physiology and Pathophysiology

editors

Florian Lang
Michael Föller
University of Tübingen, Germany

Imperial College Press

Published by

Imperial College Press
57 Shelton Street
Covent Garden
London WC2H 9HE

Distributed by

World Scientific Publishing Co. Pte. Ltd.
5 Toh Tuck Link, Singapore 596224
USA office: 27 Warren Street, Suite 401-402, Hackensack, NJ 07601
UK office: 57 Shelton Street, Covent Garden, London WC2H 9HE

British Library Cataloguing-in-Publication Data
A catalogue record for this book is available from the British Library.

ERYTHROCYTES
Physiology and Pathophysiology
Copyright © 2012 by Imperial College Press

All rights reserved. This book, or parts thereof, may not be reproduced in any form or by any means, electronic or mechanical, including photocopying, recording or any information storage and retrieval system now known or to be invented, without written permission from the Publisher.

For photocopying of material in this volume, please pay a copying fee through the Copyright Clearance Center, Inc., 222 Rosewood Drive, Danvers, MA 01923, USA. In this case permission to photocopy is not required from the publisher.

ISBN-13 978-1-84816-619-6
ISBN-10 1-84816-619-2

Typeset by Stallion Press
Email: enquiries@stallionpress.com

Printed by FuIsland Offset Printing (S) Pte Ltd Singapore

Contents

Abbreviations		vii
1	Functional Significance of Erythrocytes *Wolfgang Jelkmann*	1
2	Properties and Membrane Transport Mechanisms of Erythrocytes *Peter K Lauf and Norma C Adragna*	57
3	Erythropoiesis *Maxim Pimkin and Mitchell J Weiss*	229
4	Erythrocyte Senescence *Giel JCGM Bosman, Frans LA Willekens and Jan M Werre*	301
5	Eryptosis, the Suicidal Death of Erythrocytes *Florian Lang, Stephan Huber and Michael Föller*	327
6	Regulation of Red Cell Mass by Erythropoietin *Johannes Vogel and Max Gassmann*	351
7	Anaemia *Gordon W Stewart and Michael Watts*	367
8	Erythrocytes and Malaria *Henry M Staines, Elvira T Derbyshire, Farrah A Fatih, Amy K Bei and Manoj T Duraisingh*	387
Index		431

Abbreviations

AGM	aorta-gonad-mesonephros
AHSP	α-hemoglobin stabilizing protein
BFU-E	burst-forming unit-erythroid
CD	cluster differentiation
CFU-E	colony-forming unit-erythroid
CLP	common lymphoid progenitor
CMP	common myeloid progenitor
EPO	erythropoietin
EPO-R	erythropoietin receptor
GM-CSF	granulocyte-macrophage colony stimulating factor
GMP	granulocyte-macrophage progenitor
GR	glucocorticoid receptor
Hb	hemoglobin
HPFH	hereditary persistence of fetal hemoglobin
HSC	hematopoietic stem cells
IFN	interferon
IGF	insulin-like growth factor
IL	interleukin
MEP	megakaryocyte-erythroid progenitor
MPP	multipotential progenitor
RBC	red blood cell
SCF	stem cell factor
TGF	transforming growth factor
TNF	tumor necrosis factor

1
Functional Significance of Erythrocytes

Wolfgang Jelkmann*

Institute of Physiology, University of Lübeck, Ratzeburger Allee 160, D-23562 Lübeck, Germany

1.1 Introduction

Erythrocytes (from ancient Greek *erythrós* for "red" and *kýtos* for "cavity") are the most frequent cellular elements (95%) in blood. They develop in hematopoietic tissues (the bone marrow in adult humans) and circulate for 100–120 days until they are engulfed by macrophages. Human erythrocytes are anucleated discs full of hemoglobin (Hb), the oxygen (O_2) binding hemeprotein that causes the blood's red color. Erythrocytes transport O_2 from the lung to the peripheral tissues. In addition, erythrocytes expedite the carbon dioxide (CO_2) transport in blood, have buffer function, and can release vasoactive substances.

1.2 Historical Perspective

The Biblical phrase "for the life of the flesh is in the blood" (Moses 4th book; Leviticus 17:11) unveils that blood was considered the spirit of life in the early days. In ancient medicine, somatic and psychic diseases were often related to an unfit blood composition. Accordingly, physicians advocated bloodletting as a primary therapy. The blood of a healthy creature was believed to make a human recipient powerful and courageous if ingested or used for a bath.[1,2] Interestingly, the archaic supposition that

---

* Corresponding author: jelkmann@physio.uni-luebeck.de

blood is a medium transferring individual properties has received recent verification with the intriguing finding that the chemosensory identity through odor is altered in rodents after hematopoietic stem cell transplantation.[3] Peptides deriving from molecules encoded by the major histocompatibility complex (*MHC*) gene family are thought to function as olfactory signals.[4]

1.2.1 Characterization of erythrocytes

Red blood cells were first seen under a microscope in the 1660s, by the Dutch biologist Jan Swammerdam, who studied frog blood, and the Italian anatomist Marcello Malpighi, who studied hedgehog blood. Shortly thereafter, Antonie van Leeuwenhoek from Delft in the Netherlands provided a detailed microscopic characterization of erythrocytes. He correctly measured the diameter of human erythrocytes to be 7.5 µm. In 1675, van Leeuwenhoek noted: "I am apt to imagine, that those sanguineous globuls in a healthy Body must be very flexible and pliant, if they shall pass through the small capillary Veins and Arteries, and that in their passage they change into an oval figure, reassuming their roundness when they come into a larger room".[5] In 1843, the pathologist Gabriel Andral of Paris, sometimes referred to as the founder of scientific hematology, introduced the term "anemia" as the opposite of plethora. In 1852, the physiologist Karl von Vierordt of Tübingen presented a micrometer method for counting erythrocytes in diluted blood samples.[6] Von Vierordt estimated the number quite rightly at 5 Mio per µL blood for men. Subsequently, Georges Hayem of Paris developed "Hayem's solution" for blood cell counting. This pioneering work has been reviewed previously.[7] In 1868, Ernst Neumann of Königsberg and Giulio Bizzozero of Berlin reported the myeloid origin of mammalian erythrocytes.[8] George Gulliver[9] first provided a comprehensive description of the size and shape of the red corpuscles ("with drawings of them to a uniform scale, and extended and revised tables of measurements") of various vertebrate species. In the 1890s, Israel and Pappenheim succeeded in differentiating the stages of erythrocytic precursors, i.e. the transition from basophilic to polychromatic, hemoglobinized, erythroblasts.[10,11]

1.2.2 Hb and gas transport

In 1746, the chemist Vincenzo Menghini of Bologna reported that iron ("ferrum") is concentrated in the erythrocytes and that this causes the red color of the blood.[12] Heinrich Gustav Magnus of Berlin first demonstrated that there is more O_2 and less CO_2 in arterialized compared to venous blood.[13] The important role of erythrocytes as O_2 carriers was recognized in the second half of the 19th century (Table 1.1).

In 1862, the physiological chemist Felix Hoppe-Seyler of Tübingen isolated and crystallized the O_2 binding blood-borne protein. He presented the absorption spectra of the protein and gave it the name "hemoglobin".[14] Hoppe-Seyler also demonstrated that Hb can reversibly bind O_2, resulting in "oxyhemoglobin". The physiologist Eduard von Pflüger of Bonn recognized that the gas exchange between the blood and the tissue occurs by the process of diffusion.[15] In 1878, the English neurologist William Richard Gowers developed a device ("haemocytometer") for the measurement of the Hb concentration [Hb] in blood. In 1904, Christian Bohr, Karl Albert Hasselbalch, and August Krogh from Copenhagen described the sigmoid Hb-O_2 dissociation curve and the influence of CO_2 on the binding of O_2 to

Table 1.1 Seminal studies of the respiratory function of erythrocytes.

Year	Study	Investigators
1837	Blood gas measurements	H Gustav Magnus
1862	Discovery of reversible O_2 binding to Hb	Felix Hoppe-Seyler
1868	Role of bone marrow in hematopoiesis	Ernst Neumann, Guilio Bizzozero
1872	Tissue respiration	Alexander Schmidt, Eduard FW Pflüger
1880	Respiratory function at high altitude	Paul Bert, Denis Jourdanet
1904	Sigmoid Hb/O_2 binding curve and pH effect	Christian Bohr, Karl Hasselbalch and August Krogh
1909	Puffering of H^+ by Hb	Lawrence Henderson
1910	Hb aggregation hypothesis	Archibald Hill
1936	Geometry of O_2 binding	Linus Pauling and Charles Coryell
1938	O_2 dependent conformational change of Hb	Felix Haurowitz
1970	"T-" and "R-" structures of Hb	Max Perutz

Hb ("Bohr effect").[16] Hb-O_2 binding proved to be affected by pH, ionic strength, and temperature. Carl Gustav von Hüfner of Tübingen and associates calculated that the molecular weight of Hb is 16,700[17] — which holds true for the Hb subunit — and showed that 1 g of crystalline Hb can bind up to 1.34 mL O_2.[18] In 1910, the physiologist Archibald Vivian Hill of Cambridge, England, formulated an Hb aggregation hypothesis and proposed a simple equation for the Hb-O_2 dissociation curve: $SO_2/(1 - SO_2) \approx PO_2^n$, where SO_2 is the O_2 saturation ratio and PO_2 the O_2 partial pressure in mmHg, with slope n ("Hill coefficient") amounting to about 2.7.[19] Hill received the Nobel Prize in Physiology or Medicine 1922 (shared with Otto Fritz Meyerhof) "for his discovery of the fixed relationship between the consumption of oxygen and the metabolism of lactic acid in the muscle".

In 1925, Gilbert Smithson Adair of Cambridge, England, first recognized the tetrameric structure of the Hb molecule and that it contains four hemes.[20] In the same year, the American surgeon George Whipple discovered that iron, stored in the liver, was essential for heme synthesis and erythropoiesis. The development of a treatment for pernicious anemia — a fatal disease caused by atrophic gastritis, lack of intrinsic factor, and subsequent loss of the ability to absorb vitamin B_{12} — earned George Whipple, George Minot, and William Murphy the Nobel Prize in Physiology or Medicine 1934 "for their discoveries concerning liver therapy in cases of anaemia". The German chemist Hans Fischer characterized the heme structure and eventually synthesized it in 1929. The manner in which the O_2 is bound to the heme iron was first recognized by Linus Pauling and Charles Coryell of Pasadena in 1936, when they showed an ozone-type binding.[21] Pauling received the Nobel Prize in Chemistry 1954 "for his research into the nature of the chemical bond and its application to the elucidation of the structure of complex substances". In 1938, Felix Haurowitz of Prague observed that Hb forms different crystals in the absence and presence of O_2, indicating that the protein can undergo a conformational change.[22] This finding stimulated the Austrian chemist Max Ferdinand Perutz in his ambitious project of using X-ray crystallography to uncover the structure-function relationship of Hb, which he performed in Cambridge, England. It took Perutz almost 30 years of work, until he had unravelled the tertiary structure of the protein and demonstrated cooperativity between its four subunits. Perutz was awarded the Nobel Prize for Chemistry 1962, shared with John Cowdery Kendrew, "for their studies of the structures of globular proteins".

In 1967, the function of 2,3-bisphosphoglycerate (2,3-BPG) was recognized, an intraerythrocytic metabolite of glycolysis first described by Rapoport and Luebering in 1950 ("Rapoport–Luebering shunt").[23] 2,3-BPG binds to the β globin chains of Hb, thereby reducing the O_2 affinity of the erythrocytes and facilitating the unloading of O_2 in peripheral tissues.[24,25] Details of the fascinating story of the early research on Hb can be read elsewhere.[26,27]

1.2.3 *Acid base status and CO_2 transport*

Johanne Christiansen in the group of John Scott Haldane at Oxford showed that Hb releases CO_2 on oxygenation ("Haldane effect").[28] Lawrence Joseph Henderson, a biochemist at Harvard University, Boston, first noted that Hb becomes more acidic on oxygenation, and he described the relationship between pH, [CO_2], and [HCO_3^-] in blood.[29,30] He defined the constant "k" as the proton concentration at which half of the H_2CO_3 is dissociated: $k = [H^+][HCO_3^-]/[H_2CO_3]$. In 1917 Hasselbalch adapted Henderson's mass law for H_2CO_3 to the logarithmic form known as the Henderson–Hasselbalch equation: $pH = pK' + \log([HCO_3^-]/[CO_2])$.[31] The Henderson–Hasselbalch equation enabled it to calculate the PCO_2, when the blood pH was determined with a glass electrode and the CO_2 content by vacuum extraction and manometric measurement according to Van Slyke and Neill.[32] Subsequently, Poul Bjørndahl Astrup of Copenhagen designed a simpler electrometric technique, which allowed for the calculation of the *in vivo* PCO_2 based on the *in vitro* measurement of pH on equilibration of a blood sample at two different PCO_2.[33] This technique has been clinically most useful in evaluating the acid-base status of patients, particularly in determining the "Standard bicarbonate" respectively "Base excess" (BE).[34] The BE is defined as the amount of acid or alkali (in mmol/L) required to restore fully oxygenated blood *in vitro* to pH 7.40 under standard conditions, i.e. PCO_2 40 mmHg (5.3 kPa) and 37°C. The BE is an important measure of metabolic alkaloses (positive BE) or acidoses (negative BE).

1.2.4 Regulation of erythropoiesis

The French physiologist Paul Bert and his mentor Denis Jourdanet discovered that the concentration of erythrocytes depends on the O_2 supply to the tissues. In 1878, Bert first determined the *in vivo* relationship between the PO_2 and the O_2 content of the blood, as measured in samples from animals exposed to different barometric pressures.[35] Jourdanet made important observations on patients suffering from chronic mountain sickness in Mexico. The patients had thickened blood, while their symptoms were similar to those of anemic persons (dyspnea, tachycardia, and syncope). Jourdanet coined the term "anoxyhémie" to describe the lack of O_2 in arterial blood. Bert noted that animals living at 4,000 m altitude have higher [Hb] than lowlanders.[35] Bert and Jourdanet believed that polyglobulia — the "thick blood" of high altitude residents — was inherited. This concept was disproven by the French anatomist Francois-Gilbert Viault who had travelled from Bordeaux to the highland of Peru. After residence at 4,500 m for 23 days, his erythrocytes were increased from 5 to 8 Mio per μL blood. Viault concluded that erythropoiesis is acutely stimulated when the O_2 content of the blood is reduced.[36] His colleague Müntz reported that the descendants of a herd of rabbits he had brought to the Pic du Midi (2,877 m) in the Pyrenees exhibited a 75% increase in the blood's iron content, as a measure of [Hb].[37] In 1893, Friedrich Miescher[38] reported data of his co-worker Egger from Basel showing that the number of erythrocytes was increased in anemic patients with tuberculosis after they had spent a few weeks in an alpine health resort. The idea of a hormonal regulation of erythropoiesis was first formulated by Paul Carnot and Claude Deflandre of Paris in 1906.[39,40] These investigators subjected rabbits to a bloodletting (30 mL), took another blood sample on the next day and injected the serum (5–9 mL) into normal rabbits. The recipients' erythrocytes increased by 20–40% within 1–2 days. Although this strong reaction is irreproducible, Carnot and Deflandre rightly concluded that the donors' serum contained a hormone which they called "hémopoiétine". The more specific name "erythropoietin" (Epo) was introduced by Eva Bonsdorff and Eeva Jalavisto from Helsinki in 1948.[41] Allan Erslev of New Haven is generally credited for having first proven the existence of Epo,[42] although a closer look has shown that other investigators had reported this before.[43] Erslev infused large volumes (50–200 mL) of plasma from severely anemic rabbits into normals. The recipients responded with a significant reticulocytosis and, in the long term, an increase in hematocrit (Hct). Notably, Erslev predicted the potential therapeutic value of the erythropoietic factor: "Conceivably isolation and purification of this factor would provide an agent useful in the treatment of conditions associated with erythropoietic depression, such as chronic infection and chronic renal disease".[42] Recombinant human Epo (rhEpo) first became available as an anti-anemic therapeutic in the late 1980s, and has since blossomed into a leading biopharmaceutical. With respect to this success, credit is due to the work of Takaji Miyake and colleagues,[44] who collected and concentrated 2,550 L Epo-containing urine from patients with aplastic anemia in Kumamoto City, Japan. The material was used to purify the hormone and to partially identify its amino acid sequence. Based on this knowledge the human Epo gene could be characterized and be expressed in host cells for the industrial production of rhEpo for medical use.[45,46]

1.2.5 *Blood transfusion*

In tracing back to the time when the first blood transfusion in a human was performed, one hits upon Pope Innocent VIII. Legend says that he received the blood of three ten-year old boys a few days before he died in 1492. However, while the writings indicate that blood was indeed taken from the boys, evidence is missing that it was given to the Pope.[47] The transfusion of blood components was probably not experienced prior to William Harvey's description of the circulation of the blood, in his 1628 book *De Motu Cordis*. Harvey recognized that the blood is pumped from the right ventricle through the pulmonary circulation to the left ventricle.

The primal allogeneic (homologous) transfusion trials were conducted by Richard Lower on dogs in 1665.[1] The first blood transfusion in a human was likely performed by Jean-Baptiste Denis and Paul Emmerez in Paris in 1667. Reportedly, the anemic young patient improved greatly after he received blood from a lamb.[1] Mainly for ethical reasons the transfusion of heterologous blood into humans was interdicted after a few more trials, such as those by Lower and King in England. In 1825, James Blundell, an obstetrician in

London, performed the first successful allogeneic blood transfusion in a human, a woman suffering from postpartum hemorrhage, with her husband donating the blood.[48] Immune reactions and blood clotting remained major plagues in transfusion therapy. In 1875, Leonard Landois of Greifswald, Germany demonstrated that erythrocytes often clump and lyze when they are mixed with serum from a different species. About 25 years later, Karl Landsteiner of Vienna pointed out that a similar reaction can result when the blood of one person is transfused to another human being, thereby causing shock, jaundice, and hemoglobinuria. In 1909, Landsteiner classified the blood of humans into the A, B, AB, and 0 groups and showed that hemolysis likely occurs when blood of a different group is transfused. Subsequent work Landsteiner performed in collaboration with Alexander Wiener in New York led to the discovery of the Rhesus antigen (RhAg) of erythrocytes. Serum obtained from rabbits immunized with erythrocytes from the Rhesus macaque was found to react with about 85% of different human blood samples.[49,50] Landsteiner was awarded the Nobel Prize in Physiology or Medicine 1930 "for his discovery of human blood groups". In his Nobel lecture, published in modified form,[51] Landsteiner annotated:

> the erythrocytes evidently contain substances (isoagglutinogens) with two different structures, of which both may be absent, or one or both present, in the erythrocytes of a person. The groups are named according to the agglutinogens contained in the cells. More important to practical medicine than the subject with which we have just been dealing is the use of the blood-group reaction in transfusions. (www.nobelprize.org)

Ottenberg and Johnson[52] were among the first who took the risk of an immune reaction into account before transfusing blood. They stated[52]:

> It is not enough in transfusion to depend on identity of groups. "Cross tests" of donor's and patient's serum on each other's cells must invariably be done. By this means the danger of hemolysis due to "sub-groups" or rare anomalies can be completely avoided.

Based on the increasing knowledge of the blood group systems and through the use of anticoagulants, the allogeneic transfusion of blood or of packed red blood cells, became an important life-saving procedure in the 20th century. The leftover risks include febrile non-hemolytic transfusion reactions, graft-*versus*-host-disease, acute or delayed hemolytic reactions; and transmission of prions, viruses, protozoans, and bacteria. In addition, there is evidence for a high incidence of bedside transfusion errors that pass unnoticed. The SAnGUIS project assessing blood practice on 808 transfusion patients revealed 15 major errors (transfusion of a wrong unit) and 150 recording errors.[53] As the risks of allogeneic blood transfusion-transmitted viruses were reduced to exceedingly low levels in the United States (and likely in other countries), acute lung injury, hemolysis, and sepsis have emerged as the leading causes of allogeneic blood transfusion-related deaths.[54]

1.3 Volume, Number, Shape, and Composition

1.3.1 *Volume*

Blood accounts for 6–8% of the body mass in adults. The blood volume (V_{Blood}) can be measured by dilution of tracers injected into the circulation. Tracers can be erythrocytes labeled with radioactive chromium (^{51}Cr) for the measurement of the total red blood cell mass (V_{RBC}), and albumin labeled with radioactive iodine (^{131}I or ^{125}I) or the dye Evans blue for the measurement of the plasma volume (V_P). The ^{51}Cr method for estimating V_{RBC} ($V_{RBC,51Cr}$) is regarded as the reference method by the International Committee for Standardization in Haematology. CO-rebreathing is the most reliable technique for measuring the total Hb mass,[55] which amounts to about 1% of the body mass.

1.3.2 *Number*

Erythrocytes account for almost all the cellular elements in the blood: only 1 out of every 20 is a platelet and 1 out of 500 is a leukocyte. Cell numbers are usually determined by flow cytometry in clinical routine.[56] Women have normally 4–5 Mio and men 5–6 Mio erythrocytes per μL blood (Table 1.2). The packed red blood cells make up about 42% of the blood volume in females and 47% in males ("hematocrit"). During

Table 1.2 Normal* erythrocyte parameters (*normal values vary depending on measuring devices).

Hb concentration of the blood	about 150 g L^{-1} (2.3 mM Hb_4)
	♀ 120–160 g L^{-1}
	♂ 140–180 g L^{-1}
Oxygen concentration in arterial blood	about 200 mL L^{-1} (STPD; [Hb] dependent)
Hematocrit (Hct; packed red cell volume, PCV)	0.40–0.50
	♀ 0.37–0.47
	♂ 0.40–0.54
Red cell count in blood	4–6 × 10^{12} L^{-1}
	♀ 4.0–5.2 × 10^{12} L^{-1}
	♂ 4.6–5.9 × 10^{12} L^{-1}
Mean corpuscular volume (MCV)	87 (80–96) fL
Mean corpuscular Hb (MCH)	30 (27–33) pg
Mean corpuscular Hb concentration (MCHC)	340 (320–360) g L^{-1}
Mean erythrocyte lifespan	100–120 days
Reticulocytes	0.5–2.5% of red blood cells

childhood the erythrocyte count changes. In the newborn it is high (5.5 Mio per µL) because of the transfer of blood from the fetal placenta into the child's circulation at birth and the subsequent marked fluid loss. In the following months the production of erythrocytes does not keep pace with general body growth, resulting in a decrease in erythrocyte count to about 3.5 Mio per µL blood in the third month of life. Preschool and school children have somewhat lower erythrocyte counts than adult women. The higher erythrocyte count, Hct and [Hb] in males compared to females results from the stimulation of erythropoiesis by androgens and, possibly, its inhibition by estrogens.[57]

1.3.3 Shape

Human erythrocytes are anucleated, flexible, biconcave discs. Their greatest thickness (at the edge) is about 2 µm and their width 7.5 µm. There is a positive correlation among vertebrates between the species-specific width of the erythrocytes and the capillary diameter.[58] In humans, the mean corpuscular volume (MCV) of normocytic erythrocytes

averages 87 fL and the mean corpuscular Hb mass (MCH) of normochromic erythrocytes 30 pg. The gas exchange in the microcirculation is facilitated by the discoid shape of the erythrocytes, for the diffusion area is large and the diffusion distance small. The total surface area of the erythrocytes of an adult man amounts to about 4,000 m^2. Due to their flexibility, erythrocytes can pass through vessels with inside diameters smaller than their own width of 7.5 µm by reversibly adopting a parachute shape. Biophysical considerations have led to the concept that the discoid shape has major significance in the macrovasculature, where it promotes non-turbulent flow and minimizes platelet scattering, thereby acting in an anti-atherogenic and anti-thrombotic way.[59] The loss of flexibility is the main reason why aged erythrocytes and red blood cells with an altered form (e.g. spherocytes, elliptocytes, echinocytes) are retained in the meshwork of the spleen and subsequently destroyed there.

1.3.4 *Cytosol and metabolic processes*

Erythrocytes are tightly filled with Hb, which makes up 34% of their wet weight ("mean corpuscular Hb concentration", MCHC). Hb forms tetramers composed of four globin chains, each attached to a heme group containing ferrous iron (Fe^{2+}), to which O_2 can bind (Fig. 1.1).

Oxygenated Hb is scarlet, while deoxygenated Hb has a dark red burgundy color, appearing bluish through the venous vessel wall and the skin. The erythrocytes of an adult human contain collectively 2–3 g iron, representing about 65% of the body's total iron. The mean [Hb] is normally 140 g per L blood in females and 160 g per L in males, with 1 g Hb binding maximally 1.34 mL O_2 ("O_2 capacity"). Note that there are ethnic differences. For example, blacks have lower [Hb] than whites, which is independent of the iron and general nutritional status.[60] Hence, the universal acceptance of the lower limits of "normal" [Hb] as set by the WHO (120 g/L for adult females and 130 g/L for males) has been disputed.[61]

The sequestration of the O_2 binding Hb in the erythrocytes (rather than having it dissolved in the plasma) allows for less viscous blood and an increased O_2 capacity. The erythrocytes of most non-mammalian vertebrates have nuclei.[62] In most mammalian species erythrocytes are devoid of a nucleus and of other organelles such as mitochondria, Golgi

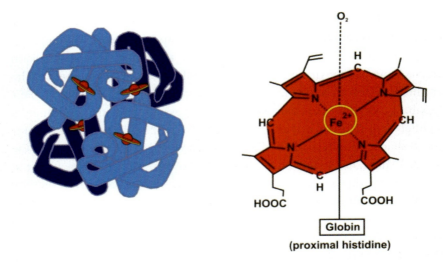

Figure 1.1 *Left:* Structure of the tetrameric hemoglobin (Hb) composed of two pairs of identical globins (blue), each attached to a heme group (red). *Right:* The heme contains ferrous iron (Fe^{2+}), to which O_2 can reversibly bind. With kind permission from Springer Science+Business Media: Physiologie des Menschen; Schmidt RF, Lang F, eds., Chapter 34: Atemgastransport; 30th ed., 2007; p. 806, W. Jelkmann, Fig. 34.2.

apparatus, and endoplasmic reticulum. Erythrocytes lacking mitochondria can carry more O_2 and they do not consume the O_2 they transport. With glucose as the chief substrate, the cells rely on anaerobic glycolysis, which ends with the formation of lactic acid.[63] The main energy source, as in other cells, is ATP; it is required in particular for the active transport of ions through the erythrocyte membrane and thus serves to maintain the transmembraneous ion-concentration gradients. When ATP is generated *via* glycolysis, NADH (reduced nicotinamide-adenine dinucleotide) and NADPH (reduced nicotinamide-adenine dinucleotide phosphate, derived from the pentose phosphate cycle) are produced. NADH is required for the reduction of the Fe^{3+} to Fe^{2+} in heme; NADPH is involved in the reduction of the glutathione in the erythrocytes. Glutathione, which is readily oxidizable, protects proteins with SH groups (especially the Hb molecules and membrane proteins) from oxidation. Proteomic studies have identified >700 proteins within the human erythrocyte.[64] Apart from Hb, of particular importance is the enzyme carbonic anhydrase that catalyzes the reversible reaction of CO_2 with H_2O yielding H_2CO_3, which dissociates to HCO_3^- and H^+, the latter being buffered by Hb.

1.3.5 *Membrane properties*

The erythrocyte membrane is a flexible coat of three layers:

i   the glycocalyx on the exterior, which is rich in carbohydrates;
ii the lipid bilayer, which — apart from cholesterol and phospholipids — also contains various transmembrane proteins; and
iii the protein skeleton that is linked to the inner surface of the lipid bilayer.

The membrane is about 10 nm thick; it is much more permeable to anions than to cations. Depending on their chemical properties, substances can pass through the membrane by diffusion (O_2, CO_2), or by means of transporter proteins.

More than 50 different membrane proteins and glycoproteins have been identified. Approximately 25 of these carry the various blood group antigens.[65] The membrane proteins exert several functions, such as transport of small molecules and ions across the membrane, and adhesion and interaction with endothelial cells. The complex of the transmembrane proteins anion exchanger 1 (AE1), Rh antigen (RhAg), CD47, and glycophorin A is linked to the β-spectrin/actin cytoskeleton by ankyrin in association with protein 4.2. The complex of AE1, glycophorin C, glucose transporter 1 (Glut 1 = insulin independent), and RhAg is linked to spectrin, actin, and protein 4.1.[66] Abnormalities in these proteins can cause hemolytic disorders.[66–68] Other red cell membrane transport proteins include aquaporin 1, L-dehydroascorbic acid transporter, Kidd antigen protein for urea, Na^+/K^+-ATPase, Ca^{2+}-ATPase, Na^+/K^+-$2Cl^-$-cotransporter, Na^+-Cl-cotransporter, Na^+-H^+-exchanger, K^+-Cl^- cotransporter, and the Gardos channel (Ca^{2+}-activated K^+ channel). Finally, there are cell adhesion molecules such as intercellular adhesion molecule-4 (ICAM-4), which interacts with integrins, and basal cell adhesion molecule (BCAM), a glycoprotein that is also known as Lu as it defines the Lutheran blood group.[69]

The intraerythrocytic osmotic pressure is slightly higher than that of the blood plasma, which suffices for the normal turgor of the erythrocyte. When the extracellular fluid is hypotonic, the erythrocytes swell and approach a spherical shape (spherocytes). In a hypertonic medium the cells lose water and become crenated.[70] The osmotic resistance of erythrocytes can be studied on suspension in progressively diluted media. Fifty percent of the erythrocytes of a healthy person are hemolyzed when the tonicity of the medium falls below ~145 mosmol/L. Osmotic resistance is reduced in certain forms of anemia.

1.4 Role in Gas Transport

1.4.1 *O_2 transport*

According to Henry's law the concentration of physically dissolved gas [G] is proportional to its partial pressure ($[G] = P_G \times \alpha_G$). The factor α is the gas-specific solubility (or Bunsen absorption) coefficient, which amounts for O_2 to 24 mL [STPD, standard temperature (0°C) and pressure (760 mmHg), dry] per L solution per atmosphere (760 mmHg). Thus, about 3 mL O_2 per L is dissolved in arterialized blood (PO_2 about 95 mmHg). In reality, however, arterialized blood contains about 200 mL O_2 per L, because O_2 is chemically bound to Hb.

Hb takes up O_2 in the lung (about 20 moles/day) and releases it in the respiring tissues. Furthermore, Hb buffers H^+ and binds CO_2 to form carbamino-Hb. Hb is a tetramer (Hb_4, molecular mass: 64.5 g/mol) of four subunits each composed of a heme (protoporphyrin III with a central Fe^{2+}) and a globin chain (Fig. 1.1). The protoporphyrin is formed by four pyrrole rings that are synthesized from δ-aminolevulinic acid (δ-ALA) and linked by methine bridges. The planar porphyrin structures are nestled in pockets of the globins. The Fe^{2+} is located in the center of the protoporphyrin, bonded to the four pyrrole nitrogen atoms. At its 5th coordination site, the Fe^{2+} is linked to the imidazole ring of the proximal histidine of the globin. At the 6th coordination site, the Fe^{2+} can reversibly bind O_2 ("oxygenation" and "deoxygenation"). In the deoxygenated state, the Fe^{2+} is located 0.4 Å outside the porphyrin. On O_2 binding, the Fe^{2+} becomes smaller and glides into the plane of the porphyrin ring.

Oxidation of Fe^{2+} to Fe^{3+} converts Hb to Met-Hb ("hemiglobin"), which cannot bind O_2. Also, Hb cannot bind O_2, when other ligands occupy the Fe^{2+}, such as CO (carboxy-Hb). Because the absorption spectra of the Hb depend on the degree of oxygenation, venous blood appears darker (bluish-red) than arterialized blood. The spectral difference is used for clinical non-invasive study of blood oxygenation by pulse oximetry.[71]

Each two of the four globins of the Hb tetramer are always identical. Adult humans have mainly HbA (98%) that consists of two α and two β globin chains ($\alpha_2\beta_2$), which are composed of 141 and 146 amino acids, respectively. In the follow-up of diabetes mellitus non-glycated (HbA_0, normally > 94%) and glycated (HbA_1, <6%) forms of HbA are distinguished. The increased percentage of HbA combined with glucose (HbA_{1c}) is indicative of abnormally high glucose concentrations.[72] Minor Hb isoforms of the adult are HbA_2 (2%, $\alpha_2\delta_2$) and HbF (<1%, $\alpha_2\gamma_2$; fetal-type Hb), which differ in amino acid sequences as compared to the β-chains present in HbA. The globins of the Hb tetramer are connected by salt bridges, hydrogen bonds, and hydrophobic contacts, which dissolve on O_2 binding.[73] This is called the transition from the T-("tense") to the R-("relaxed") structure (Fig. 1.2). One mole Hb_4 can maximally take up four moles O_2. Taken the mole volume of gas (22.4 L), one mole Hb_4 theoretically binds up to 89.6 L O_2 (1.39 mL/g Hb). Due to the presence of minor amounts of Met-Hb and CO-Hb, which cannot take up O_2, the maximal O_2 loading — the "O_2 capacity" of the blood — is in reality slightly lower, and calculations are based on "Hüfner's number" (1.34 mL O_2/g Hb). The percentage of the Hb that is oxygenated ("O_2 saturation", SO_2) depends on the prevailing PO_2. In the fully oxygenated blood of the pulmonary veins, the PO_2 is about 100 mmHg (13.3 kPa; $SO_2 = 100\%$). This fully oxygenated blood is mixed with venous blood that passes through the physiological shunt on its way to the left heart. Thus, the PO_2 is lowered to 95 mmHg (12.6 kPa) in the arterial blood of the aorta ($SO_2 = 98.5\%$).

The binding of O_2 — first occurring at the α_1 heme — causes a gliding of the Fe^{2+} into the plane of the porphyrin ring, a pulling at the proximal histidine and a rotation of the $\alpha_1\beta_1$ dimer by about 15° relative to the $\alpha_2\beta_2$ dimer.[73] Thus, the oxygenation of one Hb subunit produces a conformational change of the whole tetramer, and the other subunits gain an increased affinity for O_2 ("cooperativity" or "allostery"). This explains as to why the

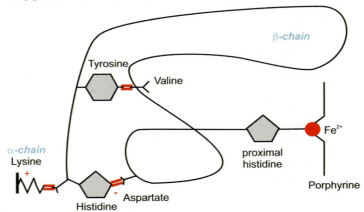

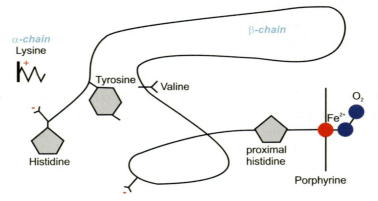

Figure 1.2 Oxygenation causes a transition from the *T*- ("tense") to the *R*- ("relaxed") structure of Hb. On O_2 binding, the Fe^{2+} moves into the plane of the porphyrin ring, thereby pulling at the proximal histidine of the globin chain. This produces a conformational change of the whole tetramer. According to Perutz.[74] With kind permission from Springer Science + Business Media: Physiologie des Menschen; Schmidt RF, Lang F, eds., Chapter 34: Atemgastransport; 30th ed., 2007; p. 809, W. Jelkmann, Fig. 34.5.

Hb-O_2 binding curve is sigmoid (S-shaped). The sigmoid shape of the Hb-O_2 binding curve is advantageous for the O_2 supply of the tissues (Fig. 1.3).

Even when the arterial PO_2 decreases (e.g. at high altitude) SO_2 is still relatively high, because the Hb-O_2 binding curve is very flat in the upper part (SO_2 is still 90% at PO_2 60 mmHg). On the other hand, the steep slope in the middle of the Hb-O_2 binding curve allows for the release of O_2 at relatively high PO_2 values in the peripheral tissues. In mixed-venous blood the PO_2 is about 40 mmHg (5.3 kPa) and SO_2 75%. The Hill coefficient — as a measure of the steepness of the Hb-O_2 binding curve — is calculated by plotting the log PO_2 *versus* the log fractional SO_2. The positon of the Hb-O_2 binding curve — as a measure of the Hb-O_2 affinity — can be quantified by the P_{50}-value, which is the PO_2 yielding 50% SO_2 (Fig. 1.3). The P_{50} of adult human blood is 27 mmHg (3.5 kPa) at standard conditions (pH 7.40, PCO_2 40 mmHg, 37°C). In addition, the tetrameric structure of the Hb allows for allosteric reactions. Hb-O_2 binding is reduced (increase in P_{50}) on increasing [H^+] and PCO_2 ("Bohr effect") or of temperature (Fig. 1.4). The binding of H^+ and CO_2 to amino acids of the globins stabilizes the T-structure and reduces the Hb-O_2 affinity, thereby facilitating the release of O_2 in the periphery.[74] At low pH, the imidazole groups of the terminal histidines of the β chains (His^{146}) are protonated and form salt bridges with a negatively charged aspartate (Asp^{94}) in the same β chain. This reaction is favored in the deoxygenated state, because a lysine (Lys^{40}) of the α chain of the other αβ dimer bridges with the β chain His^{146}, thereby stabilizing the Hb tetramer.

The T-structure of human HbA is also stabilized by 2,3-BPG, which binds in the central cavity of deoxygenated Hb. There, it interacts with three positively charged amino acid residues of the two β chains (His^2, Lys^{82}, and His^{143}), thereby reducing the Hb-O_2 affinity.[24,25] Human erythrocytes are rich in 2,3-BPG (1 mol per mol of Hb_4) due to their high 2,3-BPG mutase activity.[75] Two alternative strategies have been followed in mammalian evolution in order to facilitate the unloading of O_2 in the periphery:

i   intrinsically high Hb-O_2 affinity, which is lowered in a regulated way by 2,3-BPG (e.g. in humans); and
ii intrinsically low Hb-O_2 affinity and low sensitivity to 2,3-BPG (e.g. in ruminants).[76]

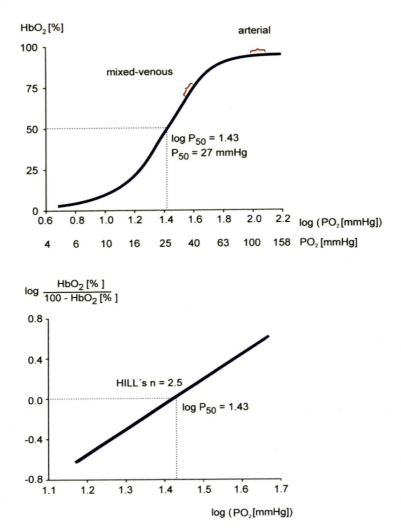

Figure 1.3 O_2 concentration curve of human blood as a function of the PO_2. *Upper part*: Sigmoid shape Hb-O_2 binding curve of human blood (blue). The O_2 saturation (SO_2) is >98% in arterialized (PO_2 90–100 mmHg) and about 75% in mixed-venous (PO_2 40 mmHg) blood (curly brackets). The PO_2 producing 50% O_2 saturation (P_{50}) of adult human blood is 27 mmHg (3.5 kPa) at standard conditions (pH 7.40, PCO_2 40 mmHg, 37°C). *Lower part*: The Hill coefficient as a measure of the steepness of the Hb-O_2 binding curve is calculated by plotting the log PO_2 *versus* the log fractional SO_2. With kind permission from Springer Science + Business Media: Physiologie des Menschen; Schmidt RF, Lang F, eds., Chapter 34: Atemgastransport; 30th ed., 2007; p. 808, W. Jelkmann, Fig. 34.4.

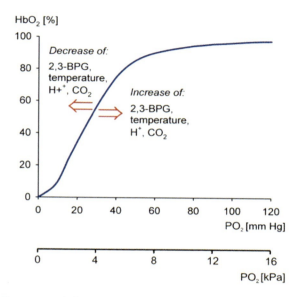

Figure 1.4 Parameters influencing the Hb-O_2 affinity. Hb-O_2 binding is reduced (increase in P_{50}) on increasing [H^+] and PCO_2 (Bohr effect) or of temperature (T), and *vice versa*. $\Delta \log P_{50}/\Delta$ pH = -0.48; $\Delta \log P_{50}/\Delta$ T = 0.023. With kind permission from Springer Science+Business Media: Physiologie des Menschen; Schmidt RF, Lang F, eds., Chapter 34: Atemgastransport; 30th ed., 2007; p. 810, W. Jelkmann, Fig. 34.6.

The former strategy is advantageous, because it allows for an adaptation of the Hb-O_2 affinity to pH changes. While the share of the 2,3-BPG bypass in human erythrocytes amounts normally to about 20% of the overall flux through glycolysis, it increases in alkalotic states.[75]

HbF contains 2 γ-chains instead of the 2 β-chains of HbA. HbF is incapable of binding 2,3-BPG, which explains the high O_2 affinity of fetal blood compared to that of adults.[77,78]

The primary embryonic Hb is Gower-1 ($\zeta_2\varepsilon_2$). In addition, Hb Gower-2 ($\alpha_2\varepsilon_2$), Hb Portland-1 ($\zeta_2\gamma_2$) and Portland-2 ($\zeta_2\beta_2$) are expressed at relatively low levels during embryonic and early fetal development.[79,80]

1.4.2 CO_2 transport and buffering

The CO_2 produced in the mitochondria of the peripheral tissue (about 16 moles CO_2/day) diffuses into the capillary blood and enters the erythrocytes.

The PCO_2 is 46 mmHg (6.1 kPa) in mixed-venous and 40 mmHg (5.3 kPa) in arterialized blood. The solubility coefficient for CO_2 is 570 mL STPD per L solution per atmosphere (760 mmHg). Thus, about 30 mL CO_2 per L is dissolved in arterialized blood. However, the total [CO_2] is about 500 mL STPD per L (22.3 mM) in mixed-venous and 460 mL STPD per L (20.7 mM) in arterialized blood, under conditions of physical rest.[81] This is because CO_2 mainly exists as HCO_3^- in blood.

The reaction of dissolved CO_2 with H_2O yielding H_2CO_3 is slow in plasma, but >10,000 times faster in the erythrocytes because of the action of carbonic anhydrase.[82] H_2CO_3 rapidly dissociates into H^+ and HCO_3^-. The H^+-ions are buffered by Hb. Note that H^+ can bind more readily to the NH-moiety of the imidazol rings of deoxygenated Hb molecules. About 70% of newly formed HCO_3^- is exchanged for Cl^- into the blood plasma, through anion exchanger AE1 (Fig. 1.5). Ninety percent of the total CO_2 in blood is transported as HCO_3^- (Fig. 1.6), with 5% being dissolved and another 5% being bound to terminal amino groups of deoxy-Hb to form

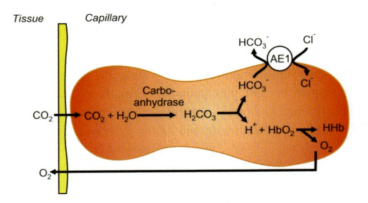

Figure 1.5 CO_2 exchange between tissue and blood. Intraerythrocytic carbonic anhydrase catalyzes the reaction of dissolved CO_2 with water yielding H_2CO_3, which dissociates into H^+ and HCO_3^-. The H^+ is buffered by Hb. About 70% of newly formed HCO_3^- is exchanged for Cl^- into the blood plasma, through anion exchanger AE1. With kind permission from Springer Science+Business Media: Physiologie des Menschen; Schmidt RF, Lang F, eds., Chapter 34: Atemgastransport; 30th ed., 2007; p. 812, W. Jelkmann, Fig. 34.7.

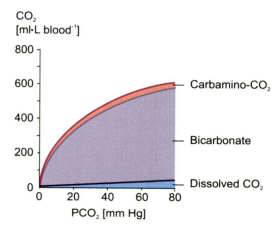

Figure 1.6 CO_2 concentration curve of human blood as a function of the PCO_2. Of the total CO_2 in blood, 90% is transported as HCO_3^-, with 5% being dissolved and another 5% being bound to terminal amino groups of deoxy-Hb to form carbamino-groups (Hb-NH_2 + CO_2 ⇌ Hb-NH-COO^- + H^+). With kind permission from Springer Science + Business Media: Physiologie des Menschen; Schmidt RF, Lang F, eds., Chapter 34: Atemgastransport; 30th ed., 2007; p. 813, W. Jelkmann, Fig. 34.8.

carbamino-groups (Hb-NH_2 + CO_2 ⇌ Hb-NH-COO^- + H^+). At rest, 85% of the CO_2 expired in the lung originates from the HCO_3^- pool, 9% from dissolved CO_2, and 6% from the carbamino compounds ("oxy-labile carbamate"). However, during heavy exercise, the contribution of dissolved CO_2 can greatly increase, making up almost one-third of the total CO_2 exchange.[81]

The total CO_2-concentration of blood rises with the PCO_2. The CO_2-dissociation curve is almost linear in the physiological PCO_2 range between 40 and 46 mmHg. In mixed-venous blood the CO_2-concentration is larger for a given PCO_2 than in arterial blood, because deoxygenated Hb binds more CO_2 than oxygenated Hb ("Haldane effect"). The Haldane effect facilitates the release of CO_2 in the lungs.[83] Due to the increased formation of HCO_3^-, the erythrocytes contain more osmotically active particles in the venous than in the arterial blood. Consequently, they take up water and swell. Thus, the Hct is higher in venous than in arterial blood.

1.4.3 Thrombus formation

The incidence of thrombus formation increases with Hct. The erythrocytes mainly occupy the center of the blood vessels, thereby expelling the small platelets into the plasma-skimming layer (Fig. 1.7). This extrusion favors the activation and adherence of platelets at sites of endothelial lesions. Large erythrocytes enhance platelet adhesion more strongly than small erythrocytes.[84] At high Hct or high shear rates, thrombus formation is promoted. Accordingly, present guidelines for the use of recombinant erythropoiesis-stimulating agents (ESA) for the treatment of renal or chemotherapy-induced anemias recommend not to raise [Hb] above 120 g/L (Hct 0.36).[85] On the other hand, the prolonged bleeding time in some anemias may be corrected by transfusion of red cells[86] or by ESA therapy.[87]

1.4.4 Control of vascular tone

Erythrocytes encountering shear stress in narrow vessels release ATP which causes vasodilation. Although the understanding of the mechanosensitive ATP release is still incomplete, evidence suggests that the spectrin-actin network and the cystic fibrosis transmembrane conductance regulator (CFTR) are involved.[88] In addition, erythrocytes have been assigned an essential role in matching blood flow to local metabolic demands during hypoxic

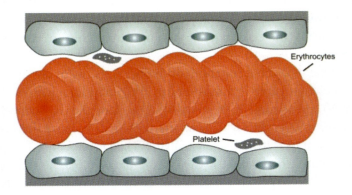

Figure 1.7 Drawing of a small blood vessel, where the erythrocytes occupy the center thus expelling the small platelets into the plasma-skimming layer. This extrusion favors the activation and adherence of platelets at sites of endothelial lesions.

stress. On deoxygenation Hb releases NO and S-nitrosothiol derivatives. The hypoxia-induced vasodilation is dependent on cGMP but largely endothelium-independent.[89] Other investigators have reported that erythrocytes express NO synthase and synthesize NO enzymatically from L-arginine.[90] Exposure of erythrocytes to physiological levels of shear stress activates NO synthase and stimulates the formation of NO, which may contribute to the regulation of vascular tone.[91]

1.5 Life Cycle

1.5.1 *Production of erythrocytes*

Erythrocytes are produced in the hematopoietic tissues: the yolk sac of the embryo (from week 2 post conception, p.c.), in the liver (from week 6 p.c.), and in the spleen (from month 4 p.c.) of the fetus; and — becoming the primary site by the 7th month p.c. — in the red marrow of the flat bones and the epiphyses of the humerus and femur throughout one's life.[79] These tissues contain the pluripotent hematopoietic stem cells, which are the common progenitors of all kinds of blood cells.[92] At the next level of differentiation are the committed progenitors, which can only give rise to restricted kinds of blood cells (erythrocytes, thrombocytes, monocytes, granulocytes). Several stages of differentiation and maturation are distinguished, until finally the young anuclear erythrocytes leave the bone marrow as reticulocytes. Apart from iron, erythropoiesis requires vitamin B_{12}, folic acid, vitamin B_6, and vitamin C in addition to basic nutrients (amino acids, lipids, carbohydrates), and Epo. About 1% of the 25×10^{12} erythrocytes of an adult are renewed every day. This implies an erythropoiesis rate of $2-3 \times 10^6$ red cells per second.

Erythrocytes are the offspring of a small pool of self-perpetuating hematopoietic stem cells called "CFU-GEMM" (colony-forming unit generating granulocytic, erythrocytic, megakaryocytic, and monocytic progeny) or $CD34^+$ cells (CD means "cluster of differentiation", with the number indicating specific membrane marker proteins). Their proliferative activity depends on the local environment and the presence of cytokines such as interleukin-3 (IL-3). A small number of stem cells emigrates from the hematopoietic tissue and circulates in blood ("peripheral blood stem cells", PBSC), which can be purified for patients undergoing hematopoietic cell transplantation.[93] In the erythrocytic line (the "erythron"), at the next level of differentiation in the bone marrow are the "BFU-Es" (burst-forming units-erythroid), which can generate several hundred erythroblasts within 10–20 days. The more differentiated "CFU-Es" (colony-forming units-erythroid) generate 8–64 erythroblasts within 7–8 days. They form erythroblastic islands with a central macrophage in the bone marrow.[94] The proerythroblasts (also called pronormoblasts) are the first cells of the erythron that are microscopically identifiable ("erythrocytic precursors"). As proliferation progresses, the nuclei of the descendants become smaller and their cytoplasm more basophilic ("basophilic erythroblasts"), which is due to the presence of ribosomes. When the descendants begin to synthesize Hb, they are called "polychromatic erythroblasts". Hb synthesis involves three major substrates: protoporphyrin, iron, and amino acids. The δ-aminolevulinic acid (δ-ALA) for protoporphyrin formation derives from glycine and succinyl-CoA in a reaction that requires pyridoxal phosphate, the active form of vitamin B_6.

Once the level of "orthochromatic erythroblasts" (also known as "normoblasts") is reached, the cells do not divide any more but extrude their nuclei to become reticulocytes.[95] Over a few days, these complete the synthesis of Hb and other proteins, and start to degrade their organelles.[96] Then the reticulocytes enter the circulation. Vital staining of the reticulocytes (e.g. with new methylene blue) reveals their filamentous structures. Circulating reticulocytes lose their polyribosomes and other organelles, as well as tubulin and cytosolic actin. Thereby, the cells diminish in size and become mature erythrocytes within 1–2 days. The membrane content of myosin, tropomyosin, ICAM-4, Glut-4, Na^+-K^+-ATPase, Na^+-H^+-exchanger 1, glycophorin A, CD47, Duffy, and Kell is now relatively low.[97] The transition into erythrocytes results in the acquisition of the biconcave shape and an increase in shear resistance.

Under normal conditions the reticulocytes account for about 1% of the red blood cells. An acceleration of erythropoiesis increases this percentage, and a retardation decreases it.[98] The assessment of the number and maturation status of the reticulocytes in blood samples is clinically useful in evaluating the rate of erythropoiesis. Immature reticulocytes are larger and have more reticular fibers than mature reticulocytes. On accelerated erythropoiesis, such as following bleeding or hemolysis, the proportion of immature reticulocytes increases in the marrow and the blood.

1.5.2 Role of Epo

Erythropoiesis maintains the steady-state concentration of erythrocytes, and it hastens red cell recovery after a loss of blood. The basal rate of erythropoiesis can increase about ten-fold after severe bleeding or when the erythrocyte life span is abnormally shortened. The effective stimulus that triggers erythropoiesis is a fall in the PO_2 in respiring tissue (an imbalance between O_2 supply and demand). Under such hypoxic conditions the concentration of the hormone Epo increases in the plasma. Human Epo is a heat-stable glycoprotein (30.4 kDa; 165 amino acids, 40% carbohydrates). The kidneys are the main organs of the production of Epo after birth, while the liver produces the hormone in the fetal stage. The plasma concentration of Epo is 6–32 International Units (IU) per L (about 10^{-11} mol/L) in healthy humans. The levels vary greatly between individuals, but there are no major gender- or age-specific differences. Plasma Epo increases exponentially, when [Hb] falls below ~125 g/L or the arterial SO_2 below 80%.[57] Epo synthesis is little stimulated, when the renal blood is lowered, because the renal O_2 demand decreases with the flow reduction.[99]

The discovery of the hypoxia-inducible transcription factors (HIFs) has provided insights into the molecular mechanisms of the hypoxia-induced *Epo* expression.[100,101] HIFs are heterodimeric proteins composed of the subunits α and β.[102] The C-terminus of HIF-α comprises O_2-dependent degradation domains (O-DDDs), in which proline residues are hydroxylated by specific HIF-α prolyl hydroxylases (PHD-1, -2, and -3) in the presence of O_2.[103–107] Prolyl hydroxylated HIF-α binds the von Hippel–Lindau tumor suppressor protein (pVHL) in complex with an E_3-ligase and undergoes immediate proteasomal degradation.[108] HIF-2 is the primary transcription factor inducing *Epo* expression.[109,110] HIF-2α is degraded on hydroxylation of Pro^{405} and Pro^{531}.[111] The transcriptional activity of the HIFs is further suppressed by O_2-dependent hydroxylation of an asparagine residue, namely Asn^{847} in HIF-2α. The HIF-α hydroxylases are enzymes containing Fe^{2+} and requiring α-ketoglutarate. α-Ketoglutarate blocking agents (clinical jargon: "HIF stabilizers") stimulate Epo production in healthy humans and in anemic patients with renal failure.[112]

The human Epo receptor (Epo-R) is a 484 amino acid membrane-spanning glycoprotein. Two Epo-R molecules form a homodimer, to which one Epo molecule can bind. In turn, Janus kinases 2 (JAK2) are activated, which are in contact with the cytoplasmic region of Epo-R[113,114] and mediate EPO-R signalling by means of tyrosine phosphorylation of intracellular proteins. Epo prevents the erythrocytic progenitors (primarily CFU-Es) from undergoing apoptosis and stimulates them to generate erythroblasts.[115] CFU-Es express the major erythroid-specific transcription factor, GATA-1.[116] The balance between GATA-1 and caspases determines the processes of apoptosis, proliferation, and differentiation of the erythrocytic progenitors.[117] Epo-R signalling is negatively regulated by hematopoietic cell-specific tyrosine phosphatases (HCP; such as the SH_2 domain-containing tyrosine phosphatases SHP-1 and SHP-2) and by members of the suppressors of cytokine signalling family (SOCS).[118]

While Epo is strictly required for erythropoiesis, it is supported by other hormones, namely testosterone, thyroid hormone, somatotropin, and insulin-like growth factor 1 (IGF-1). Stress hormones (catecholamines, cortisol) acutely enhance the release of reticulocytes from the bone marrow.

In several other mammalian species, such as dogs and horses, the spleen functions as a reservoir of erythrocytes, which are acutely expelled into the blood in response to an increased activity of the sympathetic nervous system. However, this effect is limited in humans. Maximally, erythrocytes amounting to only about 5% of the total red cell mass can be released from the human spleen.[119]

1.5.3 *Destruction of erythrocytes and degradation of Hb*

Human erythrocytes circulate for 100–120 days.[120] Over time, they undergo changes in their membrane, making them susceptible to the attack of macrophages. Senescent erythrocytes are phagocytized primarily in the bone marrow, and under pathological conditions also in the liver and the spleen. In healthy humans about 1% of all erythrocytes are subjected to hemolysis (mainly extravascularly) every day. In addition, erythrocytes undergo suicidal death. This process ("eryptosis") is stimulated by various endogenous and exogenous factors.[121] The degradation products of the Hb molecules (heme and amino acids) are partly reused. The heme is broken down into biliverdin and Fe^{3+}. The biliverdin is reduced to bilirubin, which is released and transported to the liver bound to albumin. The liver secretes the bilirubin in bile into the intestines, where it is metabolized into urobilinogen and egested as stercobilin. The iron is salvaged for reuse. In the blood plasma it is carried by the hepatic 678-amino-acid glycoprotein transferrin (79.5 kDa).[122] Transferrin keeps Fe^{3+} nonreactive in the circulation and delivers it to cells *via* specific membranous homodimeric transferrin receptors (TfR). The erythrocytic precursors use the Fe^{3+} for Hb synthesis after reduction to Fe^{2+}. The serum transferrin Fe^{3+}-saturation, and the serum levels of ferritin and of the soluble form of the TfR (sTfR) provide diagnostic information on the rate of erythropoiesis and the body iron status. sTfR is elevated under conditions of increased erythropoiesis.[123] The iron-export membrane protein ferroportin is required for iron resorption in the intestine and for iron release from storage sites.[122,124] The cellular iron availability is reduced by the 25-amino acid hepatic acute-phase protein hepcidin, which promotes the internalization and degradation of ferroportin.[125,126] Serum ferritin of 12 µg/L or less is generally indicative of iron deficiency. However, because ferritin is an acute-phase protein, its concentration may be normal or increased in iron-deficient patients with inflammation. Because reticulocytes are the youngest erythrocytes in blood, their mean Hb mass (CHr) provides a measure of iron available to red cells recently produced in the bone marrow. A low CHr is a strong indicator of iron deficiency in clinical practice.[127,128]

1.6 Erythrocytes and Physical Performance

In endurance sports — such as long-distance running, cycling, or skiing — performance relies on an adequate O_2 supply to the heart and the skeletal muscles. Hence, the rate of maximal O_2-uptake ($\dot{O}_2$ max) is an important determinant of the aerobic power. Amongst other things, $\dot{O}_2$ max depends on the total Hb mass (Hb_{mass}). In untrained lowlanders Hb_{mass} can be roughly estimated to amount to about 12 g per kg body mass (i.e. 900 g Hb in a 75 kg person) and $\dot{O}_2$ max to about 50 mL per kg body mass and min.

A change of one gram of Hb produces a change in O_2 max of approximately 4 mL min^{-1}.[129] Hb$_{mass}$ affects O_2 max in two ways:

i an accompanying increase in blood volume augments cardiac output; and
ii an increase in [Hb] at maintained blood volume augments the peripheral O_2 extraction.[129]

Forbidden procedures to increase Hb$_{mass}$ in athletes include the transfusion of erythrocytes, the infusion of Hb and the misuse of erythropoiesis-stimulating agents (ESAs). Erythrocyte transfusion maneuvers increase O_2 max and prolong the time to exhaustion on heavy workload.[130] The effect is due to increases in the O_2 transport capacity and the buffering capacity of the blood. rhEpo and its analogues enhance performance mainly by raising Hb$_{mass}$, while the plasma volume decreases.[131]

Presently, only the misuse of allogeneic — but not autologous — blood can be proven directly. rhEpo and its analogues can be detected by isoelectric focusing and immunoblotting of urine or plasma samples, because "Epo" is in reality a mixture of isoforms and the N-glycans of endogenous Epo and of the recombinant products differ.[132] Endogenous Epo (10 isoforms in the isoelectric point (pI) range 3.77–4.70) is more acidic[133,134] than rhEpo (4–6 isoforms in the pI range 4.42–5.21). However, there is a plethora of novel ESAs, including peptidic as well as non-peptidic Epo mimetics. Furthermore, chemical compounds may be misused in sports that can be taken orally and stimulate endogenous Epo production by activating the *Epo* promoter ("GATA-inhibitors": diazepane derivatives) or the *Epo* enhancer ("HIF-stabilizers": α-ketoglutarate blocking agents, see above).[135] In December 2009, the World Anti-Doping Agency (WADA) implemented operating guidelines for the monitoring of selected red blood cell variables ("Athlete Biological Passport"), which indirectly uncover doping, as opposed to the traditional direct detection of substances.

1.7 Pathophysiology of Anemias and Laboratory Analyses

Anemia means, literally, bloodlessness. In clinical usage, the term refers primarily to a lowered [Hb] (Fig. 1.8).

Functional Significance of Erythrocytes

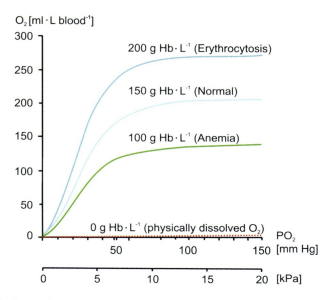

Figure 1.8 Dependence of the O₂ capacity on the Hb concentration of the blood (calculated on Hüfner's number: 1.34 mL O₂/g Hb). With kind permission from Springer Science+Business Media: Physiologie des Menschen; Schmidt RF, Lang F, eds., Chapter 34: Atemgastransport; 30th ed., 2007; p. 807, W. Jelkmann, Fig. 34.3.

In this state, there can be a reduction in the number of erythrocytes as compared with the norm and/or in the mean Hb mass (MCH) of the individual erythrocytes (Table 1.3). The WHO has defined anemia as [Hb] <120 g/L in women and <130 g/L in men.[136]

Note that these values originally referred to nutritional anemias.[136] The lower limits set by the WHO are not generally accepted because of methodological (measurement inaccuracies, epidemiologic weaknesses) and biological (ethnic differences) deficiencies.[61] The symptoms in anemic subjects depend on the severity, abruptness of onset, age, and cardiopulmonary reserve (Fig. 1.9). Pallor, hyperdynamic precordium, and flow murmurs are typical signs. However, the questions regarding the "optimal [Hb]" and the "critical lowest [Hb]" (requiring therapeutic intervention) have remained controversial since the early days of anemia research.[137]

Table 1.3 Pathophysiology of erythropoiesis (main causes).

A) Decrease → Anemia	
Defect in erythropoiesis	Iron deficiency (insufficient resorption, chronic bleeding)
	Vitamin B_{12} deficiency (gastric disorder)
	Folic acid deficiency
	Epo deficiency (chronic kidney disease)
	Primary bone marrow failure (cytotoxic drugs, radiotherapy)
Increased loss of erythrocytes	Acute bleeding
	Hemolytic diseases (spheroid cells, sickle cells, thalassemia)
B) Increase → Erythrocytosis	
Polycythemia vera	Genetic disorders (JAK2 mutation)
Secondary polycythemia	Epo excess (altitude residence)

1.7.1 *Primary bone marrow hypoplasia*

In aplastic anemia only the erythrocytes are reduced (pure red cell aplasia) whereas in pancytopenia all bone marrow-derived blood cells are affected. Among the aplastic anemias are both hereditary (Diamond–Blackfan, Fanconi) and acquired idiopathic forms. The inhibition of blood cell production in the pancytopenias can be caused by bone-marrow damage due to ionizing radiation (X-rays or exposure to radioactive elements), cell toxins (cytostatics, benzene etc.), or tumor metastases, which take the place of normal bone marrow tissue.

1.7.2 *Iron deficiency*

Worldwide, the most common cause of anemia is iron deficiency. This can be produced by a diet with inadequate iron content (especially common among infants), by impaired iron absorption from the duodenum and upper jejunum (for example, in the so-called malabsorption syndrome), or by chronic loss of blood due to ulcers, carcinomas, and polyps in the gastrointestinal tract, esophageal varicosities, hookworm infestation (common in the tropics), and heavy menstrual bleeding. A normal diet

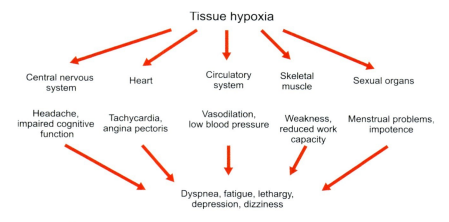

Figure 1.9 Pathophysiological consequences of anemia.

contains about 15 mg of iron per day, with about 15% of this being absorbed in the small intestine. The uptake requirements amount to 1–2 mg per day in menstruating females, and to 0.5–1 mg in non-menstruating females, males, and children. In iron-deficiency anemia the blood contains small erythrocytes with a subnormal Hb mass (hypochromic microcytic anemia). There is also increased anisocytosis (variation in size) and poikilocytosis (variation in shape). Diagnostically, serum ferritin, serum iron, and percentage transferrin saturation are reduced, while the total iron-binding capacity (TIBC) is increased.[138]

1.7.3 *Vitamin B_{12} and folate deficiency*

The most important common characteristic of these anemias is an enlargement of the erythrocytes (megalocytes) and their precursors in the bone marrow (megaloblasts). Production of the large cells is caused by a deficiency of vitamin B_{12} and/or folic acid. When these vitamins are lacking, DNA synthesis is impaired and cell division is delayed. Vitamin B_{12} is required for the synthesis of pyrimidine, and, thus, of nucleobases. Megalocytes have a shorter life span than normal erythrocytes and this, together with the limited generation of reticulocytes, leads to anemia. Note, however, that vitamin B_{12} deficiency is primarily a neurologic disease with demyelination of corticospinal tracts and posterior columns of the spinal cord. Clinically relevant vitamin B_{12} deficiency is more likely to occur when intrinsic factor-driven absorption fails (pernicious anemia) than when diet is poor (vitamin B_{12} is missing in vegetable foods). The clinical course is insidious. Most cases take years to produce obvious deficiency signs.[139] The hepatic vitamin B_{12} stores suffice to supply daily needs for 2–3 years.

1.7.4 *Epo deficiency*

The anemia in chronic kidney disease (CKD) is usually normochromic and normocytic, and its severity increases with the impairment of renal function. Significant anemia develops in general when the glomerular filtration rate falls below ca. 30 mL per min and 1.73 m² body surface area.[140–142] Basically, however, renal anemia is an endocrine disorder. Apart from insufficient Epo production, the factors causing the anemia in CKD include shortening of erythrocyte life-span, bleeding, malnutrition, and inhibition of erythropoiesis by inflammatory cytokines[143] and uremia toxins.[144,145] The therapy with rhEpo raises Hct and [Hb] in a dose-dependent and predictable way.[146,147] It abolishes the need for erythrocyte transfusions, prevents the hyperdynamic cardiac state, and improves brain function. As noted above, the target [Hb] is presently set in the range 100–120 g/L.[85]

Apart from CKD, the anemias associated with chemotherapy in cancer patients and with zidovudine-treatment in HIV-infected patients can be alleviated by the administration of rhEpo or its analogs. The primary goals of ESA therapy are to maintain [Hb] above the transfusion trigger, increase the exercise tolerance, prevent fatigue, and improve quality-of-life parameters. In the surgical setting, Epo may be administered pre-operatively in order to stimulate erythropoiesis in phlebotomy programs for autologous re-donation or for correction of a pre-existing anemia, and post-operatively for recovery of erythrocyte mass.[148]

1.7.5 *Hemolytic disorders*

Inherited hemolytic disorders may be caused by abnormalities of the erythrocyte membrane or of erythrocytic enzymes. In severe cases anemia develops as the production of erythrocytes cannot keep pace with the accelerated destruction. Apart from anemia symptoms, the patients present with jaundice and splenomegaly. Erythrocyte cell fragments (schistocytes) may be observed in blood. Because Hb released into the plasma is captured by the plasma protein haptoglobin and removed by macrophages *via* endocytosis,[149,150] a decrease in the level of circulating haptoglobin can support the diagnosis of hemolytic anemia,[151] especially when accompanied by an increased reticulocyte count.

1.7.5.1 *Red cell membrane defects*

The deficiency in one of the membrane proteins involved in the "vertical" interaction (anion exchanger AE1, Rh Ag, ankyrin, protein 4.2, and spectrin) anchoring the membrane skeleton to the lipid bilayer leads to hereditary spherocytosis (HS).[67] HS is an autosomal dominant disorder. Anemia due to HS is the most common inherited anemia in Northern European descendants, affecting approximately 1 in 1,000–2,500 individuals depending on the diagnostic criteria.[66,68] In HS, the erythrocytes are small and spherical. The clinical features include increased hemolysis, splenomegaly, and reticulocytosis. MCHC is often above normal.[152] However, most of the afflicted persons are asymptomatic, as they compensate for their hemolysis with increased erythropoiesis.[68]

Hereditary elliptocytosis (HE) results from abnormal "lateral" interactions in the membrane skeleton and is a rare autosomal dominant disorder. HE is due to defects in either α- or β-spectrin, resulting in reduced avidity to spectrin dimer–dimer interaction or defects in protein 4.1.[66]

1.7.5.2 *Red cell enzyme deficiencies*

Most of the mutations occur sporadically; some, such as glucose-6-phosphate-dehydrogenase (G6PD) mutants, are endemic. The deficiency is sex-linked and affects males more than females. More than 300 variants have been defined.[153] G6PD accomplishes the reduction of NADP to NADPH. On oxidative stress cells containing denatured Hb ("Heinz bodies") may be seen on blood smears.[153] During hemolytic episodes nonspecific abnormalities (anisocytosis, polychromasia) occur. In response to hemolysis, the number of reticulocytes increases. Therefore, the diagnosis is difficult to assess in cases with milder forms, because G6PD activities are relatively high in young erythrocytes.

Pyruvate kinase (PK) deficiency is the most common congenital red cell enzyme deficiency, particularly in Northern Europe. PK deficiency is an autosomal recessive disorder causing poikilocytosis with echinocytes (crenated erythrocytes) and acanthocytes. ATP formation is reduced while 2,3-BPG levels are increased. Persons homozygote for PK deficiency show severe anemia and are usually discovered in childhood. Splenomegaly, cholelithiasis, and jaundice are frequent.[153]

1.7.5.3 *Sickle cell disease*

A point mutation in the β globin gene results in the replacement of glutamic acid in position 6 by valine (HbS). When HbS is deoxygenated it becomes insoluble and forms fibrils, leading to sickle-shaped erythrocytes, which are rigid and cause blood vessel occlusion. The disease is most common in Africans, in particular in those living in areas where malaria is endemic (see Chapter 8).

1.7.5.4 *Thalassemias*

The disease is generally due to an autosomal recessive genetic defect of at least one of the 4 α and 2 β globin genes, producing an abnormal ratio of the Hb subunits.[80] Thalassemias are particularly common in people of Mediterranean, African, and Southeast Asian ancestry. The typical symptoms include anemia, hemolysis, splenomegaly, bone marrow hyperplasia, and iron overload due to multiple transfusions. The diagnosis is based on Hb electrophoresis and DNA testing. Serum bilirubin, iron, and ferritin levels are generally increased. Treatment for severe forms may include red blood cell transfusion, splenectomy, iron chelation, and stem cell transplantation.

In α-thalassemia, heterozygotes for a single gene defect (α-thalassemia-2 [silent]) are not usually apparent. Heterozygotes with defects in 2 of the 4 genes (α-thalassemia-1 [trait]) tend to develop mild microcytic anemia. Defects in 3 of the 4 genes result in severely reduced γ-globin chain levels and the formation of tetramers of β-globin chains (HbH) or, in fetuses, γ-globin chains (Hb Bart), which can be assessed by Hb electrophoresis. Recombinant DNA approaches of gene mapping have proved useful for prenatal diagnosis. Defects in all four β-globin genes lead to prenatal death.

In β-thalassemia, heterozygotes have mild microcytic anemia and normal life expectancy (β-thalassemia minor). Homozygotes (β-thalassemia major) develop severe anemia (Cooley's anemia) and bone marrow hyperplasia.[154] HbF is increased, sometimes to as much as 90%, and HbA$_2$ to >3%. Splenomegaly is common. Bone marrow hyperactivity causes thickening of the cranial bones. Long bone involvement predisposes to pathologic fractures and impairs growth. Life expectancy is decreased in β-thalassemia major; only some patients live to puberty or beyond. The afflicted children should receive as few transfusions as possible to avoid iron overload. Excess (transfusional) iron must be removed by iron-chelation therapy. Hematopoietic stem cell gene transfer may become a therapeutic option in the future.[155]

1.7.5.5 Acquired abnormalities

There are various non-immune and immune-mediated forms. Non-immune forms may be associated with trauma, infections, or toxins. Immune mediated hemolysis mostly occurs extravascularly and is associated with IgG antibodies on the red cell surfaces. The immune reactions may be stimulated by chemicals.[156]

1.7.6 Chronic inflammation

Anemia of chronic disease (ACD) is also called anemia of inflammation, as it frequently develops in association with malignancies, autoimmune diseases, and infections. The etiology of ACD is multifactorial but is basically due to an increased formation of proinflammatory cytokines such as IL-1 and tumor necrosis factor α, which cause hemolysis, impaired responsiveness of erythrocytic progenitors, reduced iron availability, and lowered Epo expression (Fig. 1.10).[126,157,158] The laboratory findings include hypochromic or normochromic erythrocytes, increased hepcidin levels, reduced serum iron, and increased iron stores in macrophages.[126]

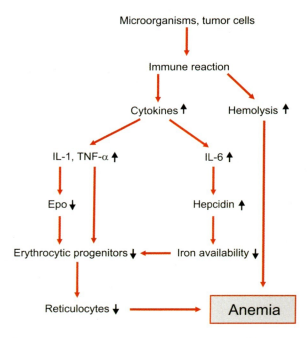

Figure 1.10. Factors causing the anemia of chronic disease (ACD). IL: interleukin, TNF: tumor necrosis factor, Epo: erythropoietin.

1.8 Erythrocytoses

Erythrocytoses are characterized by an abnormally high concentration of erythrocytes. The high viscosity of the blood increases cardiac afterload and the risk of developing heart failure, myocardial infarction, peripheral thromboses, pulmonary embolism, and seizures. An absolute erythrocytosis is present when [Hb] is >165 g/L in females and >185 g/L in males, and Hct and the total red cell mass (RCM; normal value about 30 mL/kg body weight) are elevated.[159,160]

Erythrocytoses can be congenital or acquired, and have primary or secondary causes.[160,161] In primary erythrocytosis there is an intrinsic bone marrow hyperactivity, while the concentration of circulating Epo is low. In contrast, secondary erythrocytosis results from increased Epo production. This increase may be associated with a mutation in one of the genes involved in O_2 sensing,[162] or be due to tissue hypoxia. The most prevalent cause of secondary erythrocytosis is residence at high altitude.

1.8.1 *Primary erythrocytosis*

Relevant diagnostic tools are the study of erythrocyte parameters including RCM, screens of Epo-R and JAK2 mutation, assay of serum Epo level (should be low)[163], and bone marrow histopathology.[160]

1.8.1.1 *Epo-R mutations*

Congenital primary erythrocytosis can be due to an *Epo-R* mutation where the Epo-R signals constitutively. In contrast to polycythemia vera (PV), only the erythrocytic line is affected and the cardiovascular system adapts to the increased RCM from the beginning of life. The afflicted persons do not have an increased risk of thrombosis and bleeding. In the first mutation identified the Epo-R was truncated and lacked deactivation by the tyrosine phosphatase SHP-1.[164] Within the affected Finnish family, one famous member is the cross-country skier Eero Mäntyranta who likely benefited — with respect to sports — from his increased RCM (59 mL/kg; [Hb] ~200 g/L, Hct >0.60).[165] Subsequently, several other cases of primary familial and congenital erythrocytosis due to a mutation in the cytoplasmic region of the Epo-R have been detected.[161,166]

1.8.1.2 *JAK2 mutations (PV)*

Polycythemia vera (PV) is the most prevalent acquired genetic disorder leading to primary erythrocytosis. For example, the yearly incidence for PV was reported to amount to 2.8 per 100,000 persons in Sweden in 2001.[167] The disease is generally diagnosed after the age of 60 years. A loss-of-function mutation of the suppressor of cytokine signalling (SOCS3) may play a role in disease onset.[118] PV is a clonal disease, whereby the erythrocytic progenitors proliferate in the absence of Epo.[168] In the majority of patients with PV — and at high frequency in patients with essential thrombocythemia and chronic idiopathic myelofibrosis — the disease is due to the JAK2V617F mutation in exon 14 of JAK2.[169]

More recently, mutations were discovered in exon 12 of JAK2 in patients with PV lacking the V617F mutation.[170-172] Patients with exon 12 mutations have higher erythrocyte but lower leukocyte and platelet numbers than patients with the exon 14 mutation.[171]

According to the WHO, diagnosis can rely on raised [Hb] (>165 g/L in women and >185 g/L in men) or raised RCM and a JAK2 mutation, in combination with a minor criterion (bone marrow biopsy abnormality, low serum Epo, or reduced *in vitro* endogenous erythroid colony formation).[173] The prognosis of PV without treatment is poor, because the incidence of thromboembolic events increases with rising Hct. Long-term risks of the disease include transition to acute leukemia or myelofibrosis.[174] The primary therapeutic goal is to permanently reduce Hct to <0.40 in female and to <0.45 in male patients. Phlebotomy is the treatment of choice. Hydroxyurea and interferon-α2b are alternate treatment choices.[175]

1.8.2 *Secondary erythrocytosis*

Here, Epo production is abnormally increased. There are congenital and acquired forms. Relevant diagnostic tools are Hb electrophoresis; determination of the P_{50} value; assay of 2,3-BPG mutase activity; analyses of pVHL, PHD-2, and HIF-2α genes; pulse oximetry; measurement of CO-Hb; radiology of head, chest and abdomen; and sleep studies.[160]

1.8.2.1 *Congenital forms*

Heart defects associated with cyanosis (such as septal defects, left-right shunt) may lead to chronic hypoxemia, and thus to increased Epo production.[176] High O_2 affinity Hb may also cause erythrocytosis as first reported in 1966.[177] Since then over 90 Hb variants of α-globin and β-globin genes have been described.[178] High O_2 affinity of the blood due to a deficiency of 2,3-BPG mutase leading to erythrocytosis was first described in 1978.[179] A few more cases have since been discovered with either autosomal dominant[180] or recessive inheritance.[181]

Mutations in the genes of the HIF pathway may also cause erythrocytoses.[162] Of note is the activation of HIF in normoxia which can result in the expression of genes and proteins that promote tumor growth.[182] The so-called Chuvash polycythemia was first discovered in Eastern Russia. This disease is based on an autosomal recessive inherited mutation in VHL, which results in the formation of pVHL with reduced activity. Hence, the HIF-α proteins are not degraded even under normoxic conditions and Epo transcription is increased.[183] The afflicted persons present with an increased incidence of thromboses and bleeding.[184] Outside the Chuvash area, the mutation was more recently discovered in some families of Bangladeshi or Pakistani origin,[185] and in habitants of the island of Ischia, Italy.[186] All these subjects may have had a single founder some 10,000 years ago.[187] Percy et al.[188] have first described a family in which erythrocytosis is associated with an inherited mutation in PHD-2. Further cases of PHD-2 mutations in familial erythrocytoses have been reported recently.[189–191] A family with erythrocytosis and elevated Epo levels in three generations was found to have a mutation of the O_2-dependent degradation domain of HIF-2α.[192] Further HIF-2α mutations have been described.[191,193]

1.8.2.2 Acquired forms

Mountain sickness is caused by the lack of O_2 as the PO_2 declines at altitude proportional with the barometric pressure as a whole (e.g. the PO_2 of dry air is 100 mmHg at 4,750 m compared to 160 mmHg at sea level). While "acute mountain sickness" (AMS; symptoms: headache, nausea, dizziness, dyspnea, tachycardia, pulmonary and cerebral edema) occurs immediately at high altitude, "chronic mountain sickness" (CMS; also known as "Monge's Disease" after its first description by Carlos Monge in 1925) can develop after years of altitude residence, both in natives or in sojourners. Most cases of CMS are seen at over 3,000 m altitude. Severe CMS is characterized by dyspnea, erythrocytosis, pulmonary hypertension, right heart hypertrophy and, eventually, congestive heart failure. Although CMS is associated with relatively high Epo levels, its pathogenesis is more complicated. Pre-existing pulmonary insufficiency, cardiovascular or other disorders that are worsened at low inspiratory PO_2 are obviously involved, because only a certain percentage of people living at altitude develop CMS. South American high altitude natives often suffer from erythrocytosis and CMS,[194] while Tibetans living at about 4,000 m altitude have low [Hb] and do not develop CMS.[195,196] This difference suggests that there has been natural selection and genetic adaptation in the evolution of tolerance of humans to hypoxia.[197–199]

In addition to erythrocytosis, the low inspiratory PO_2 causes pulmonary arteriolar constriction, which is the main reason for pulmonary hypertension. In the long term, there is a malorganization of the segments of the lung arteries. Clinical laboratory findings are [Hb] >200 g/L, Hct >0.65 and arterial O_2 saturation <85%. The symptoms diminish upon descent from altitude and Hct returns to normal within about one month. Bloodletting is proposed as an acute therapeutic option. Pharmacological treatment trials were performed with female sex hormone (medroxyprogesterone), angiotensin converting enzyme (ACE) inhibitor (enalapril),[200] and carbonic anhydrase inhibitor (acetazolamide).[201]

Reactive erythrocytosis can develop at sea level as a physiological compensatory reaction in smokers. Chronic obstructive pulmonary disease or sleep apnea are rarely associated with increased Epo production and erythrocytosis.[202]

The prevalence of erythrocytosis is low in humans with renal artery stenosis. A flow reduction of at least 50% below normal is necessary to elicit some Epo production in experimental animals.[99] Erythrocytosis may occur in association with polycystic kidney disease.[203] Furthermore, following renal transplantation, postrenal transplant erythrocytosis (PTE) develops in 5–15% of the recipients.[204,205] The pathogenesis of PTE is not fully clear. Epo and angiotensin II, as well as IGF-1 and IGF binding proteins have been implicated.[206,207] PTE has been treated with ACE inhibitors[206,208] or selective angiotensin II type 1 receptor antagonists.[209]

Certain malignant and nonmalignant tumors cause erythrocytosis secondary to the production of Epo by the neoplastic cells.[57] Erythrocytosis develops in 3% of the patients with renal carcinomas and (with lower prevalence) in patients with Wilms' tumor. Perhaps arising from pressure-induced ischemia, benign renal tumors may also induce Epo production and erythrocytosis. Erythrocytosis is observed in 4–12% of liver carcinoma patients. Of the ectopic sites of Epo production, first to be noted are cerebellar hemangioblastomas, present in 20% of all patients with tumor-associated erythrocytosis. Von Hippel–Lindau syndrome typically presents with renal carcinoma or cysts, cerebellar hemangioblastoma, and pheochromocytoma. pVHL mutations have been associated with paranaoplastic Epo production and erythrocytosis.[210] Other sites of paraneoplastic Epo synthesis have included the adrenal gland and smooth muscle tumors such as uterine fibromyomas and leiomyomas.

Bibliography

1. Maluf NS. History of blood transfusion. *J Hist Med Allied Sci.* 1954;9:59–107.
2. Bauer AW. From blood transfusion to haemotherapy — the anniversary of the German society for transfusion medicine and immunology (DGTI) from a medicinal-historical and bioethical perspective. *Transfus Med Hemother.* 2004;31:1–6.
3. Sobottka B, Eggert F, Ferstl R, Muller-Ruchholtz W. Veränderte chemosensorische Identität nach experimenteller Knochenmarktransplantation: Erkennung durch eine andere Spezies. *Zeitschrift für experimentelle und angewandte Psychologie.* 1989;36:654–664.
4. Leinders-Zufall T, Ishii T, Mombaerts P, Zufall F, Boehm T. Structural requirements for the activation of vomeronasal sensory neurons by MHC peptides. *Nat Neuroscience.* 2009;12:1551–1558.
5. Rolleston H. The history of haematology: (Section of the History of Medicine). *Proc R Soc Med.* 1934;27:1161–1178.
6. Von Vierordt K. Neue Methode der chemischen Analyse des Blutes. *Arch physiol Heilkunde.* 1852;11:47–72.
7. Verso ML. Some nineteenth-century pioneers of haematology. *Med Hist.* 1971;15:55–67.
8. Tavassoli M. Bone Marrow: the seedbed of blood. In: Wintrobe MM, (ed.). *Blood, Pure and Eloquent.* McGraw-Hill Book Company; 1980, pp. 57–79.
9. Gulliver G. On the size and shape of red corpuscles of the blood of vertebrates, with drawings of them to a uniform scale, and extended and revised tables of measurements. *Proc Royal Soc London.* 1875; 474–495.
10. Israel O, Pappenheim A. Ueber die Entkernung der Säugethiererythroblasten. *Arch Physiol Anat.* 1896;143:419–447.
11. Pappenheim A. Abstammung und Entstehung der rothen Blutzelle. Eine cytologisch-mikroskopische Studie. *Virchows Archiv.* 1898;151:89–158.

12. Menghini V, Zanottum FM. De ferrearum particularum sede in sanguine. De bononiensi scientiarum et artium Instituto atque Academia commentarii. *Catalogo Articoli.* 1746;2:244–266.
13. Magnus HG. Über die im Blute enthaltenen Gase, Sauerstoff, Stickstoff und Kohlensäure. *Poggend Annal.* 1837;41:583–606.
14. Hoppe-Seyler F. Über das Verhalten des Blutfarbstoffes im Spektrum des Sonnenlichtes. *Virchows Archiv.* 1862;23:446–449.
15. Pflüger EFW. Über die Diffusion des Sauerstoffs, den Ort und die Gesetze der Oxydationsprocesse im thierischen Organismus. *Arch Gesamte Physiol.* 1872;6:43–64.
16. Bohr C, Hasselbalch KA, Krogh A. Ueber einen in biologischer Beziehung wichtigen Einfluss, den die Kohlensaeurespannung des Blutes auf dessen Sauerstoffbindung uebt. *Skand Arch Physiol.* 1904;16:401–412.
17. Hüfner G, Gansser E. Über das Molekulargewicht des Oxyhämoglobins. *Arch Physiol.* 1907;314:209–216.
18. Hüfner CG. Neue Versuche zur Bestimmung der Sauerstoffcapacität des Blutfarbstoffs. *Arch Pathol Anat Physiol.* 1894; 130–176.
19. Hill AV. The possible effects of the aggregation of the molecules of haemoglobin on its dissociation curves. *J Physiol.* 1910;40:4–7.
20. Adair GS. A critical study of the direct method of measuring the osmotic pressure of hemoglobin. *Proc R Soc London Ser A.* 1925; 108A:627–637.
21. Pauling L, Coryell CD. The magnetic properties and structure of hemoglobin, oxyhemoglobin and carbonmonoxyhemoglobin. *Proc Natl Acad Sci USA.* 1936;22:210–216.
22. Haurowitz F. Das Gleichgewicht zwischen Hämoglobin und Sauerstoff. *Zeitschrift für Physikalische Chemie.* 1938;254:266–274.
23. Rapoport S, Luebering J. The formation of 2,3-diphosphoglycerate in rabbit erythrocytes: the existence of a diphosphoglycerate mutase. *J Biol Chem.* 1950;183:507–516.
24. Benesch R, Benesch RE. The effect of organic phosphates from the human erythrocyte on the allosteric properties of hemoglobin. *Biochem Biophys Res Commun.* 1967;26:162–167.
25. Chanutin A, Curnish RR. Effect of organic and inorganic phosphates on the oxygen equilibrium of human erythrocytes. *Arch Biochem Biophys.* 1967;121:96–102.

26. Antonini E, Brunori M. Hemoglobin. *Annu Rev Biochem*. 1970;39:977–1042.
27. Edsall JT. Hemoglobin and the origins of the concept of allosterism. *Fed Proc*. 1980;39:226–235.
28. Christiansen J, Douglas CG, Haldane JS. The absorption and dissociation of carbon dioxide by human blood. *J Physiol*. 1914;48:244–271.
29. Henderson LJ. Das Gleichgewicht zwischen Basen und Säuren im tierischen Organismus. *Ergebnisse der Physiologie*. 1909;55:254–325.
30. Henderson LJ. The equilibrium between oxygen and carbonic acid in blood. *J Biol Chem*. 1920;41:401–430.
31. Hasselbalch H. Die Berechnung der Wasserstoffzahlen des Blutes aus der freien und gebundenen Kohlensäure desselben und die Sauerstoffbindung des Blutes als Funktion der Wasserstoffzahl. *Biochemische Zeitschrift*. 1917;78:112–144.
32. Van Slyke DD, Neill JM. The determination of gases in blood and other solutions by vacuum extraction and manometric measurement. *J Biol Chem*. 1924;61:523–543.
33. Astrup P. A simple electrometric technique for the determination of carbon dioxide tension in blood and plasma, total content of carbon dioxide in plasma and bicarbonate content in "separated" plasma at a fixed carbon dioxide tension. *Scand J Clin Lab Invest*. 1956;8:33–43.
34. Siggard Andersen O, Engel K, Jorgensen K, Astrup P. A micro method for determination of pH, carbon dioxide tension, base excess and standard bicarbonate in capillary blood. *Scand J Clin Lab Invest*. 1960;12:172–176.
35. Bert P. La Pression Barométrique. *Recherches de Physiologie Expérimentale*. Paris: Libraire de l'Académie de Médicine; 1878.
36. Viault F. Sur l'augmentation considérable du nombre des globules rouges dans le sang chez les habitants des hauts plateaux de l'Amérique du Sud. *Comptes Rendus Hebdomadaires des Séances de L'Académie des Sciences* 1890;111:917–918.
37. Müntz A. De l'enrichissement du sang en hémoglobine, suivant les conditions d'existence. *Comptes Rendus Hebdomadaires des Séances de L'Académie des Sciences* 1891;112:298–301.
38. Miescher F. Über die Beziehungen zwischen Meereshöhe und Beschaffenheit des Blutes. *Correspondenz-Blatt Schweizer Aerzte*. 1893;23:809–830.
39. Carnot MP, Deflandre C. Sur l'activité hémopoiétique des différents organes au cours de la régénération du sang. *Comptes Rendus Hebdomadaires des Séances de L'Académie des Sciences*. 1906;143:432–435.

40. Carnot P, Deflandre C. Sur l'activité hémopoiétique du sérum au cours de la régénération du sang. *Comptes Rendus Hebdomadaires des Séances de L'Académie des Sciences.* 1906;143:384–386.
41. Bonsdorff E, Jalavisto E. A humoral mechanism in anoxic erythrocytosis. *Acta Physiol Scand.* 1948;16:150–170.
42. Erslev A. Humoral regulation of red cell production. *Blood.* 1953;8: 349–357.
43. Fisher JW. Landmark advances in the development of erythropoietin. *Exp Biol Med (Maywood).* 2010;235:1398–1411.
44. Miyake T, Kung CK, Goldwasser E. Purification of human erythropoietin. *J Biol Chem.* 1977;252:5558–5564.
45. Jacobs K, Shoemaker C, Rudersdorf R, Neill SD, Kaufman RJ, Mufson A, *et al.* Isolation and characterization of genomic and cDNA clones of human erythropoietin. *Nature.* 1985;313:806–810.
46. Lin FK, Suggs S, Lin CH, Browne JK, Smalling R, Egrie JC, *et al.* Cloning and expression of the human erythropoietin gene. *Proc Natl Acad Sci USA.* 1985;82:7580–7584.
47. Lindeboom G. The story of a blood transfusion to a Pope. *J Hist Med Allied Sci.* 1954;9:455–459.
48. Blundell J. Successful case of transfusion. *Lancet.* 1828;i:431–432.
49. Landsteiner K, Wiener AS. An agglutinable factor in human blood recognized by immune sera for rhesus blood. *Proc Soc Exp Biol Med.* 1940;43:223–224.
50. Landsteiner K, Wiener AS. Studies on an agglutinogen (Rh) in human blood reacting with anti-rhesus sera and with human isoantibodies. *J Exp Med.* 1941;74:309–320.
51. Landsteiner K. On individual differences in human blood. *Science.* 1931; 73:234–245.
52. Ottenberg R, Johnson A. A hitherto undescribed anomaly in blood groups. *J Immunol.* 1926;12:35–44.
53. Baele PL, De Bruyere M, Deneys V, Flament J, Lambermont M, Latinne D, *et al.* Bedside transfusion errors. A prospective survey by the Belgium SAnGUIS Group. *Vox Sang.* 1994;66:117–121.
54. Vamvakas EC, Blajchman MA. Transfusion-related mortality: the ongoing risks of allogeneic blood transfusion and the available strategies for their prevention. *Blood.* 2009;113:3406–3417.

55. Gore CJ, Hopkins WG, Burge CM. Errors of measurement for blood volume parameters: a meta-analysis. *J Appl Physiol*. 2005;99:1745–1758.
56. Kickler TS. Clinical analyzers. Advances in automated cell counting. *Anal Chem*. 1999;71:363R–365R.
57. Jelkmann W. Erythropoietin: structure, control of production, and function. *Physiol Rev*. 1992;72:449–489.
58. Snyder GK, Sheafor BA. Red blood cells: centerpiece in the evolution of the vertebrate circulatory system. *Amer Zool*. 1999;39:189–198.
59. Uzoigwe C. The human erythrocyte has developed the biconcave disc shape to optimise the flow properties of the blood in the large vessels. *Med Hypotheses*. 2006;67:1159–1163.
60. Perry SP, Byers T, Yip R, Margen S. Iron nutrition does not account for the hemoglobin differences between blacks and whites. *J Nutr*. 1992;122:1417–1424.
61. Beutler E, Waalen J. The definition of anemia: what is the lower limit of normal of the blood hemoglobin concentration? *Blood*. 2006;107:1747–1750.
62. Cohen WD. The cytomorphic system of anucleate non-mammalian erythrocytes. *Protoplasma*. 1982;113:23–32.
63. Rapoport S. The regulation of glycolysis in mammalian erythrocytes. *Essays Biochem*. 1968;4:69–103.
64. Goodman SR, Kurdia A, Ammann L, Kakhniashvili D, Daescu O. The human red blood cell proteome and interactome. *Exp Biol Med (Maywood)*. 2007;232:1391–1408.
65. Denomme GA. The structure and function of the molecules that carry human red blood cell and platelet antigens. *Transfus Med Rev*. 2004;18:203–231.
66. An X, Mohandas N. Disorders of red cell membrane. *Br J Haematol*. 2008;141:367–375.
67. Iolascon A, Perrotta S, Stewart GW. Red blood cell membrane defects. *Rev Clin Exp Hematol*. 2003;7:22–56.
68. Perrotta S, Gallagher PG, Mohandas N. Hereditary spherocytosis. *Lancet*. 2008;372:1411–1426.
69. Cartron JP, Colin Y. Structural and functional diversity of blood group antigens. *Transfus Clin Biol*. 2001;8:163–199.
70. Brecher G, Bessis M. Present status of spiculed red cells and their relationship to the discocyte-echinocyte transformation: a critical review. *Blood*. 1972;40:333–344.

71. Severinghaus JW, Astrup P, Murray JF. Blood gas analysis and critical care medicine. *Am J Respir Crit Care Med*. 1998;157:114–122.
72. Bunn HF, Gabbay KH, Gallop PM. The glycosylation of hemoglobin: relevance to diabetes mellitus. *Science*. 1978;200:21–27.
73. Perutz MF, Wilkinson AJ, Paoli M, Dodson GG. The stereochemical mechanism of the cooperative effects in hemoglobin revisited. *Annu Rev Biophys Biomol Struct*. 1998;27:1–34.
74. Perutz MF. Stereochemistry of cooperative effects in haemoglobin. *Nature*. 1970;228:726–739.
75. Chiba H, Sasaki R. Functions of 2,3-bisphosphoglycerate and its metabolism. *Curr Top Cell Regul*. 1978;14:75–116.
76. Perutz MF, Imai K. Regulation of oxygen affinity of mammalian haemoglobins. *J Mol Biol*. 1980;136:183–191.
77. Bauer C, Ludwig I, Ludwig M. Different effects of 2.3 diphosphoglycerate and adenosine triphosphate on the oxygen affinity of adult and foetal human haemoglobin. *Life Sci*. 1968;7:1339–1343.
78. Tyuma I, Shimizu K. Different response to organic phosphates of human fetal and adult hemoglobins. *Arch Biochem Biophys*. 1969;129:404–405.
79. Wood WG. Haemoglobin synthesis during human fetal development. *Br Med Bull*. 1976;32:282–287.
80. Bank A, Mears JG, Ramirez F. Disorders of human hemoglobin. *Science*. 1980;207:486–493.
81. Geers C, Gros G. Carbon dioxide transport and carbonic anhydrase in blood and muscle. *Physiol Rev*. 2000;80:681–715.
82. Meldrum NU, Roughton FJW. The state of carbon dioxide in blood. *J Physiol*. 1933;80:143–170.
83. Wagner PD. Diffusion and chemical reaction in pulmonary gas exchange. *Physiol Rev*. 1977;57:257–312.
84. Aarts PA, Bolhuis PA, Sakariassen KS, Heethaar RM, Sixma JJ. Red blood cell size is important for adherence of blood platelets to artery subendothelium. *Blood*. 1983;62:214–217.
85. Locatelli F, Gascon P. Is nephrology more at ease than oncology with erythropoiesis-stimulating agents? Treatment guidelines and an update on benefits and risks. *Oncologist*. 2009;14:57–62.
86. Turitto VT, Weiss HJ. Red blood cells: their dual role in thrombus formation. *Science*. 1980;207:541–543.

87. Moia M, Mannucci PM, Vizzotto L, Casati S, Cattaneo M, Ponticelli C. Improvement in the haemostatic defect of uraemia after treatment with recombinant human erythropoietin. *Lancet*. 1987;2:1227–1229.
88. Wan J, Ristenpart WD, Stone HA. Dynamics of shear-induced ATP release from red blood cells. *Proc Natl Acad Sci USA*. 2008;105:16432–16437.
89. Diesen DL, Hess DT, Stamler JS. Hypoxic vasodilation by red blood cells: evidence for an s-nitrosothiol-based signal. *Circ Res*. 2008;103:545–553.
90. Kleinbongard P, Schulz R, Rassaf T, Lauer T, Dejam A, Jax T, et al. Red blood cells express a functional endothelial nitric oxide synthase. *Blood*. 2006;107:2943–2951.
91. Ulker P, Sati L, Celik-Ozenci C, Meiselman HJ, Baskurt OK. Mechanical stimulation of nitric oxide synthesizing mechanisms in erythrocytes. *Biorheology*. 2009;46:121–132.
92. Till JE, McCulloch EA. A direct measurement of the radiation sensitivity of normal mouse bone marrow cells. *Radiat Res*. 1961;14:213–222.
93. Weissman IL, Shizuru JA. The origins of the identification and isolation of hematopoietic stem cells, and their capability to induce donor-specific transplantation tolerance and treat autoimmune diseases. *Blood*. 2008;112:3543–3553.
94. Chasis JA, Mohandas N. Erythroblastic Islands: niches for erythropoiesis. *Blood*. 2008;112:470–478.
95. Hebiguchi M, Hirokawa M, Guo YM, Saito K, Wakui H, Komatsuda A, et al. Dynamics of human erythroblast enucleation. *Int J Hematol*. 2008;88:498–507.
96. Koury MJ, Koury ST, Kopsombut P, Bondurant MC. *In vitro* maturation of nascent reticulocytes to erythrocytes. *Blood*. 2005;105:2168–2174.
97. Liu J, Guo X, Mohandas N, Chasis JA, An X. Membrane remodeling during reticulocyte maturation. *Blood*. 2010;115:2021–2027.
98. Riley RS, Ben-Ezra JM, Tidwell A, Romagnoli G. Reticulocyte analysis by flow cytometry and other techniques. *Hematol Oncol Clin N Am*. 2002;16:373–420.
99. Pagel H, Jelkmann W, Weiss C. O_2-supply to the kidneys and the production of erythropoietin. *Respir Physiol*. 1989;77:111–117.
100. Semenza GL, Nejfelt MK, Chi SM, Antonarakis SE. Hypoxia-inducible nuclear factors bind to an enhancer element located 3′ to the human erythropoietin gene. *Proc Natl Acad Sci USA*. 1991;88:5680–5684.

101. Semenza GL, Wang GL. A nuclear factor induced by hypoxia via de novo protein synthesis binds to the human erythropoietin gene enhancer at a site required for transcriptional activation. *Mol Cell Biol.* 1992;12:5447–5454.
102. Wang GL, Semenza GL. Purification and characterization of hypoxia-inducible factor 1. *J Biol Chem.* 1995;270:1230–1237.
103. Epstein ACR, Gleadle JM, McNeill LA, Hewitson KS, O'Rourke J, Mole DR, *et al.* C. Elegans EGL-9 and mammalian homologs define a family of dioxygenases that regulate HIF by prolyl hydroxylation. *Cell.* 2001;107:43–54.
104. Bruick RK, McKnight SL. A conserved family of prolyl-4-hydroxylases that modify HIF. *Science.* 2001;294:1337–1340.
105. Jaakkola P, Mole DR, Tian YM, Wilson MI, Gielbert J, Gaskell SJ, *et al.* Targeting of HIF-α to the von Hippel-Lindau ubiquitylation complex by O_2-regulated prolyl hydroxylation. *Science.* 2001;292:468–472.
106. Ivan M, Kondo K, Yang H, Kim W, Valiando J, Ohh M, *et al.* HIFα targeted for VHL-mediated destruction by proline hydroxylation: implications for O_2 sensing. *Science.* 2001;292:464–468.
107. Yu F, White SB, Zhao Q, Lee FS. HIF-1α binding to VHL is regulated by stimulus-sensitive proline hydroxylation. *Proc Natl Acad Sci USA.* 2001;98:9630–9635.
108. Ohh M, Park CW, Ivan M, Hoffman MA, Kim TY, Huang LE, *et al.* Ubiquitination of hypoxia-inducible factor requires direct binding to the β-domain of the von Hippel-Lindau protein. *Nat Cell Biol.* 2000;2:423–427.
109. Warnecke C, Zaborowska Z, Kurreck J, Erdmann VA, Frei U, Wiesener M, *et al.* Differentiating the functional role of hypoxia-inducible factor (HIF)-1alpha and HIF-2α (EPAS-1) by the use of RNA interference: erythropoietin is a HIF-2alpha target gene in Hep3B and Kelly cells. *FASEB J.* 2004;18:1462–1464.
110. Haase VH. Hypoxic regulation of erythropoiesis and iron metabolism. *Am J Physiol Renal Physiol.* 2010;299:F1–13.
111. Erbel PJ, Card PB, Karakuzu O, Bruick RK, Gardner KH. Structural basis for PAS domain heterodimerization in the basic helix — loop — helix-PAS transcription factor hypoxia-inducible factor. *Proc Natl Acad Sci USA.* 2003;100:15504–15509.
112. Bernhardt WM, Wiesener MS, Scigalla P, Chou J, Schmieder RE, Günzler V, *et al.* Inhibition of prolyl hydroxylases increases erythropoietin production in ESRD. *J Am Soc Nephrol.* 2010;21:2151–2156.

113. Remy I, Wilson IA, Michnick SW. Erythropoietin receptor activation by a ligand-induced conformation change. *Science*. 1999;283:990–993.
114. Witthuhn BA, Quelle FW, Silvennoinen O, Yi T, Tang B, Miura O, *et al*. JAK2 associates with the erythropoietin receptor and is tyrosine phosphorylated and activated following stimulation with erythropoietin. *Cell*. 1993;74:227–236.
115. Koury MJ, Bondurant MC. Erythropoietin retards DNA breakdown and prevents programmed death in erythroid progenitor cells. *Science*. 1990;248:378–381.
116. Suzuki N, Suwabe N, Ohneda O, Obara N, Imagawa S, Pan X, *et al*. Identification and characterization of 2 types of erythroid progenitors that express GATA-1 at distinct levels. *Blood*. 2003;102:3575–3583.
117. De Maria R, Zeuner A, Eramo A, Domenichelli C, Bonci D, Grignani F, *et al*. Negative regulation of erythropoiesis by caspase-mediated cleavage of GATA-1. *Nature*. 1999;401:489–493.
118. Suessmuth Y, Elliott J, Percy MJ, Inami M, Attal H, Harrison CN, *et al*. A new polycythaemia vera-associated SOCS3 SH2 mutant (SOCS3F136L) cannot regulate erythropoietin responses. *Br J Haematol*. 2009;147: 450–458.
119. Stewart IB, McKenzie DC. The human spleen during physiological stress. *Sports Med*. 2002;32:361–369.
120. Berlin NI, Waldmann TA, Weissman SM. Life span of red blood cell. *Physiol Rev*. 1959;39:577–616.
121. Lang F, Gulbins E, Lerche H, Huber SM, Kempe DS, Foller M. Eryptosis, a window to systemic disease. *Cell Physiol Biochem*. 2008;22:373–380.
122. Andrews NC. Forging a field: the golden age of iron biology. *Blood*. 2008;112:219–230.
123. Raya G, Henny J, Steinmetz J, Herbeth B, Siest G. Soluble transferrin receptor (sTfR): biological variations and reference limits. *Clin Chem Lab Med*. 2001;39:1162–1168.
124. Lee PL, Beutler E. Regulation of hepcidin and iron-overload disease. *Annu Rev Pathol*. 2009;4:489–515.
125. Nemeth E, Tuttle MS, Powelson J, Vaughn MB, Donovan A, Ward DM, *et al*. Hepcidin regulates cellular iron efflux by binding to ferroportin and inducing its internalization. *Science*. 2004;306:2090–2093.
126. Ganz T, Nemeth E. Hepcidin and disorders of iron metabolism. *Annu Rev Med*. 2011;62:347–360.

127. Brugnara C. Use of reticulocyte cellular indices in the diagnosis and treatment of hematological disorders. *Int J Clin Lab Res*. 1998; 28:1–11.
128. Mast AE, Blinder MA, Lu Q, Flax S, Dietzen DJ. Clinical utility of the reticulocyte hemoglobin content in the diagnosis of iron deficiency. *Blood*. 2002;99:1489–1491.
129. Schmidt W, Prommer N. Impact of alterations in total hemoglobin mass on VO_2 max. *Exerc Sport Sci Rev*. 2010;38:68–75.
130. Gaudard A, Varlet-Marie E, Bressolle F, Audran M. Drugs for increasing oxygen and their potential use in doping: a review. *Sports Med*. 2003;33:187–212.
131. Lundby C, Thomsen JJ, Boushel R, Koskolou M, Warberg J, Calbet JA, *et al*. Erythropoietin treatment elevates haemoglobin concentration by increasing red cell volume and depressing plasma volume. *J Physiol*. 2007;578:309–314.
132. Reichel C, Gmeiner G. Erythropoietin and analogs. *Handb Exp Pharmacol*. 2010; 251–294.
133. Lasne F, de Ceaurriz J. Recombinant erythropoietin in urine. *Nature*. 2000;405:635.
134. Catlin DH, Breidbach A, Elliott S, Glaspy J. Comparison of the isoelectric focusing patterns of darbepoetin alfa, recombinant human erythropoietin, and endogenous erythropoietin from human urine. *Clin Chem*. 2002;48: 2057–2059.
135. Jelkmann W. Erythropoiesis stimulating agents and techniques: a challenge for doping analysts. *Curr Med Chem*. 2009;16:1236–1247.
136. World Health Organization. Nutritional anaemias. Report of a WHO scientific group. *World Health Organ Tech Rep Ser*. 1968;405:5–37.
137. Elwood PC. Evaluation of the clinical importance of anemia. *Am J Clin Nutr*. 1973;26:958–964.
138. Thomas C, Thomas L. Biochemical markers and hematologic indices in the diagnosis of functional iron deficiency. *Clin Chem*. 2002;48:1066–1076.
139. Carmel R. How I treat cobalamin (vitamin B12) deficiency. *Blood*. 2008;112:2214–2221.
140. Radtke HW, Claussner A, Erbes PM, Scheuermann EH, Schoeppe W, Koch KM. Serum erythropoietin concentration in chronic renal failure: relationship to degree of anemia and excretory renal function. *Blood*. 1979;54:877–884.

141. McGonigle RJ, Boineau FG, Beckman B, Ohene-Frempong K, Lewy JE, Shadduck RK, et al. Erythropoietin and inhibitors of *in vitro* erythropoiesis in the development of anemia in children with renal disease. *J Lab Clin Med*. 1985;105:449–458.
142. Chandra M, Clemons GK, McVicar MI. Relation of serum erythropoietin levels to renal excretory function: evidence for lowered set point for erythropoietin production in chronic renal failure. *J Pediatr*. 1988;113:1015–1021.
143. Cooper AC, Mikhail A, Lethbridge MW, Kemeny DM, MacDougall IC. Increased expression of erythropoiesis inhibiting cytokines (IFN-γ, TNF-α, IL-10, and IL-13) by T cells in patients exhibiting a poor response to erythropoietin therapy. *J Am Soc Nephrol*. 2003;14:1776–1784.
144. Wallner SF, Vautrin RM. Evidence that inhibition of erythropoiesis is important in the anemia of chronic renal failure. *J Lab Clin Med*. 1981;97:170–178.
145. McGonigle RJ, Husserl F, Wallin JD, Fisher JW. Hemodialysis and continuous ambulatory peritoneal dialysis effects on erythropoiesis in renal failure. *Kidney Int*. 1984;25:430–436.
146. Winearls CG, Oliver DO, Pippard MJ, Reid C, Downing MR, Cotes PM. Effect of human erythropoietin derived from recombinant DNA on the anaemia of patients maintained by chronic haemodialysis. *Lancet*. 1986;2:1175–1178.
147. Eschbach JW, Egrie JC, Downing MR, Browne JK, Adamson JW. Correction of the anemia of end-stage renal disease with recombinant human erythropoietin. Results of a combined phase I and II clinical trial. *N Engl J Med*. 1987;316:73–78.
148. Goodnough LT, Monk TG, Andriole GL. Erythropoietin therapy. *N Engl J Med*. 1997;336:933–938.
149. Kristiansen M, Graversen JH, Jacobsen C, Sonne O, Hoffman HJ, Law SK, et al. Identification of the haemoglobin scavenger receptor. *Nature*. 2001;409:198–201.
150. Nielsen MJ, Moestrup SK. Receptor targeting of hemoglobin mediated by the haptoglobins: roles beyond heme scavenging. *Blood*. 2009;114:764–771.
151. Kormoczi GF, Saemann MD, Buchta C, Peck-Radosavljevic M, Mayr WR, Schwartz DW, et al. Influence of clinical factors on the haemolysis marker haptoglobin. *Eur J Clin Invest*. 2006;36:202–209.

152. Bolton-Maggs PH, Stevens RF, Dodd NJ, Lamont G, Tittensor P, King MJ. Guidelines for the diagnosis and management of hereditary spherocytosis. *Br J Haematol.* 2004;126:455–474.
153. Prchal JT, Gregg XT. Red cell enzymes. *Hematology Am Soc Hematol Educ Program.* 2005; 19–23.
154. Cunningham MJ. Update on thalassemia: clinical care and complications. *Hematol Oncol Clin North Am.* 2010;24:215–227.
155. Persons DA. Hematopoietic stem cell gene transfer for the treatment of hemoglobin disorders. *Hematol J.* 2009; 690–697.
156. Gertz MA. Cold hemolytic syndrome. *Hematology Am Soc Hematol Educ Program.* 2006; 19–23.
157. Means RT, Krantz SB. Progress in understanding the pathogenesis of the anemia of chronic disease. *Blood.* 1992;80:1639–1647.
158. Jelkmann W. Proinflammatory cytokines lowering erythropoietin production. *J Interferon Cytokine Res.* 1998;18:555–559.
159. Pearson TC, Guthrie DL, Simpson J, Chinn S, Barosi G, Ferrant A, et al. Interpretation of measured red cell mass and plasma volume in adults: expert panel on radionuclides of the international council for standardization in haematology. *Br J Haematol.* 1995;89:748–756.
160. McMullin MF. The classification and diagnosis of erythrocytosis. *Int J Lab Hematol.* 2008;30:447–459.
161. Percy MJ, Rumi E. Genetic origins and clinical phenotype of familial and acquired erythrocytosis and thrombocytosis. *Am J Hematol.* 2008;84: 46–54.
162. Semenza GL. Involvement of oxygen-sensing pathways in physiologic and pathologic erythropoiesis. *Blood.* 2009;114:2015–2019.
163. Messinezy M, Westwood NB, Woodcock SP, Strong RM, Pearson TC. Low serum erythropoietin — a strong diagnostic criterion of primary polycythaemia even at normal haemoglobin levels. *Clin Lab Haematol.* 1995;17: 217–220.
164. de la Chapelle A, Träskelin A-L, Juvonen E. Truncated erythropoietin receptor causes dominantly inherited benign human erythrocytosis. *Proc Natl Acad Sci USA.* 1993;90:4495–4499.
165. Juvonen E, Ikkala E, Fyhrquist F, Ruutu T. Autosomal dominant erythrocytosis caused by increased sensitivity to erythropoietin. *Blood.* 1991;78:3066–3069.

166. Kralovics R, Indrak K, Stopka T, Berman BW, Prchal JF, Prchal JT. Two new EPO receptor mutations: truncated EPO receptors are most frequently associated with primary familial and congenital polycythemias. *Blood*. 1997;90:2057–2061.
167. Kutti J, Ridell B. Epidemiology of the myeloproliferative disorders: essential thrombocythaemia, polycythaemia vera and idiopathic myelofibrosis. *Pathol Biol (Paris)*. 2001;49:164–166.
168. Campbell PJ, Green AR. The myeloproliferative disorders. *N Engl J Med*. 2006;355:2452–2466.
169. Baxter EJ, Scott LM, Campbell PJ, East C, Fourouclas N, Swanton S, *et al*. Acquired mutation of the tyrosine kinase JAK2 in human myeloproliferative disorders. *Lancet*. 2005;366:1054–1061.
170. Percy MJ, Scott LM, Erber WN, Harrison CN, Reilly JT, Jones FGC, *et al*. The frequency of JAK2 exon 12 mutations in idiopathic erythrocytosis patients with low serum erythropoietin levels. *Haematologica*. 2007;92:1607–1614.
171. Scott LM, Tong W, Levine RL, Scott MA, Beer PA, Stratton MR, *et al*. JAK2 exon 12 mutations in polycythemia vera and idiopathic erythrocytosis. *N Engl J Med*. 2007;356:459–468.
172. Pietra D, Li S, Brisci A, Passamonti F, Rumi E, Theocharides A, *et al*. Somatic mutations of JAK2 exon 12 in patients with JAK2 (V617F)-negative myeloproliferative disorders. *Blood*. 2008;111:1686–1689.
173. Tefferi A, Thiele J, Orazi A, Kvasnicka HM, Barbui T, Hanson CA, *et al*. Proposals and rationale for revision of the World Health Organization diagnostic criteria for polycythemia vera, essential thrombocythemia, and primary myelofibrosis: recommendations from an ad hoc international expert panel. *Blood*. 2007;110:1092–1097.
174. Passamonti F, Rumi E, Pungolino E, Malabarba L, Bertazzoni P, Valentini M, *et al*. Life expectancy and prognostic factors for survival in patients with polycythemia vera and essential thrombocythemia. *Am J Med*. 2004;117:755–761.
175. Samuelsson J, Hasselbalch H, Bruserud O, Temerinac S, Brandberg Y, Merup M, *et al*. A phase II trial of pegylated interferon alpha-2b therapy for polycythemia vera and essential thrombocythemia: feasibility, clinical and biologic effects, and impact on quality of life. *Cancer*. 2006;106:2397–2405.
176. Thorne SA. Management of polycythaemia in adults with cyanotic congenital heart disease. *Heart* 1998;79:315–316.

177. Charache S, Weatherall DJ, Clegg JB. Polycythemia associated with a hemoglobinopathy. *J Clin Invest.* 1966;45:813–822.
178. Means RT. Hepcidin and cytokines in anaemia. *Hematology.* 2004;9:357–362.
179. Rosa R, Prehu MO, Beuzard Y, Rosa J. The first case of a complete deficiency of diphosphoglycerate mutase in human erythrocytes. *J Clin Invest.* 1978;62:907–915.
180. Galacteros F, Rosa R, Prehu MO, Najean Y, Calvin MC. Diphosphoglyceromutase deficiency: new cases associated with erythrocytosis. *Nouv Rev Fr Hematol.* 1984;26:69–74.
181. Hoyer JD, Allen SL, Beutler E, Kubik K, West C, Fairbanks VF. Erythrocytosis due to biphosphoglycerate mutase deficiency with concurrent glucose-6-phosphate dehydrogenase (G-6-PD) deficiency. *Am J Hematol.* 2004;75:205–208.
182. Semenza GL. Regulation of cancer cell metabolism by hypoxia-inducible factor 1. *Semin Cancer Biol.* 2009;19:12–16.
183. Ang SO, Chen H, Hirota K, Gordeuk VR, Jelinek J, Guan Y, *et al.* Disruption of oxygen homeostasis underlies congenital Chuvash polycythemia. *Nat Genet.* 2002;32:614–621.
184. Gordeuk VR, Prchal JT. Vascular complications in Chuvash polycythemia. *Semin Thromb Hemost.* 2006;32:289–294.
185. Percy MJ, McMullin MF, Jowitt SN, Potter M, Treacy M, Watson WH, *et al.* Chuvash-type congenital polycythemia in 4 families of Asian and Western European ancestry. *Blood.* 2003;102:1097–1099.
186. Perrotta S, Nobili B, Ferraro M, Migliaccio C, Borriello A, Cucciolla V, *et al.* Von Hippel-Landau-dependent polycythemia is endemic on the island of Ischia: identification of a novel cluster. *Blood.* 2006;107:514–519.
187. Liu E, Percy MJ, Amos CI, Guan Y, Shete S, Stockton DW, *et al.* The worldwide distribution of the VHL 598C>T mutation indicates a single founding event. *Blood.* 2004;103:1937–1940.
188. Percy MJ, Zhao Q, Flores A, Harrison C, Lappin TR, Maxwell PH, *et al.* A family with erythrocytosis establishes a role for prolyl hydroxylase domain protein 2 in oxygen homeostasis. *Proc Natl Acad Sci USA.* 2006;103:654–659.
189. Al-Sheikh M, Moradkhani K, Lopez M, Wajcman H, Préhu C. Disturbance in the HIF-1α pathway associated with erythrocytosis: further evidences brought by frameshift and nonsense mutations in the prolyl

hydroxylase domain protein 2 (PHD2) gene. *Blood Cells Mol Dis.* 2008;40:160–165.
190. Percy MJ, Furlow PW, Beer PA, Lappin TR, McMullin MF, Lee FS. A novel erythrocytosis-associated PHD2 mutation suggests the location of a HIF binding groove. *Blood.* 2007;110:2193–2196.
191. Furlow PW, Percy MJ, Sutherland S, Bierl C, McMullin MF, Master SR, *et al.* Erythrocytosis-associated HIf-2alpha mutations demonstrate a critical role for residues C-terminal to the hydroxylacceptor proline. *J Biol Chem.* 2009;284:9050–9058.
192. Percy MJ, Furlow PW, Lucas GS, Li X, Lappin TR, McMullin MF, *et al.* A gain-of-function mutation in the HIF2A gene in familial erythrocytosis. *N Engl J Med.* 2008;358:162–168.
193. van Wijk R, Sutherland S, Van Wesel ACW, Huizinga EG, Percy MJ, Bierings M, *et al.* Erythrocytosis associated with a novel missense mutation in the *HIF2A* gene. *Haematologica.* 2010;95:829–832.
194. Leon-Velarde F, Monge CC, Vidal A, Carcagno M, Criscuolo M, Bozzini CE. Serum immunoreactive erythropoietin in high altitude natives with and without excessive erythrocytosis. *Exp Hematol.* 1991; 19:257–260.
195. Beall CM, Brittenham GM, Strohl KP, Blangero J, Williams BS, Goldstein MC, *et al.* Hemoglobin concentration of high-altitude Tibetans and Bolivian Aymara. *Am J Phys Anthropol.* 1998;106:385–400.
196. Winslow RM, Chapman KW, Gibson CC, Samaja M, Monge CC, Goldwasser E, *et al.* Different hematologic responses to hypoxia in Sherpas and Quechua Indians. *J Appl Physiol.* 1989;66:1561–1569.
197. Hochachka PW, Rupert JL, Monge C. Adaptation and conservation of physiological systems in the evolution of human hypoxia tolerance. *Comp Biochem Physiol A Mol Integr Physiol.* 1999;124:1–17.
198. Beall CM. Two routes to functional adaptation: Tibetan and Andean high-altitude natives. *Proc Natl Acad Sci USA.* 2007;104(Suppl 1):8655–8660.
199. Beall CM, Cavalleri GL, Deng L, Elston RC, Gao Y, Knight J, *et al.* Natural selection on EPAS1 (HIF2alpha) associated with low hemoglobin concentration in Tibetan highlanders. *Proc Natl Acad Sci USA.* 2010;107:11459–11464.
200. Plata R, Cornejo A, Arratia C, Anabaya A, Perna A, Dimitrov BD, *et al.* Angiotensin-converting-enzyme inhibition therapy in altitude polycythaemia: a prospective randomised trial. *Lancet.* 2002;359:663–666.

201. Richalet JP, Rivera-Ch M, Maignan M, Privat C, Pham I, Macarlupu JL, et al. Acetazolamide for Monge's disease: efficiency and tolerance of 6-month treatment. *Am J Respir Crit Care Med.* 2008;177:1370–1376.
202. Pokala P, Llanera M, Sherwood J, Scharf S, Steinberg H. Erythropoietin response in subjects with obstructive sleep apnea. *Am J Respir Crit Care Med.* 1995;151:1862–1865.
203. Chandra M, Miller ME, Garcia JF, Mossey RT, McVicar M. Serum immunoreactive erythropoietin levels in patients with polycystic kidney disease as compared with other hemodialysis patients. *Nephron.* 1985;39:26–29.
204. Sun CH, Ward HJ, Paul WL, Koyle MA, Yanagawa N, Lee DBN. Serum erythropoietin levels after renal transplantation. *N Engl J Med.* 1989;321:151–157.
205. Vlahakos DV, Marathias KP, Agroyannis B, Madia NE. Posttransplant erythrocytosis. *Kidney Int.* 2003;63:1187–1194.
206. Gaston RS, Julian BA, Curtis JJ. Posttransplant erythrocytosis: an enigma revisited. *Am J Kidney Dis.* 1994;24:1–11.
207. Brox AG, Mangel J, Hanley JA, St. Louis G, Mongrain S, Gagnon RF. Erythrocytosis after renal transplantation represents an abnormality of insulin-like growth factor-1 and its binding proteins. *Transplantation* 1998;66:1053–1058.
208. Glicklich D, Kapoian T, Mian H, Gilman J, Tellis V, Croizat H. Effects of erythropoietin, angiotensin II, and angiotensin-converting enzyme inhibitor on erythroid precursors in patients with posttransplantation erythrocytosis. *Transplantation* 1999;68:62–66.
209. Julian BA, Brantley RR, Barker CV, Stopka T, Gaston RS, Curtis JJ, et al. Losartan, an angiotensin II type 1 receptor antagonist, lowers hematocrit in posttransplant erythrocytosis. *J Am Soc Nephrol.* 1998;9:1104–1106.
210. Kaelin W-GJ. The von Hippel-Lindau tumour suppressor protein: O_2 sensing and cancer. *Nat Rev Cancer.* 2008;8:865–873.

2
Properties and Membrane Transport Mechanisms of Erythrocytes

Peter K Lauf*,†,‡ and Norma C Adragna†,§

†*Cell Biophysics Group;* ‡*Department of Pathology;* §*Department of Pharmacology & Toxicology; Boonshoft School of Medicine, Wright State University, Dayton, OH 45435, USA*

2.1 General Introduction: Historical Aspects and the Significance of Erythrocytes

This chapter is an overview of the properties and membrane transport mechanisms of volume and ion homeostasis in mammalian erythrocytes (synonyms: red blood cells, RBCs). In face of the overwhelming progress of the past 50 years on this topic, the strategy chosen here was, in part and wherever possible, to review past important discoveries of RBC membrane transport systems and then to guide the reader to their current status. The significance of the erythrocyte lies in the fact that, together with the early work on artificial phospholipid bilayer (black lipid) membranes,[1] it served for decades as the dominant cellular model to study cell homeostasis and transport. Historically, this research contributed seminally to the evolution of our current concepts of the detailed molecular mechanisms of ion pumps, channels, co-transporters, and exchangers, later developed in nucleated mammalian cell systems. In

---

* Corresponding author: peter.lauf@wright.edu

time, this appeared to be a unidirectional flux of information culminating in the discovery of the crystal structure of potassium $(K)^2$; and water channels[3]; of the Na/K pump[4,5]; or their transport subunits or fragments, by investigators — two of whom received the Nobel Prize in Chemistry (Roderick MacKinnon 2003, Peter Agre 2003). Yet there is emerging evidence for reverse information transfer: the structural knowledge of major functional components originally found or suspected to be present in RBCs, such as the Rhesus (Rh) antigen[6-8] or the calcium-activated intermediate K (IK or $K_{Ca}3.1$) channels, were often found first elsewhere, reactivating RBC research.[9] The molecular basis of such phenomena as transport of ammonia[10] or the Gardos effect[11-13] were thus explained in satisfying detail.

Historically, several major phases of research on the nature of RBC transport functions can be distinguished. The original discoveries of frog and human erythrocytes were made by Jan Swammerdam (1658) and Antonie van Leeuwenhoek in 1674, respectively. Two centuries passed before the most important carrier function of the erythrocyte, i.e. oxygen (O_2) transport, was established by Felix Hoppe-Seyler (1865) and the modulation of the sigmoid O_2 association curve by carbon dioxide (CO_2) by Niels Bohr (1904). The finding by Hartog Jacob Hamburger (1890), that CO_2 left the RBC associated with bicarbonate (HCO_3) entry and chloride (Cl) exit — the Hamburger shift — as part of the carbonic anhydrase (CA)-driven Jacobs–Parpart–Stewart cycle,[14] heralded the arrival of the membrane exchange principle that only 70 years later was kinetically characterized[15] and attributed in subsequent decades to the now well-known anion exchange protein isoform 1 (AE1) (Section 2.7). Given Charles Ernest Overton's theorem[16] that cell membranes were for all practical purposes cation-"impermeable" and the fixed charge hypothesis of Mond (1929) to explain anion binding,[17] the CO_2-driven anion movements were thus historically addressed before the small cation "leaks" of the RBC membrane.

Meanwhile, the arrival of more advanced microscopic and biochemical techniques permitted detailed RBC shape and membrane composition analyses such as lipids and proteins of several species, leading to the early lipid bilayer membrane models, with beta-sheeted protein attachments, of Gorter and Grendel (1925), and of Davson and

Danielli (1936),[18] and, based on permanganate staining of the phospholipid bilayer, called the "unit membrane", by Robertson in 1959.[19] The arrival of electrophoresis techniques, first with urea starch (Oliver Smithies, Nobel Prize 2007) and then with polyacrylamide[20] and biphasic extraction techniques to separate the relatively insoluble major membrane proteins from lipids revealed that the former, with an initially large molecular weight estimated in the mega daltons,[21,22] were composed of greatly heterogeneous and smaller molecular weight components,[23] thus paving the way for the fluid mosaic model of Singer.[24] The conventional lipid bilayer membrane model was replaced with one that permitted proteins and thus pores to cross and move within the hydrophobic fatty acid core of the plasma membrane overcoming the "hydrophobic effect".[25,26] Still, the carbohydrate diversity of the RBC surface lost its enigma as the structural basis of the major human blood groups ABO, MN, and Rhesus was elucidated.[27–29] Parallel developments occurred in RBCs of ungulates[30,31] suggesting relationships between surface antigens and membrane transport function.[32]

Thus, toward the 1960s, the stage was set to characterize the cation and anion permeabilities of the RBC membrane. At that time, the RBC, lacking the Krebs citric acid cycle, was recognized to have the glycolytic pathway or Embden–Meyerhoff hexose-monophosphate pathway leading to the 1922 and 1931 Nobel Prizes for Meyerhoff and Warburg, respectively.[33] This pathway produces anaerobically 2 adenosine triphosphates (ATP)/glucose, whereas the pentose-phosphate shunt generates the reduced redox partners glutathione (GSH) and nicotinamide adenine dinucleotide phosphate (NADPH). ATP was the energy behind the conserving cation transport processes counteracting the very small dissipative cation "leaks" that made the erythrocyte and muscle membranes apparently impermeable to cations. However, the discovery by Skou (Nobel Prize in Chemistry 1997) of a (Na + K)-dependent ATPase in crab nerve preparations[34,35] also jolted the red cell field. Soon, several groups in the United States led by Hoffman, Tosteson, Post, and Sachs, and by Nakao in Japan, began to explore the kinetic details of the Na, K, and MgATP dependencies of what became the Na/K pump and its quantitative relationship to the cation "leaks". This culminated in the now classic "pump-leak hypothesis"[36] and in the Post–Albers canonical cycle of the

Na/K pump or Na/K ATPase intermediates[37,38] (Section 2.4.2). Crucial was the discovery of the action of the digitalis ouabain, the only specific inhibitor known today.[39] Transforming RBCs by two step hemolytic procedures[40] into hemoglobin (Hb) free "ghosts" provided a new tool to study the forward and reverse partial reactions of the Na/K pump, providing the context for dialogue between those supporting the concept of membrane bound, ATP-dependent Na/K pumps and those explaining active transport by adsorption and desorption to the cellular gel/sol boundaries.[41,42] With the technological advances greatly benefitting from fallouts of the US Space programs, it took no more than two decades to unravel in other non-red cell systems the intricacies of the molecular mechanism of the Na/K pump, and the biophysical basis of its electrogenicity, inhibited by cardiotonic steroids.[5,43]

The seemingly intractable small erythrocyte "cation leaks" were explained/approached by the Nernst–Planck–Goldman–Hodgkin–Huxley constant field equation, and thus considered to be of an electrogenic nature and moving through the fluctuating phospholipid bilayer's fatty acid chains. In contrast to nerve and muscle, these cation leaks were not rate-limited because of higher Cl permeability in RBCs which determined the Cl distribution and membrane potential (Section 2.3.2). A superb summary of the behavior of ions and their passive permeabilities under constant field assumptions has been published.[44] Thus, the dramatic transition from the static Robertson–Davson–Danielli membrane concept to the fluid mosaic model of Nicholson and Singer was another jolt intensifying the quest to understand the underlying molecular basis of the passive, energy-independent cation leaks of RBCs. Interestingly, these leaks were amplified in RBC ghost preparations leading to systematic attempts to separate by ion flux studies so called "leaky from tight" ghost populations.[45] The prediction that the erythrocyte's high anion permeability was in part due to the existence of an electroneutral Cl exchange flux component,[46] was some 15 years later assigned to an anion exchanger inhibited by 4,4'-diisothiocyano-2,2'-stillbene-disulfonic acid (DIDS).[15,47] This anion exchanger (now AE1) first discovered in the red cell was also called band 3 protein, based on the relative mobilities of RBC membrane proteins separated by sodium dodecyl lauryl sulfate (SDS) polyacrylamide gel electrophoresis.[20,48]

Thus the time was ripe for a new look at the passive cation "leak" fluxes. Indeed, early work with cation replacement studies revealed the presence of Na-dependent K movements[49,50] and Na-independent K movement with pH optima around 7.[51] Band 3 was found to exchange Cl for HCO_3 by the Hamburger shift and other anions such as SO_4, NO_3 and I[52] with small changes in steady state volumes that can be corrected with sucrose. With this discovery studies on the effect of Cl replacement by these anions on Na-dependent and independent K movements were carried out. These studies in nucleated bird and enucleate human RBCs led to the discovery of the shrinkage-activated, electroneutral bumetanide-inhibited coupled transport of Na, K and Cl,[53–55] whose operation in a 1:1:2 stoichiometry was shown in mouse Ehrlich ascites tumor cells.[56] It did not take long to show that the Na-uncoupled K movements[49,51] were due to another electroneutral system, the furosemide-inhibited K-Cl cotransporter which was swelling-stimulated in sheep,[57] duck[53] and fish RBCs[58–61] and thiol-activated in all red cell systems studied thus far.[62–64]

In the 1980s, with the arrival of myriads of cell culture techniques and special tissue preparations such as the shark rectal gland, and the refinement of molecular biology tools, the major transport systems such as the Na/K pump, the Cl/HCO_3 exchanger, the Na-K-2Cl and K-Cl cotransporters and others such as the Na/Li or Na/H exchangers (Sections 2.4.2 and 2.5.2), found to be present in RBC, were molecularly characterized in cells and tissues derived from the central nervous system (for the Cl-dependent cation transporters see[65]) to the cardiovascular, intestinal, and renal organ systems in many species. These transporters were cloned, their chromosomal location established, and their properties such as isoform-specific ion selectivity studied after transfection of their cDNAs into cell lines with suitably low backgrounds confirming in most cases the biophysical properties anticipated from the "black box studies" in RBCs. An example is the calcium (Ca)-activated K (K_{Ca}3.1 or IK) channel that was predicted in RBCs from the K loss induced by ATP depletion,[12,13,66] and which was molecularly cloned and characterized within the last decade (Section 2.6). Its inhibition by clotrimazole and other drugs is now tested in clinical trials to attenuate its role in RBC dehydration associated with HbS disease. More recent developments are the characterization of NH_3/NH_4 transport through the Rh-associated glycoproteins (Section 2.8), a variant of similar proteins occurring in the kidney and playing a role in its nitrogen metabolism. Finally, RBCs have been recently shown to undergo apoptosis, coined "eryptosis" by Florian Lang and colleagues (see Chapter 5) thus becoming once more a simplified model to better understand complex mechanisms in nucleated cells.

From this short and obviously biased historical sweep, that by all means could not be fair to many more players and discoveries in the field of red cell ion transport, it is nevertheless evident that the erythrocyte's significance was to serve as "the" major cellular model to uncover and understand biological transport mechanisms in homeostatic control.

2.2 Shape, Surface Area (SA), Volume (V), and Membrane Constituents

The shape, size, surface area (SA) and volume (V) of normal nucleated and enucleate RBCs vary widely across the animal kingdom. Both SA and V determine the osmotic fragility of all RBCs, their susceptibility to hemolyze in solutions of lesser osmolarity, i.e. greater chemical water potential, because the plasma membrane can barely withstand small hydrostatic pressure differentials of more than a few dyne/cm. Both SA and V also strongly determine how fast gases equilibrate across the plasma membrane and are transported by Hb contained within the cells. The Stoke–Einstein equation predicts that small red cells confer upon their mammalian hosts a distinct advantage of rapid water, O_2, CO_2, and anion equilibration. Compared to current clinical data on mammalian RBCs, the early works of Ponder[67] (see also an extensive data compilation in Ref. 68) and others were done on air-dried samples. Bird, fish, amphibian, and reptile RBCs are nucleated. The ratio of the length (L) and width (W) in bird RBCs, used in exploring the Na-K-2Cl cotransporters, is closer to unity. The L/W ratio of elasmobranch RBCs, such as those from sharks and skates, exceeds by a factor of two that of teleosts (*Oncorhynchus mykiss*, trout) also used in studies of volume regulation[58–59,69–71] and with lengths up to 20 μ.[72] Amphibians have the largest RBCs, with lengths of up to 65 μ and a L/W ratio of near 2 in Amphiumidae, and have become excellent study objects for volume regulation[73] where the first patch clamps were performed to measure the membrane potential.[74,75] Among the reptiles, RBCs of crocodilians are similar in size and L/W ratio to elasmobranches, and the RBCs of snakes and turtles follow in size with an L/W ratio in most cases between 1.9 and 1.3. Thus the sizes of the nucleated RBCs are not correlated with that of the host animal. In fact, estimated cell volumes of different animal species vary from 10,700 μ^3 in the *Necturus*, to 450 μ^3 in the alligator, and ~120 μ^3 in the chicken and the catfish.[68] The osmotic fragility of nucleated RBCs from amphibians and reptiles is lower than in enucleate RBCs from several mammals[76] suggesting that other factors play a role, in addition to the SA/V ratio. Given their large cell size relative to mammalian RBCs, the cell counts in these ectotherms are low.[72]

As in nucleated RBCs, the size of enucleate mammalian red cell is not correlated with that of the host animal.[67] Thus, the mean erythrocyte diameters (ED, in μ) are 6.6 in mice, 4–4.8 in sheep, 7.5 in man, 9.2 in elephants, and 9.2 in anteaters, respectively.[68] The SA/V ratio and the osmotic fragility vary between species and within a species due to pathological disorders, primarily loss of membrane SA due to mostly genetic instabilities of the cytoskeleton (see below). A typical interspecies variation is an SA/V ratio of ~1.6 in normal human RBCs (SA = ~163 μ^2, V = ~88–96 μ^3),[67–77] and ~2 for sheep red cells (SA = 67 μ^2, V = 30 μ^3).[67] The mean osmotic fragility (50% hemolysis, MOF) is not similar in man and sheep. In man, the MOF is around 150 mOsM measured in saline[78] and ~185 mOsM in high potassium (HK), as opposed to 195 mOsM in low potassium (LK) sheep RBCs.[79] Based on the van't Hoff–Morse equation, this translates into a differential of almost 3 Atm compared to human RBCs. Most likely, the SA of sheep RBCs is an overestimate because under the microscope they appear more spherical than the biconcave human RBCs, or, as in amphibian RBCs,[72] other factors yet to be determined are at play in defining their higher MOF. The difference between the HK and LK sheep RBC is explained by the smaller volume of the former (30 μ^3) compared with the latter (33 μ^3),[80,81] and remains at variance with an earlier report.[36] That it is not solely the SA/V ratio determining the MOF can be realized from a comparison of a decreased osmotic fragility coupled to a reduced K content and twice higher K-Cl cotransport (KCC) activity in BXD-31 mouse RBCs as opposed to their parent DBA/2J strain, which are under apparent control of a single genetic locus.[82] This work confirmed earlier observations of the control of cell volume by the "rol" (resistance to osmotic lysis) gene in mice.[83] However, in sheep, where KCC is up to ten-fold higher in LK than in HK RBCs,[84] this correlation was inverted, i.e. the former had a higher MOF than the latter.[81] Determination of the quantitative trait loci that assess the important SA and V parameters and the tightness of their correlation with the MOF are certainly in the right direction.[85]

Exposure of normal human RBCs to moderate anisosmotic solutions causes transitions from the normal discocyte (~88–96 μ^3) to hyposmotically swollen spherocytes (~140 μ^3) with subsequent hemolysis, or hyperosmotically shrunken echinocytes (~70–80 μ^3). Due to the extra SA, the red cell volume attained just before hemolysis is larger than the prehemolytic volume V_h[67] calculated from its diameter. Shape transformations at constant RBC volume have been ascribed to changes in relative surface energies between the two halves of the lipid bilayer as predicted from the bilayer couple model.[86] The erythrocyte shape changes not only in anisosmotic media, i.e. as a function of initially different water chemical activities across the plasma membrane, but also as a response to media pH variations and plasma composition. Two excellent dispositions of the structural and medium determinants of erythrocyte shape have been published.[87,88] Permanent spherocyte-discocyte-echinocyte transitions to shapes with decreased SA/V ratios such as spherocytes, stomatocytes, and elliptocytes and their combination, or shapes with increased SA/V ratios such as codocytes, echinocytes, and xerocytes are pathologic and occur in various clinical disorders (see Chapter 7 on anemias).

Human RBC membranes are composed of some 40–50% protein and the remainder is lipid partitioned into about 25% cholesterol wedged in between the fatty acid acyl chains of the diacyl phospholipids phosphatidyl choline, phosphatidyl ethanolamine, and sphingomyeline (75%). The nature and distribution of membrane lipids, especially the asymmetrization of the membrane due to unequal and inwardly favored partition of negative phospholipids by phospholipid flippases countering scramblases, is an important current research area for our understanding not only of normal volume maintenance but also of destruction by eryptosis, i.e. apoptosis (as discussed in Chapter 5 by Lang, Huber and Föller). An explanation for the lower passive ionic permeability of ruminant RBCs is that sphingomyelin, binding cholesterol tighter than the remainder phospholipids, is by far in excess in their membranes. The lipid physicochemical basis for this phenomenon has been summarized by two excellent chapters.[87–89]

The pathologies underlying the permanent shape transitions mentioned in the preceding paragraph are primarily at the membrane level, i.e. genetic deficiencies in its cytoskeletal and transmembrane proteins (see fluid mosaic models in Section 2.1). Figure 2.1 depicts a convenient and relatively timeless cartoon of the sub-phospholipid bilayer membrane protein assembly in the erythrocyte,[90] and the proteins appearing in the cartoon are listed in Table 2.1 with respect to their nomenclature, mass, amino acids, chromosomal assignment, and role (resource input from Ref. 90, 91) and others. For several other minor membrane proteins supporting the bilayer structure, please consult.[92–94] In support of the erythrocyte membrane bilayer and shape, the spectrins heteroligomerize, interact with the internal phospholipid-charged head groups and crosslink *via* 4.1 proteins, 4.2 proteins, and

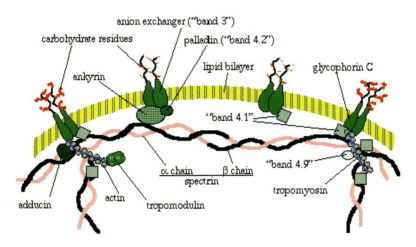

Figure 2.1 Cartoon depicting the sub-phospholipid bilayer membrane protein assembly (with permission to republish from Ref. 90). The nature of the individual protein components and their role in the context of this figure are given in Table 2.1.

Table 2.1 Major human erythrocyte membrane proteins with current nomenclature, size, amino acids, chromosomal (chr) assignment and role. Data assembled based on resource input from Ref. 90, 270, 882, 903) and others (n.a. = not available).

SDS-Pagel "band"	Common name	Molecular mass (kDa)	Amino acids	Chr	Copies 10^4/cell	Role in erythrocytes
1	A Spectrin	240	2429	1	24	Bilayer support
2	β Spectrin	220	2137	14	24	Bilayer support
2.1	Ankyrin	215	1881	8	n.a.	Cross linker
2.9	α/β Adducin	103/97	737/726	4/2	3	Stabilizer
3	Anion Exchanger (AE1)	95	911	17	100	Cl/HCO$_3$ exchange
4.1	Protein 4.1	78	588	1	20	Stabilizer/cross linker
4.2	Palladin	72	691	15	n.a.	Stabilizer/cross linker
4.9	Dematin	50	383	8	4	Cross linker
5a	β-Actin	43	375	7	50	Stabilizer/microfilament
5b	Tropomodulin	41	359	9	3	Binds tropomyosin
6	GAPDH	35	335	12	n.a.	1,3 DPG production
7a	Stomatin	31	288	9	10	Adaptor protein?
7b	Tropomyosin	28	239	1	8	Microfilament with actin
7c	Aquaporin	28	269	7	100	Water transport
PAS1/2/3	Glycophorin A/B/C	36/32/20	131/128/72	4/2/4	n.a.	Blood groups M,N, & Ss

β actins to the glycophorins, and *via* ankyrin to the AE1 or band 3 proteins, the latter constituting 10% of the protein mass.

Other proteins such as protein 4.2 (palladin), 4.9 (dematin), tropomyosin, and β-actin through tropomodulin, congregate and attach at key crossover points with the αβ-spectrins and membrane-inserted glycophorins. Thus, this sub-bilayer protein network, with its principally non-covalent Ca- and phosphorylation-mediated interaction, guarantees the pliability of the membrane that is necessary during the constant erythrocyte motion across the vascular bed. This requires instantaneous shape changes at constant SA/V ratio from the discocyte to bullet-like cell shapes during single file movement through capillaries.[95] The functionally important ~174 kDa large Rhesus complex (not shown here), consisting of two ~30 kDa RhD/CcEe peptides and two ~50 kDa Rh-associated glycoproteins (RhAG), associating with minor proteins such as intracellular adhesion molecules (ICAM, earlier called the Landsteiner–Wiener substance), and the CD47 or integrin-associated protein (IAP), in addition to AE1, will be detailed in Section 2.8.3.

Mutations in these membrane-supporting proteins show their importance for the functional stability of the erythrocyte that undergoes permanent pathological discocyte-spherocyte transition in hereditary spherocytosis or elliptocytosis (deficiencies in spectrin, ankyrin, AE1, 4.1, and 4.2 proteins and the glycophorins A and C). Chapter 7 addresses the clinical details of these erythrocyte membrane disorders.

2.3 Water, Anions, Cations, and Other Solutes

2.3.1 *Water*

The RBC water content varies between 60 and 70%, the remainder made up primarily of the large impermeable solute hemoglobin (Hb). In human RBCs, most of the water is osmotically active, i.e. a plot of the water content in terms of kg/kg dry solids *versus* the inverse of the osmolarity is a linear function with the slope yielding the osmotically active cellular ion and other solute content. Electron probe analysis shows that the RBC Hb concentration is an inverse function of the Na+K content.[96] In RBCs permeabilized by the nystatin method,[97] the acid-base titration is not dependent on cell volume suggesting at constant pH the Hb net charge is independent of the protein concentration, and a model has been proposed to account for the cell volume dependency on osmotic pressure.[98]

For osmotically balanced RBCs, water has been always assumed to be at equilibrium, i.e. its chemical activity, μ_{H2O}, is equal on both sides of the plasma membrane which cannot sustain hydrostatic pressure gradients. Indeed, the diffusional water permeability, P_d, of RBCs is 3×10^{-3} cm/sec[99] whereas the filtrational permeability coefficient, P_f, is 2×10^{-2} cm/sec.[100] With the discovery that mercury (Hg) inhibited P_f to a value closer to P_d,[101] the long-held notion that water equilibration was purely bilayer-mediated was modified to one including > 60% water flux through a Hg-sensitive protein, aquaporin 1 (Section 2.8.1).

2.3.2 Anions

A general paradigm is that Cl, the major RBC anion, appears to occur at similar concentrations throughout the vertebrate species. For example, data (in meq/L) from Ref.(68) indicate: man (78), cat (84), dog (87), dolphin (58–83), horse (85), monkey (78), rabbit (80), rat (82), sheep (78), and cow (84). The plasma or serum Cl concentrations of man and the animals quoted vary between 108 and 112 meq/L. RBCs from reptiles exhibit slightly higher (114–133) and amphibians and fish lower (70–85), $[Cl]_i$ values (the only exception found is the goose fish with 4 meq/L). From the $[Cl]_i/[Cl]_o$ ratio of close to 1.4, i.e. close to equilibrium, one can make at least four predictions:

1. From the Donnan distribution of small ions, intra-erythrocyte macromolecules, i.e. mainly hemoglobin, for all these species must have average of negative charge equivalents equal to the missing Cl electrons.
2. The Nernst–Planck equation permits calculation of a Cl equilibrium potential (E_{Cl}) of about -11 mV, which is close to the membrane potential, V_m, measured for RBCs of many species throughout the animal kingdom, a value that was experimentally confirmed by elegant direct electrode measurements in *Amphiuma* RBCs.[74]
3. A high membrane Cl conductance must exist. However, in his 1959 classic study, Tosteson[46] showed by [36]Cl measurements, the partitioning of Cl movements into a fast performing electrically neutral Cl exchange and a conductive Cl pathway. This double paradigm was confirmed in *Amphiuma* red cells[74] where >89% of the Cl tracer flux was electrically silent[75] and theoretically and structurally verified by the AE1 anion flux work.[15]

4. Commensurate with the Donnan ratio of Cl anion and proton concentrations (commonly given in []), $[Cl]_i/[Cl]_o = [H]_o/[H]_i$, acid titration of the negative charges of Hb within a RBC suspension leads to a $[Cl]_i/[Cl]_o \cong 1$, V_m depolarization, water entry, cell swelling, and hemolysis whereas alkalinization results in a $[Cl]_i/[Cl]_o < 1$, with cell shrinkage and V_m hyperpolarization. This fact has been widely utilized to pH-clamp RBCs with anion exchange inhibitors to study the effect of intracellular pH on transport processes, for example in the case of K-Cl cotransport.[102]

Assignment of the anion exchange function to band 3 protein, later named AE1, followed.[103–105] Today, after the concept of anion slippage, it is accepted that the Cl conductance is mediated by the same protein *via* a conductive channel.[106] The distribution of other non-Cl anions such as HCO_3, and mono- and divalent phosphates follows that of Cl. There is less than 1 mM inorganic phosphate in the erythrocyte. Most acid soluble phosphate is present in the form of ATP and the various diphosphoglycerates which, compared to the ~50 meq monovalent negative Hb charges/ (L cell water), play yet a minor role in the outward distribution of Cl. Separately, phosphate utilizes cotransport with Na to enter the cell.[107] Moreover, Tosteson's second paradigm of an anion exchanger has been extended to the experimental application of anions of the Hofmeister series not normally present in blood such as sulfates (SO_4, SO_3), nitrates (NO_3), iodide (I), and bromide (Br), and thus became later the basis for detecting the Cl-dependencies of the electroneutral cation cotransporters (CCC), since unlike AE1, these systems are only transporting Cl[64,108] (see Sections 2.5.3 and 2.5.4).

2.3.3 *Monovalent cations*

Because of the paradigm that all nucleated mammalian cells have a high K and low Na concentration due to the presence of the Na/K pump, the surprising inter-species variability shown by the RBC cation composition is easily overlooked. Obviously, this fact is related to widely different active uphill transport of K by the Na/K pump, details of which will be discussed further in Section 2.4.2. More recent concentrations[109] in man for K and Na are $[K]_i = 102$ and $[Na]_i = 6.2$ meq/(L red blood cells) which, based on the provided 72% water value, translates into 142 meq $[K]_i$ and 8.3 meq $[Na]_i$/(L cell H_2O), respectively. Thus the $[K]_i/[K]_o$ ratio is ~28 and the $[Na]_o/[Na]_i$ ratio ~18, highlighting the direction of the electrochemical driving forces for the two ions. In the extended Nernst–Planck equation — the Goldman–Hodgkin equation — ions and their permeability coefficients form a product. As discussed in Section 2.3.2, the anion permeability is high and the red cell is close to Cl equilibrium, therefore these high cation ratios do not affect the membrane potential measured by electrodes or by merocyanin dye,[110] commensurate with RBC membranes being considered cation impermeable. Thus major changes are only evoked by introducing the highly K-selective valinomycin,[111] or Na-selective hemisodium,[112] in essence converting the membrane from a Cl toward a K or Na electrode. The Ca ionophore A23187 is often used to force Ca to enter the red cell by far in excess of the Ca pump capacity. Under these conditions, ATP is depleted, and a Ca-activated K channel opens thus hyperpolarizing V_m. In essence, this is the effect Gardos reported when depleting human red cells of their ATP by metabolic poisoning with iodoacetamide.[110] The significance of this mechanism will be discussed in Section 2.6. In membrane disorders such as hereditary spherocytosis, the passive cation permeability is elevated reducing both the $[K]_i/[K]_o$ and the $[Na]_o/[Na]_i$ ratios. This aspect will be dealt with in Chapter 7.

The paradigm of a high cellular $[K]_i/[Na]_i$ ratio (68), correlated with high Na/K pump activity, in man[18] applies to many mammals, such as horse,[9] rabbit,[7] mouse,[9] monkey,[6] raccoon,[5] and dolphin.[8] However, at least since the reports of Evans of a genetic cation polymorphism in sheep,[113,114] it is known that there are animals from the same species with $[K]_i/[Na]_i$ ratios of 9 as well as 0.1. A similar dimorphism, though not as sharply demarcated, is seen in other ungulates such as cattle and goats.[115,116] A genetic association exists between the HK/LK cation polymorphism and blood group surface antigens in sheep.[30,117] HK red cells contain the functionally silent M antigen, and LK red cells the functionally significant L antigen. Based on the action of specific allo-antibodies, a subset of the L antigen, L_p, is a functional repressor substance of the Na/K pump in LK sheep and goat RBCs and thus explains mechanistically the low K/Na ratio in the cells of these animals.[118] The other subset, L_1, is an activator of K-Cl cotransport[119] (details in Section 2.5.3.2.4). The low K/Na ratio is not at all rare and also occurs in all *Felidae* with 0.06 and *Canidae* with 0.07 (species which rarely have high K/Na ratios due to apparent man-induced breeding selections); in the elephant seal with 0.07; in the Japanese[120] and North American black bears[121,122]; and in others. Erythrocytes from these species with diverse K/Na ratios have been extensively studied in terms of their predictably lower or even absent Na/K pump activities, and their volume regulation in general, a topic to be dealt with in Section 2.5.

2.3.4 *Divalent cations*

In comparison to the defining presence of the alkaline metals K and Na, the next major cation in erythrocytes is the earth alkaline metal magnesium, Mg (~3 mM/L cells, dependent on the species). As cofactor with ATP, it is essential for the canonical cycle of the Na/K pump and regulation of many glycolytic enzyme functions such as hexokinase, phosphofructokinase, aldolase, phosphoglycerate kinase, and pyruvate kinase.[123] Given a negative V_m, a driving force always exists for Mg to passively enter RBCs. The free (ionized) Mg concentration in human red cells is ~0.4 mM.[124] About 0.3–0.5 mM ionized Mg has been determined by the divalent ionophore A23187 equilibration technique and NMR in sheep erythrocytes.[125] The free Mg concentration oscillates with oxygenation,[126] and appears to be maintained by an ATP hydrolysis-dependent electrogenic Mg/Na exchange.[127,128] Due to its complex with its major buffers ATP (K_d = 0.08 mM), 2,3 DPG (K_d = 3.6 mM), and their binding to and de-binding from deoxygenated and oxygenated Hb, respectively, the free Mg concentration oscillates constantly.[129] Mg levels, decreased by an elevated membrane permeability induced by drepanocyte (HbS or sickle cell) formation,[130] have been shown in HbS cells.[131] An increased Mg/ATP ratio reduces Na/K pump activity and hence causes cellular Na load[130]; and Mg elevation by A23187 reduces KCC activity in low K sheep red blood cells.[132,133] Conversely, as a consequence of deoxygenation-induced permeabilization of HbS cells to and loss of Mg,[130] KCC activity increases.[134] An excellent review of the

Mg metabolism in different species and pathophysiologies has been published.[126]

In contrast to Mg, only a very small fraction (~10 nEq/L) of the total available earth alkaline metal calcium, Ca (0.018 mEq/L) occurs in ionized form REF. 135). Ca ions play an important role in the dynamic deformation of the red cell membrane.[136] In turn, membrane deformation may open a transient Ca permeability[137] with Ca-calmodulin-induced activation of $K_{Ca}3.1$ and cellular dehydration. The red cell's Ca buffering has been partitioned into at least two fractions to be further identified: an "α" component with high Ca binding capacity and low affinity, and a low Ca binding capacity component of intermediate Ca affinity[138] kept at its low level by the Ca pump (Section 2.4.2). Elevation of free Ca beyond the "magic" 10 nM level causes ATP depletion by the Ca pump, inhibition of the Na/K pump[139] and activation of K loss *via* $K_{Ca}3.1$ channels.[9,66] Because of the latter's high K conductance, loss of K, Cl and water should affect the MOF. This effect may be offset by elevated Ca itself increasing progressively the MOF.[140] A concise review of Ca homeostasis in red cells has been published.[141]

The human erythrocyte contains nanomolar traces of other elements, such as zinc, manganese (elevated in sheep due to high Mn content of pasture grass, and present in Asian teas), and rubidium. For details on RBCs transport of trace metals (Mn, Zn, Fe, Cr, Cd, Co & Cu) see Ref. 142.

Basic biochemical and physiological aspects of hemoglobin, the main erythrocyte solute, are dealt with in Chapter 1. In Sections 2.4–2.8, nine major erythrocyte transport systems will be reviewed. For the reader's convenience, Table 2.2 summarizes some of the key properties of these transporters and will be referred to throughout these sections.

2.4 Chemi-Osmotic Transduction: The Na/K and Ca Pumps

2.4.1 *Introduction*

The Na/K pump distinguishes all moving animals from plants, their roots relying on osmosis only. By the end of the fifth decade of the past century the search for the principle underlying conservative processes maintaining a high K/Na ratio in erythrocytes and other cells had intensified

Table 2.2 Major human erythrocyte transport pathways with current nomenclature, mass, amino acids, chromosomal (chr) assignment, and role. Data collected from references quoted in Sections 2.4–2.8.

Transporter	Short name (isoforms)	Molecular mass, kDa	Amino acids	Chr	Oligomers	Copies 10³/cell	Role
Na/K ATPase	Na/K Pump α (1,3), β(1-3)	112 + 55	~1020	α1,1 α3,19	($\alpha + \beta$)-Heterodimer	250–400	Na efflux/K influx
Ca/H ATPase	PMCA (4)	110–130	1000–1200	12/3/X/1	Homodimer	700?	Ca efflux/H influx
Na/H exchanger	NHE (4)	99	800?	1/2/5/2	Homodimer	n.a.	Na influx/H efflux
Na-K-2Cl Cotransporter	NKCC1	~132	1212	15	Likely	n.a.	Na-K-2Cl influx
K-Cl Cotransporter	KCC (1,3 & 4)	120/130/135	1085/1083/1150	16/15/5	Likely	n.a.	K-Cl efflux
Gardos (IK) Channel	IK, K_{Ca}3.1, sK4	~50	425	19	Unknown	120	K efflux
Anion exchanger	AE1	95	911	17	Homodimer	106	Cl/HCO_3 Exchange
Water channel	Aquaporin 1, 3, 9	28	269	7	Tetramer	105	H_2O, CO_2 glycerol
Rhesus Proteins	RhAG	~50	409	6	Dimers	105	NH_3 or NH_4, CO_2

when Jens Christian Skou (Nobel Prize 1997) discovered in crab nerves an enzyme that hydrolyzed ATP in the presence of Na and K.[34,35] In electric eel tissue the phospho-intermediate was not phosphatidic acid as originally thought.[143,144] Indeed, the free energy available in ATP is just sufficient to oppose the electrochemical Na and K gradients across the plasma membrane, and to keep Ca ions out of the cell *via* the Ca pump.[145,146] The significance of the erythrocyte as a model to study the sidedness of this extraordinary discovery became quickly apparent for five major reasons:

1. The two-step red cell hemolysis procedure[40] provided large quantities of Hb-free ghosts.
2. The cellular ionic composition could be changed with reversibly binding mercury benzoate derivatives[147] or with peptide ionophores such as nystatin,[97] developed simultaneously in artificial lipid membranes.
3. Based on experiments to reconstitute ghosts to volume commensurate with RBCs,[148] ghosts could be kept open at low temperature in the presence of chelators and loaded with ATP, ions, and solutes and resealed at elevated temperature to restore ionic permeability. Though not identical, these ghosts are close enough to the intact red cell to obtain good signal/noise ratios of conservative ion pump/ dissipative ion leaks.[45,149]
4. Inside-out vesicles as opposed to outside-out ghosts[150,151] provided access to the membrane side that contained the phosphorylation sites for ATP derived from glycolytic metabolism of glucose. Thus ATP became the *deus ex machina* that converts scalar chemical energy into vectorial transport of Na and K through the Na/K pump on one hand, and of Ca, through the Ca/H pump, on the other.
5. Microsomal Na/K ATPase was fused into red cell ghosts.[152,153]

These five tools, together with the availability of the specific inhibitor ouabain, permitted the demonstration of the various Na/K pump modes of the Post–Albers sequential canonical model.[37,38] Absence of such an inhibitor (vanadate or lanthenum inhibit the PMCA by different mechanisms but are highly non-selective compounds not touching the power of cardiac glycosides in the Na/K pump) explains why work on the Ca pump followed much later. Less than 20 years thereafter, the fundamental understanding of the operation of the erythrocyte Na/K pump was further expanded by research on Na/K ATPase-rich microsomal kidney preparations from pig, sheep, and rabbit, and from brain tissue and the rectal gland of the shark, allowing further dissection and application of the partial reactions seen in RBCs.

2.4.2 The Na/K pump

2.4.2.1 Structure

The Na/K pump (Table 2.2, and a cartoon enhancing visually our understanding of its subcomponents is given in Fig. 2.2) or in its detergent-solubilized form, the plasma membrane Na/K ATPase (ATP phosphohydrolase, enzyme classification EC3.6.3.9), is a heterotrimer containing:

1. A 112 kDa α subunit[154] with the catalytic site;
2. A 55 kDa glycosylated β subunit[155] that is required for proper membrane insertion and earlier thought to play a role in the binding of and inhibition by ouabain[156]; and
3. A γ peptide which serves as a pump regulator.[157]

In man, there are 4 α subunits, α1 and 2 encoded by genes on chromosome 1, α3 on chromosome 19, and α4 on chromosome 13.[158,159] These alpha subunits were originally and partially uncovered by Sweadner as α + (now α1) and α, now α2.[160] The human α1 and α3, occurring in the red cell, map to chromosome 1 and 19, and β to chromosome 1.[158] Human RBCs, like those of sheep, have the α1 subunit-β1 dimer[161] and more recently, the additional presence of α3 (but not α2), and of the β2 and 3, and the modulatory γ subunit have been shown by molecular techniques.[162] There are three β subunits (β1, 2, and 3)[163] and two γ (γa and γb) isoforms.[164] Depending on the source and isoforms, the Na/K ATPase α subunit has 1016,[165] 1023[166] or 1028[167] amino acids, is non-glycosylated, and crosses the membrane ten times with the N and C termini in the cytosol. The γ subunit or FCYD2, composed of 66 or 68 amino acids and

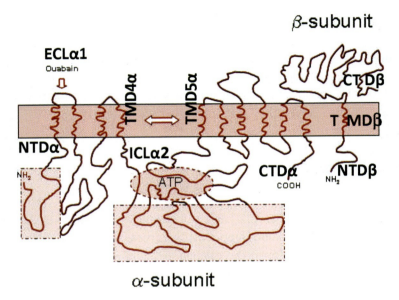

Figure 2.2 Cartoon of the molecular structure of α and β subunits, the αβ heterodimer or protomer of the Na/K-ATPase pump. The large catalytic α subunit with some 1020 amino acids (see Table 2.2) has ten conserved transmembrane domains (TMDαs) with intracellular amino-terminal domain (NTDα) and carboxy-terminal domain (CTDα). TMDα 4 and 5 (separated by double-headed arrow) are involved in ion binding. The large second intracellular loop (ICLα) 2 between TMDα 4 and 5 contains the sequence for ATP binding and hydrolysis (shaded oval). Most α isoform-specific sequences are found at the NTD and in parts of ICLα2 (shaded rectangles). Cardiac glycoside binding is determined by the sequence of first extracellular loop (ECLα1) of the enzyme (arrow). The β subunit glycoprotein, important for intracellular transport and membrane assembly,[914] has only one TMD(α) with the NTDβ in the cytosol and the CTDβ extra-cellularly. The more recently elucidated γ subunit, making up the prototrimer, has been omitted in this cartoon (figure and legend modified from Ref. 915).

localized to chromosome 11q23, is a member of the FXYD family of transmembrane peptides[157,168] and occurs in the kidney.[169]

The Na/K ATPase belongs to the P-type ATPases and is phosphorylated by ATP at asp 369 during its catalytic cycle.[170,171] The cation binding sites are located within or close to the first transmembrane domains (TMDs) and extracellular loops[172] with ser 775 located in the cation binding pocket.[173] Likewise, substitution experiments suggest ouabain binding involves amino acids in the "H1–H2 domains" and the "H5–H6" and "H7-H8" membrane spanning regions[174,175] with the amino acids making up the ouabain binding site much conserved throughout evolution.[43] While Skou's group elucidated early much of the biochemical mechanisms of the non-red cell Na/K ATPase commensurate with the Post–Albers operational scheme,[34,35,176–185] this portion of the review focuses on the evolution of our concepts of the Na/K pump and its partial reactions in RBCs. This is followed by a section on its current biophysical operational realities.

2.4.2.2 *Function in human erythrocytes*

Human RBCs, depending on the measurements, possess some 250–400 ^{3}H-ouabain binding sites and hence pumps per cell.[186–188] At an average K influx rate of ~2 mmol/(L. original cells × h)[187] the pump cycles close to 100 Hz, and exchanges 3 Na for 2 K per ATP/cycle, i.e. is electrogenic. Indeed, the electrogenic contribution of the red cell Na/K pump to V_m has been estimated as 0.1–0.2 mV.[74] There is evidence for a yet undisclosed red cell membrane ATP pool to which the Na/K pump or the glycolytic pathway have access.[189–191] At the exofacial aspect of the Na/K pump, Na inhibits,[192] whereas rubidium, cesium, ammonium, and lithium stimulate K influx as predicted for 2 K sites.[193,194] Thus the $K_{0.5}$ in the presence of 128 mM $[Na]_o$ is 1.4 mM $[K]_o$ in human,[192,195] and in sheep RBCs 3 mM and 0.6 mM in the presence and absence of $[Na]_o$ respectively.[196] At the endofacial aspect, 3 Na bind with the same affinity to three sites[197] and can be replaced by 3 Li.[198]

The Post–Albers canonical cycle (Fig. 2.3) predicts the sequential binding of 3 cytosolic Na to E1-ATP, the intermediate enzyme form that delivered K at the cytoplasmic side, followed by phosphorylation at asp 369 to the high energy phosphate intermediate (E1~P), a step during which the 3 Na are occluded for the electrogenic transport to the external face (E2-P). More precisely, the Na_3E1~P to Na_2E2-P transition releasing the first Na ion is the rate-limiting step.[199] Subsequent binding of two external K ions causes autohydrolysis of E2-P, K-occlusion, translocation

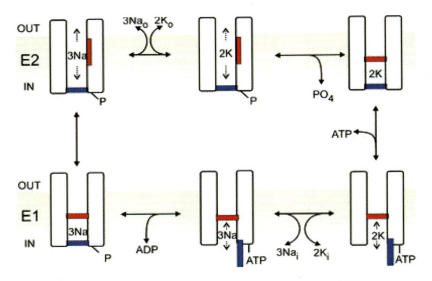

Figure 2.3 Cartoon of the alternating-gate model of the Post–Albers[37,38] transport cycle of the Na/K pump represented as an ion channel with two gates, a red extracellular-side gate (labelled "out"), and blue cytoplasmic-side gate (labelled "in"), which, consistent with the ping-pong model,[197,201,203] open alternating, but not simultaneously. The E2 (upper row; extracellular-side gate may open) and E1 (lower row; cytoplasmic-side gate may open) states are shown. Occluded states, with both gates shut, follow binding of two external K (top right) and of three internal Na and subsequent phosphorylation (bottom left). ATP acts with low affinity to accelerate the opening of the cytoplasmic-side gate and concomitant K deocclusion, and with high affinity to phosphorylate the pump. (Figure with permission and legend adapted from Ref. 275).

to, and release at the cytoplasmic side upon low affinity ATP binding. The tacit proposition of this model is the existence of five modes in which the Na/K pump operates. Indeed, the work with RBCs utilizing the five techniques listed in Section 2.4.1 led to the initial concepts that the Na/K pump can operate in full, and in two half cycles depending on the thermodynamic driving forces, i.e. that ion occlusion and delivery at the opposing membrane side alternates. After considering a flip-flop model of the Na/K pump in which two protomers would be 180° out of phase,[200] a more recent model is that the Na/K pump as protodimer operates by a ping-pong mechanism[197,201] and not by simultaneous site occupation.[202] An excellent up-to-date review comparing the kinetics of the various operational mechanisms of the Na/K pump has been published by John Sachs.[203]

The landmark red cell paper by Garrahan and Glynn in 1967[204] showed for the first time the presence of one 3Na/2K exchange cycle per one ATP hydrolyzed, and the existence of a Na/Na exchange without ATP hydrolysis in K free media inhibited by oligomycin which, like N-ethylmaleimide, slows down or blocks the canon in the E1~P to E2-P conformer transition.[205] This 1/1 Na/Na exchange did not exist in the presence of K_o.[206] When the external monovalent alkali cations were replaced by choline, a ouabain-sensitive Na efflux was discovered, the uncoupled Na efflux as the third mode of the Na/K pump[207] that was later shown to be coupled to the hydrolysis of ATP.[208–210] This uncoupled Na efflux is electroneutral and has been associated with anion cotransport[211] as well as with P_i efflux deriving from the hydrolysis of ATP.[212] In another follow-up paper, ghosts were used with high K, low Na and low [ATP]/[ADP][P] ratios but with high Na_o and $32P_i$ to show, through net ATP synthesis from ADP[210] and hence increased [$-\Delta G$], the reversal of the Na/K pump.[213] Following this lead work, a ouabain-sensitive K efflux mode was found associated with Na/K pump reversal.[214,215]

In IOVs, the ATP produced by the glycolytic enzymes appears in a compartment or membrane pool close and readily available to the Na/K pump.[189–191,216,217] Besides Na or K, the Na/K pump transports Li.[198,218] Side-specific effects of Na and K were studied in IOVs at very low ATP concentrations.[219,220] Based on a rapid non-hydrolytic disappearance of the phosphoenzyme in human ghosts, Blostein independently concluded there was a reversal of phosphorylation due to ATP resynthesis.[219] Furthermore, in IOVs from human RBCs, a proton-activated ATP-dependent net K efflux (using Rb) was observed in the absence of Na, suggesting uncoupling of inward K transport from the obligatory Na efflux occurring in the intact cell.[221]

Thus, by the 1970s, the following five operational modes of the RBC Na/K pump were known: a Na/K exchange, a Na/Na exchange, an uncoupled Na efflux, a K/Na reversal mode, and an uncoupled K efflux or K/K exchange mode. The implications are that the erythrocyte Na/K pump is primarily in the Na/K exchange mode under physiological conditions, and at any time and condition may operate in one of its other canonical modes.

For the normal Na/K exchange mode, the apparent external affinity is 1–2 mM $[K]_o$,[204] while that for internal Na binding to three identical non-interacting sites[197] was close to 10 mM Na_i in the absence of K_i which alters the Na-affinity and shifts the internal Na activation curve from a hyperbolic rectangular to a sigmoid function.[197] The maximal turnover of the Na/K pump usually refers to its activity at $V_{max,max}$, i.e. when both external and internal cation binding sites are saturated.

After the groundbreaking work on red cells, more detailed information on the order of addition and release of the occluded K and Na ions was reported not in the red cell with its few hundred pump sites, but rather in high Na/K ATPase activity preparations such as the pig and dog kidneys.[222–224] However, Patricio Garrahan and colleagues continued to explore active transport in red cells by studying the K-loading aspect of the Na/K pump with p-nitrophenylphosphate generating a lower energy E-P supporting the K and Na half cycles but not the full Na/K cycle of the pump.[225] Further work with this compound is limited since it also changed the passive monovalent cation permeability.[225] Using pH-sensitive dye-loaded and DIDS-pH clamped human red cell IOVs, a strophanthidin-sensitive increase in pH was observed after addition of both ATP and Na,[226,227] pointing to a Na/K-ATPase-mediated Na/H exchange in the absence of K_o, and a pH decrease following addition of ATP, provided the vesicles contained K (i.e. K_o).[228] These Na/H and K/H exchanges are reminiscent of the uncoupled Na and K flux modes of the Na/K pump seen earlier in ion-permeabilized red cells, and of Skou's coupling of the ATP-3Na-form with deprotonation and the P_i-2K-form with protonation and thus of a proton-dependent conformational change in the enzyme.[229]

2.4.2.3 *Function in cation dimorphic ungulate red cells*

In Section 2.3.3, a large group of animal families was shown to have RBCs with a low K/Na ratio raising the question about the mechanism by which their cells maintain steady state volumes. Using high (HK) and low K (LK) sheep RBCs, Tosteson and Hoffman in their seminal 1960 paper established for the first time a model ("the pump-leak model") correlating in parallel Na/K pump fluxes with cation leaks.[36] The model, predicting that HK cells are leakier to Na than LK cells and LK cells more to K ions than the former and that the Na/K pump works in parallel to achieve this steady state, defined membrane physiology for years. In both HK and LK sheep RBCs, the observed Na/K pump stoichiometry was 1:1 and thus deviated from the 3Na:2K ratio found later in other red cells.[230] Twenty years later, it was shown that the increased K leak in LK cells was due to K-Cl cotransport (KCC, see Section 2.5.3.2) which, in contrast to HK cells, was active even under isosmotic conditions.[57,62] The prediction of an increased Na leak in HK rather than in LK red cells is not readily commensurate with the finding that LK RBCs are ~10% larger than HK RBCs and that their MOF is about 10% higher than in HK cells.[80,81] The model obviously needs revisiting, perhaps using a comprehensive approach similar to that developed by Lew and Bookchin in 1991 to explain volume maintenance in human RBCs.[231] Nevertheless, the "pump-leak" concept was part of a larger effort to explain the physiological basis of the cation dimorphism in ungulate red cells. In a follow-up study, Philip Hoffman and Tosteson[196] showed different internal activation kinetics for both HK and LK RBCs in reversibly para-chloromercuribenzoate (PCMBS)-treated, cation-equilibrated sheep red cells. Whilst HK exhibited a gradual Na/K pump increase with Na loading, like human RBCs studied earlier by Garrahan and Glynn,[204] this was not the case for LK cells whose ouabain-sensitive K influx only appeared at >70 mM $[Na]_i$ and behaved exponentially, reaching ~16% of the levels seen in HK cells only at 100% internal Na. In contrast, in the presence of normal $[Na]_o$, the external K activation curve was sigmoid and the apparent affinities for external K for both cells were about 3 mM. When Na_o was removed, the $K_{0.5}$ for HK was 0.6 and that for LK 0.2 mM $[K]_o$. This now classic paper established that HK and LK cells were qualitatively different. Since HK and LK sheep RBCs have ~100 and 40 ^{3}H-ouabain binding sites and hence active pumps, respectively, the turnover of each pump phenotype remained different thus affirming the Hoffman and Tosteson findings.[188] This conclusion also applies to an earlier report of a five-fold difference in the ^{3}H-ouabain binding sites and seven-fold Na/K pump activity difference between HK and LK red cells.[232]

In 1966, Rasmusen and Hall[30] reported that HK and LK RBCs from some sheep possessed the M antigen, which was of a proteinaceous nature and was not correlated with different K/Na ratios in the LK cell.[233] Ellory and Tucker[31] first showed the presence of another antigen, L, on all LK sheep red cells against which was produced an antiserum that stimulated the Na/K pump. Whilst the LK gene is dominant, the HK gene is autosomal recessive[113] with the M and L antigens behaving in a codominant pattern. Antibodies against the M antigen had no effect on pump and leak fluxes in either HK or LK cell.[118] However, anti-L modified the internal Na activation kinetics of the Na/K pump by shifting the sensitivity of the LK pump to lower $[Na]_i$,[118] and its effect was greatest at higher K_i (replacing Na) in LK RBCs.[116] Since the number of Na/K pumps was not changed,[188] their turnover at its highest $[Na]_i$ was not only corrected but was also higher than in HK cells, again affirming the fundamental difference between HK and LK pumps. Several publications thereafter from other laboratories basically confirmed this finding.[232–234] The L antigen, obviously increasing the affinity of the internal Na loading site for K over Na, is now called the L_p antigen in contrast to the L_l antigen that is associated with K-Cl cotransport in LK cells (see Section 2.5.2). There are $\sim 10^3$ M-, and $<10^3$ L_p- and L_l-antigens per HK and LK red cell, respectively, thus exceeding the number of Na/K pumps in both cells.[235,236] To affect the Na/K pump, the antibody anti-L_p requires the bivalent $(Fab)_2$ fragment,[237] although the monovalent Fab-IgG will also work.[238]

Cattle and goat also possess RBCs exhibiting a cation polymorphism although the distinction between HK and LK cells is not as clear as in sheep. In these species, the L_l antigen is absent, whereas the L_p antibody increases the rate of inhibition of the LK pumps by ouabain to that characteristic of HK pumps and this effect is amplified by reduction of K_i. In contrast to sheep red cells, the calculated number of ^{3}H-ouabain molecules bound/cell at 100% pump inhibition is the same in both phenotypes.[239] Also different from sheep is the bimodality of the K_i effect, as anti-L_p increased the apparent number of ^{3}H-ouabain binding sites in LK goat RBCs at normal K_i, but failed to do so at reduced K_i.[116] The coupling of the L_p antigen to the Na/K pump must be at the protein level, since trypsin destroys the L_p-antibody-mediated de-repression of the pump.[118] Despite numerous attempts[159,240–242] to isolate either the M or the L_p antigen, there is simply no solid information as to the molecular identity of this most interesting Na/K pump modulator. Thus it was natural to look elsewhere for modulators of the Na/K pump. Indeed, in the kidney, the γ_a

and γ_b subunits modify the Na/K ATPase, lowering the apparent Na affinity by increased K antagonism and increased ATP affinity,[243,244] i.e. shifting the E2 conformation toward E1.[245] A similar equilibrium shift occurs in the α2 isoform as compared to α1 and α3[246] thus effecting apparent different Na, K and ouabain affinities in the rat as compared to the human enzyme. However, as this chapter is written, the sheep erythrocyte has only the α1 and β1 subunit, and the γ component has not been found yet.

The HK-LK red cell Na/K pump dimorphism has been explained in terms of maturational differences in the non-dividing cell at the reticulocyte to mature enucleate transition.[247] This work was later complemented in LK sheep after massive bleeding, and in newborn phenotypically LK sheep which display three red cell populations, the first fetal high K, the second small population of intermediate K, and the final one of LK character appearing three months later.[248] This was confirmed by a report of a transient small cell population with lesser Na/K pump and increased L_p antigen sites.[249] In HK sheep, such cellular transitions in the K/Na ratios have not been studied. Whilst the reticulocytes of HK and LK RBCs have similar Na/K pump properties, in the LK phenotype the transition to mature cells with LK pumps occurs in parallel with the repression of the Na/K pump by the L_p antigens by an elevated L_p/Na/K pump ratio.[119,250,251] The maturational Na/K ATPase activity changes in RBCs of LK phenotype sheep have not been explained by appearance of a different α isoform.[161]

So far, the nature of the L antigen acting as a repressor of the Na/K pump at the membrane level, and the mechanism of its action or of the L_p antibody relieving the inhibition remains elusive. Heterologous antibodies raised in mice against HK and LK red cells stimulated the Na/K pumps in both.[252] Rabbit anti-sheep HK or LK RBC antibodies were without effect (Lauf, unpublished). These two examples show the capricious nature of finding an antibody that modifies the Na/K pump as does the sheep HK alloanti-LK antibody. Attempts to use modern-day phosphatase and kinase inhibitors to shed light on this unique mechanism in physiology have failed thus far.[253,254] However, based on kinetic work[118,255] and that L_p dissociates as an independent moiety from endogenous and interacts with exogenous rat microsomal Na/K pumps fused into sheep red cells,[152] we can at least simplistically formalize the changes in the Na/K pump in the following relationship:

$$[K * LK\text{ Pump} * L_p] + Na + anti-L_p \Rightarrow [Na * LK_{pump}] + [L_p * anti-L_p] + K$$

The first term on the left is the K-inhibited LK*pump*L_p complex, and the second term on the right the L_p-antigen-antibody complex functionally separated from the now Na-responsive LK pump.

2.4.2.4 *Function in cation dimorphic red cells of other species*

The Na/K pump activity defined by ouabain inhibition is absent in RBCs of most dogs.[256] RBCs from all dogs in Europe, North America, Taiwan, Indonesia, Mongolia, and Russia have a low K/Na ratio. This has been attributed to human selection because HK dogs have RBCs that are osmotically more fragile than LK dogs and exhibit hematuria (Fujise, personal communication). In Japan and Korea, however, there are limited numbers of dogs with autosomally recessive HK RBCs.[257-259] Thus only RBCs from Shiba dogs have a ouabain-sensitive Na/K pump, sometimes coupled to a recessive low glutamine transport and low GSH trait,[259] and Na/K-ATPase activity.[260] In contrast, LK dog red cells lack the Na/K pump[256] and control their volume by coordinated actions of K-Cl cotransport[261] and Na/Ca exchange[262,263] when swollen by amiloride-sensitive, and by N-phenylmaleimide-inhibited Na/H exchange[261,264,265] when shrunken, and they maintain their intracellular Ca levels by a Ca ATPase activity.[256] Similar to LK dogs, RBCs of Japanese black bears lack the Na/K pump, volume regulating in response to hyposmotic challenge by Na/Ca exchange.[120] For a comparative account dealing with Na/K pump and secondary active transporters in cation dimorphic cells, see also Ref. 266.

2.4.2.5 *Current operational realities of the Na/K pump based on RBC work*

The electrogenic step of the Na/K pump is the translocation of Na[267] whereas the K translocation is voltage-insensitive.[268] In normal and

CTD-truncated α1 subunits expressed in *Xenopus* oocytes, evidence was recently brought forward for a strictly sequential release of the three Na ions before K was bound,[269] reviving the earlier debate between sequential and simultaneous ion binding and de-binding (see Section 2.4.2.1). When K_o is removed and the Na/K pump is in Na/Na exchange mode, its electrogenic current can be measured in guinea pig heart tissue.[270]

Some 20 years ago, Habermann reported that 1 pM of Palytoxin, PTX (a marine toxin from *Palythoa tuberculosa*) activated human RBC K efflux and Na influx, both of which required ATP and were inhibited by ouabain.[271] The finding was confirmed[272,273]: 1 nM PTX opened a ouabain-sensitive cation channel with the ion selectivity of K > Rb > Cs > Na > Li > choline > triethylamonium >> Mg and a conductance of ~10 pS. Since PTX displaced ^{3}H-ouabain from the pump without ATP hydrolysis it was concluded that PTX binds to the Na/K pump, opening a non-selective cation channel.[273] PTX action was proposed to perturb the Na/K pump by uncoupling the Na-dependent ATPase activity and increasing the rate of deocclusion of the 2KE-2-intermediate and thus dissociating it from the external (K) site.[274] These seminal RBC studies were followed by work showing that PTX uncouples the Na/K pump's gates in excised guinea pig ventricular myocyte membrane patches, changing it in the presence of ATP into a Na ion channel that transports Na with a conductance of 7 pS,[275] inhibited by K. This Na/K pump ion channel has been the center of renewed investigation into the Post–Albers canonical modes of the Na/K pump and ultimately the mystery of the structural basis of the Na and K loading sites. The PTX-opened Na pathway comprises TMD 1,2,4, and 6[276] and its width is 7.5 Å occupied by one or two Na ions based on an Ussing flux ratio of near unity.[277] In outside-out patches Na-channels permeabilized by PTX, pre-bound ouabain protected against methane thiol sulfonate modification of two CYS within TMD1 of the Na/K ATPase,[278] pointing to the close proximity between the ouabain binding site and the Na site/channel.

Thus PTX, applied first in erythrocytes, converts the Na/K pump with a turnover of some 100 Hz to a cation channel with 10^6 Hz, posing an energetically interesting situation to maintain the structural and charged constrains of the Na/K pump's gates. A recent pharmacological review on the history and use of PTX has been published.[279]

2.4.3 The Ca pump

2.4.3.1 Structure

It is the role of the plasma membrane Ca pump (PMCA) or ATPase (EC 3.6.3.8) (Table 2.2) to prevent Ca entrance in the erythrocyte through various, apparently non-selective cation channels. Although calcium plays an important role in multiple cellular functions, it is detrimental to the normal volume maintenance of the red cell when its concentration rises above tightly PCMA-controlled levels, as it activates the IK channel leading to K, Cl, and water loss and hence cell shrinkage (as discussed in Section 2.6). First shown to be the ATP-dependent principle extruding Ca from erythrocytes,[280] a distinct electrophoretic separation of the Na/K ATPase from the Ca ATPase was reported.[281] The PMCA is a ~110–130 kDa protein with some 1000–1200 amino acids depending on the isotype, with ten TMDs, and both NTD and CTD within the cytoplasm.[282] A cartoon displaying visually current concepts of the PMCA structural outlay is given in Fig. 2.4.

The PMCA occurs in four isoforms encoded by four genes localized to chromosomes 12q21–23 (PMCA1), 3p25–26 (PMCA2), Xq28 (PMCA3), and 1q25–32 (PMCA4) (reviewed in Ref. 283). The PMCA isoforms 1 and 4 are found throughout the organism,[282,284] isoform 2 in brain and heart, and isoform 3 in brain and skeletal muscle.[283,285] All isoforms exist in numerous alternative splice variants involving primarily the CTD.[283,286] Regardless of the isoform, and like the Na/K pump, the Ca-pump is an E1-E2 conformer aspartyl P-type ATPase and catalyzes the electrogenic exchange of 1 Ca/1 H per ATP.[287,288] Further conclusive proof that the PMCA was not electroneutral,[289] was provided by Inesi's group showing both unitary ion stoichiometry and hence electrogenicity, and calmodulin stimulation of the human RBC PMCA incorporated into proteolipid vesicles.[290] Conserved in all isoforms, the PMCA has two large intracellular loops, ICL1-2 containing a phospholipid-responsive region (ICL1), and ICL2 between TMD 4 and 5 possessing the aspartyl site for phosphorylation by ATP. The large CTD carrying the Ca-calmodulin-(CAM) binding domain differs between isoforms.[283] CAM bridges between ICL1 and ICL5 (see Fig. 2.4). The PMCA differs from the sarcoplasmic endoplasmatic reticulum

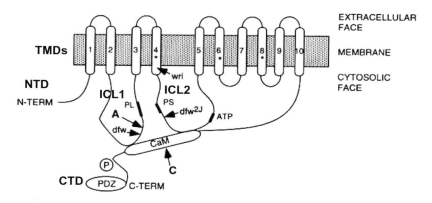

Figure 2.4 Plasma membrane calcium ATPase (PMCA) topology. As in the Na/K ATPase, a major portion of the protein is distributed toward the cytosolic face of the membrane with two large intracellular loops (ICL1 and 2). The ten TMDs are clustered in two blocks (numbered 1–4, and 5–10) and both the N-terminal (NTD) and C-terminal (CTD) domains are in the cytosol. The ATP binding site (ATP) and the phosphorylation site (PS) are opposing each other in the ECL2. The CTD contains the calmodulin- (CAM) binding domain. In the auto-inhibited state shown here, CaM is bridging two intramolecular sites on ICL1 and ICL2. TMDs 4, 6, and 8 contain amino acids ligating Ca. Arrows at A and C show regions where isoform diversity originates from alternative RNA splicing. PDZ indicates that the CTD of some PMCA splice variants can bind to the PDZ domains in several proteins involved in clustering and anchoring of membrane receptors and transporters. The positions where the "deafwaddler" (dfw and dfw^{2J}) and "wriggle mouse Sagami" (wri) mutations alter the PMCA sequence are not relevant for details in chapter Section 2.4.3. (Figure and legend modified from Ref. 301).

CaATPase (SERCA) in that it is regulated only by calmodulin whilst the latter is regulated by both calmodulin and phospholamban.[291]

2.4.3.2 *PMCA function in erythrocytes*

Due to the technically more challenging calcium-binding compartments,[138] isotope fluxes, and the absence of a specific inhibitor such as ouabain for the Na/K pump, most of the early work concentrated on the biochemical purification of the Mg-dependent PMCA from human RBC membranes.[292] The PMCA is different from the basal ouabain-insensitive

Mg-ATPase in red cell membranes[292] serving other functions such as the ATP-dependent and thiol-sensitive aminophospholipid flippase within the RBC lipid membrane.[293–298] The RBC PMCA is considered to be a high affinity and high capacity pump. Like the RBC Na/K pump, the PMCA feeds on the same membrane ATP pool whose location is not yet defined.[217] Under physiological conditions the Ca pump-leak V_m is 50 μmol/(L.cells × h) but "maximum" pump rates $V_{m,m}$ may be 200–500 times higher,[141] i.e. at least 10 mmol/(L cells × h) which was estimated in RBCs, Ca-loaded by the A23187 Ca ionophore method using the Cobalt stop technique to obtain initial velocities.[299] This value agrees with estimates based on earlier data from Schatzmann's laboratory of ca 9 mmol/(L.cells × h).[300] Estimates for the PCMA's $K_{0.5}$ range between 4×10^{-6} M in RBCs[300] and $2-5 \times 10^{-7}$ M.[301] There is no high affinity inhibitor available for the PMCA. However, 700 acyl phosphate sites and hence potential Ca pumps have been estimated per human red cell.[302] Using this number and the estimated V_m and $V_{m,m}$ values given above, we can estimate a molecular Ca pump turnover as low as 10 Hz or as high as 2000 Hz, illustrating the extraordinary adaptability of the PMCA to handle Ca overload of the red cell.

The red cell Ca pump is inhibited by vanadate with a $K_{0.5}$ of 3 μM,[138,303] an intervention causing Ca entrance *via* verapamil-sensitive channels[304] already suspected before 1982 that this activated K efflux (now known *via* the IK channel). An E1-E2-type enzyme was supported by the finding that, unlike the K-supported enzyme mode in the Na/K pump, para-nitrophenyl phosphate (pNPP) is not hydrolyzed by the PMCA to drive Ca ions.[305] There are two ATP binding sites, with K_m values of 2.5 and 145 μM ATP, respectively, both occupied under V_{max} of Ca transport (306). Mg binding ($K_d = 44$ μM) stabilizes E ~ P to bind Ca at the transport site, and Ca prevents Mg binding before E2 formation.[307] The water autohydrolysis of E-P needs binding of the second ATP[306] or a change in the K_d for ATP of the same site.[308] Proceeding from E1-P state to E2 + P requires protonation driving off Ca or Mg ions and thus completing the cycle.[309] With a K_d of 50 μM apparently through the same single site, Mg accelerates the E2-E1 transition,[310] increases the E1-P formation and together with ATP accelerates E2-P breakdown.[311] The MgATP-phosphorylated PMCA performs net ATP synthesis in the presence of high (2 mM) Ca, high P_i, and low ATP.[312,313] ATP synthesis does not occur with the Ca gradient collapsed by the Ca ionophore A23187, or by $LaCl_3$, an irreversible PCMA inhibitor.[312]

In human RBCs, calmodulin increases V_{max} and the Ca affinity, and acidic phospholipids also activate PCMA.[292,314] Ca-calmodulin effectively activates the PMCA by accelerating E2-P hydrolysis,[311] advancing the E1-E2 equilibrium toward E2.[315] Thus, like the Na/K pump reversal discussed previously and in analogy with conclusions mentioned earlier,[201] the PMCA displays the intermediate E1-E2 reaction steps that follow the Post–Albers canonical cycle of the Na/K pump, with Ca replacing Na and protons the K ions, leading to a ping-pong alternate opening of a Ca gate and proton gate. Moreover, net ATP synthesis was accomplished by running the PMCA in RBC IOVs backward.[316]

Most of these data were obtained in RBCs of man and pig. However, one cannot generalize the rationale that where there is a strong PMCA there must be an equal entrance pathway for Ca ions. Mature sheep RBCs probably lack a vigorous PMCA activity[317] since they do not undergo crenation in Ca media in the presence of A23187. They also must lack major Ca entry pathways, since, even in the presence of A23187, Ca did not activate the IK channel-mediated K loss and cell shrinkage seen in human red cells.[318,319] Consistent with these previous observations is our finding that etoposide or staurosporine failed to induce eryptosis in mature LK or HK RBCs of sheep as measured by Annexin V binding, illustrating the teleological function of the PMCA to protect the erythrocyte from potential Ca-induced K, Cl, and water loss, provided there are significant Ca inward fluxes.

2.5 Cation Exchangers and Cotransporters

2.5.1 *Introduction*

In contrast to the Na/K and Ca pumps, the cation exchangers and cotransporters are a relatively new addition to the long list of active players involved in maintaining erythrocyte homeostasis. Cation exchangers and cotransporters constitute a diverse family with some of them bearing genetic homology whereas others share certain analogies or have no structural or functional resemblance at all. We omitted consideration of the Na/Ca exchangers, and of the Na–Cl cotransporter, even though the latter has now been shown to be present outside the kidney.[320] Whenever information is available, it will be organized as much as possible according to the structural, functional, and regulatory properties of the transporters and of their impact in health and disease. Due to space limitations, the reader will be guided to other sources to obtain information on some of the transporters either not found or studied in the erythrocyte or not chosen for review in the present section.

2.5.2 Na-based exchangers for protons and lithium

This group of transporters has a common substrate, Na, whereas the ion transported in the opposite direction can be H and Li, thus constituting an electroneutral transport, or Ca, phosphate, and other polyvalent ions, leading to electrogenic transport. The chemical gradients of the transported species will affect the flux through the electroneutral transporters and the electrochemical gradients will affect that of the electrogenic type. This section will concentrate only on the erythrocyte Na/H and Na/Li exchangers (X). An evolutionary mapping of the Na/H exchangers has been published.[321]

2.5.2.1 *Structure*

The Na/HX (Table 2.2), also known as NHE or Na/H antiporter, is a plasma membrane glycoprotein involved in the regulation of intracellular pH, cell volume, and cellular proliferation. Consequently, protonation, volume sensors, and phosphorylation modulate NHE. In common with other membrane transporters, structure, function, and regulation are tightly associated and, as we will see, a clear distinction between these topics is somewhat blurred, and an overlap is more evident for this exchanger.

The NHE belongs to the SLC9A solute transporter family and, under physiological conditions, transports Na from the outside to the inside in a 1:1 exchange for H from the inside to the outside of the cell,[322] with an apparent external K_d of 13–54 mM $[Na]_o$ and for H of 35 nM, i.e. with a pK$_a$ of 7.5.[323] The Na/HX is inhibited by amiloride with a K$_i$ between 7–25 µM.[322] Intracellular protons, binding to a cytoplasmic site, are not only the driving force but also have been long recognized as allosteric modulators.[324] At the external Na site, NH$_4$>Li>K = Cs as well as amiloride compete for binding.[325]

Cloning of the NHE 1 was first reported by Sardet in 1989.[326] Four genes — SLC9A1 2, 3, and 4, encoding their respective human NHE proteins *via* 12 exons — have been allocated to human (h) chr1 (NHE1), chr2 (NHE2 and 4), and chr5 (NHE 3). Animal isoforms may localize to different chromosomes.[327,328] At the time of writing, nine isoforms have been identified in humans.[321,329] First predicted to have ten TMDs, an Mr of 99,354 daltons and some 800 amino acids, intracellular N- and C-termini and two potential glycosylation sites,[326] the now accepted model (Fig. 2.5) shows 12 TMD for isoforms NHE 1, 2, and 3.[329,330]

Key differences between these isoforms are that NHE1 responds to shrinkage and intracellular acidification with activation *via* a volume sensor in the CTD and by a proton sensor provided by TMDs in the center of the molecule, whereas NHE2 lacks the pH sensor, and NHE3

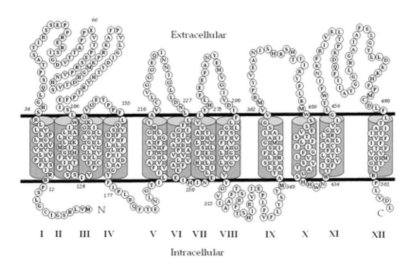

Figure 2.5 NHE1 membrane topology with 12 TMDs (I–XII) and short intracellular NTD (N) and CTD (C) (figure with permission from Ref. 329, 916.

both characteristics. Mostly located at the CTD, there are multiple phosphorylation sites for PKA, PKC, calmodulin kinase, and other serine/threonine kinases, for example mitogen-activated kinases, that are required for the multiple functions of NHE such as pH-control, sensing,[331] and cell proliferation, and hence crucial for hormonal regulation.[330,332,333] A Ca-calmodulin-dependent kinase regulates the NHE activity through phosphorylation at the C-terminus[334,335] as it occurs in so many membrane channels. Finally, it has been shown that the activation of NHE1 by hyperosmotic stress is not associated with additional phosphorylation.[336]

Assembled in the membrane as homodimers that appear to operate independently,[337] the pK of 7.3–7.5 for intracellular protons was first suggested to be a diethyl-pyrocarbonate-sensitive histidine moiety,[338] but was soon found to be a carbodiimide-sensitive carboxyl group.[339] More recent work (Fig. 2.6b), based on detailed sequence analysis, shows a cluster of titratable residues within NHE1 offered by TMDs 4–11 with participation of TMD 8.[340] Figure 2.6a represents a dynamic model of the involvement of these residues in several conformational equilibria defining the alternate gate availability for the exchange of cytoplasmic proton and extracellular Na ions.[340] Based on removal of all endogenous cysteines and their step-wise replacements with subsequent introduction of methane thio-sulfonate spin labels in the NHE's TMDs, Cala and collaborators, long known for their functional and structural work on NHE in *Amphiuma* and flounder RBCs, recently also proposed a detailed model of the Na/H exchange gate in human NHE1.[341] In this model, TMD4–TMD11 form a dynamic molecular crossover construct for the alternate binding and passage of cytosolic protons and extracellular Na. Although there is macroscopically reasonable agreement between these two models, the actually charged moieties participating in this process appear to be somewhat different, suggesting model-dependent variations.

2.5.2.2 *Function*

In the erythrocyte, the first Na/HX cloned was in trout red blood cells, called the βNHE. This distinctive name was given to differentiate this isoform from those insensitive to cAMP activation.[333] In addition to the

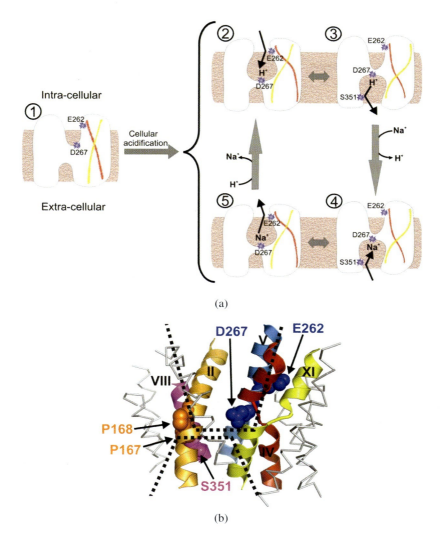

Figure 2.6 A possible molecular mechanism of NHE1. (a) State 1 represents an inactive conformation, and the exchange cycle (states 2–5) illustrates putative conformational changes in the TM domain that follow activation by cellular acidification. The cycle involves a dynamic equilibrium between alternating access conformations 2 and 5 with the cation-binding site at the cytoplasm, and conformations 3 and 4 at the extracellular matrix. The changes are mediated by the TM4-TM11 assembly and may also involve rotation of TM8 and exposure of Ser351 to the extracellular funnel (indicated in states 3 and 4). In state 2, low pH promotes the entrance of a proton to the cytoplasmic funnel, probably

cAMP signaling pathway, the βNHE is also regulated by the PKC pathway due to phosphorylation sites for both PKA and PKC,[342] which are localized in the cytoplasm.[343] βNHE contains three putative N-glycosylation consensus sites,[344] with a unique glycosylation site on the first extracellular loop (Asn49).[344]

A salient characteristic of the trout βNHE is its activation by cAMP and desensitization after stimulation, and its independent regulation by PKA and PKC.[345] In contrast to the other NHE isoforms (see above), βNHE is not correlated with cell volume or pH control but with O_2 transport under lower O_2 tension conditions.[343] Arrestin, a cytosolic protein, regulates the exchanger.[343] Also, βNHE appears to be controlled by the β3b-adrenoceptor, recently found by Nickerson and colleagues in trout RBCs.[346]

Injection of adrenaline into trout causes cell swelling and a shift of the O_2 Hb association curve to the left, i.e. increases the O_2 affinity of Hb.[347] When carps are exposed to hypoxic stress or propanolol, their RBCs accumulate Na and Cl and the pH increases, whilst plasma Na and Cl concentrations decrease.[348] Since alkaline cellular pH increases the Root effect, i.e. the ability to destabilize the oxygenated Hb resting (R) state,[349] these changes demonstrate the presumed role of NHE in positively influencing the oxygen-carrying capacity of blood under stress.[350] Adrenergic alkalinization depends on the Hb buffering capacity, which is high in teleosts and low in elasmobranches,[349] and is prevented by carbonic anhydrase inhibitors,[351] suggesting the thermodynamic coupling between NHE

---

attracted by the acidic Glu262, and the protonation of Asp267. The low pH also induces conformational changes, leading to the transfer of the proton from the cytoplasmic funnel to the extracellular funnel (state 3). Based on their chemical gradients, H is exchanged for Na in the extracellular matrix, perhaps *via* Ser351 (state 4) and Na by H at the cytoplasmic side (state 2). The continuance of the cycle is controlled by cellular pH. (b) The model structure of the TM domain of NHE1 in the inactive conformation of state 1 viewed from the membrane. The intracellular side is facing upward. TM segments important for function are represented by the colored ribbons. Other segments are represented by a gray trace. TM1 was omitted for clarity. Residues involved in the cation transport path are represented by space-filled atoms. The funnels laying the transport path are indicated by dashed lines (figure with permission from and legend modified from Ref. 340).

activation and Cl/HCO$_3$ exchange.[71] Finally, hydroxyl radicals inhibit the adrenergic NHE exchange.[352]

The *Amphiuma tridactylum* RBCs possess a Na/HX (atNHE1) that is highly homologous to the mammalian NHE1 in its structure whereas it is insensitive to HOE-694, a potent inhibitor of NHE1.[353] Likewise, the RBCs of the winter flounder *Pseudopleuronectes americanus* possess an NHE1 isoform, paNHE1, with 74% similarity to that of trout RBCs, βNHE, and 65% to the human NHE1.[354] Interestingly, paNHE1 can be activated by PKA, as in *Amphiuma*, and by shrinkage, as in human NHE1. These effects are mediated by two independent regulatory pathways. Furthermore, calyculin A, a potent serine/threonine phosphatase inhibitor, also activates paNHE1 by an additive mechanism to that of PKA and shrinkage.[354] Studies with chimeras obtained from at NHE1 and that of the winter flounder *Pleuronectes americanus* (paNHE1) indicate that the transmembrane domains (TM) 4 and 10–11 are involved in the partial inhibition by amiloride and HOE.[355]

Rabbit and rat erythrocytes express NHE1 and NEH3, as determined by immunoblot analysis with specific antisera for five different NHE isoforms (NHE1-5).[356] Studies on the topology of the NHE1 C-terminus from rat and mouse RBC membrane vesicles indicated presence of one or more epitopes in the extracellular aspect of resealed and separated ROVs in these species.[357] In support of these findings, human erythroid cells were found to possess the NHE1 isoform, as determined by Northern blot and RT-PCR analyses.[358] However, adult RBCs possess an amiloride-insensitive Na/HX and Na/Li countertransport (CT)[329,359] under physiological conditions, although manipulations of Ca, Na, and H gradients brings about amiloride-sensitivity.[360,361]

Since the finding of an increased Na/LiCT in erythrocytes from essential hypertensive and diabetic nephropathy patients[362–368] in the 1980s there has been a long-standing controversy about its relationship to the NHE isoforms due to lack of inhibition by amiloride of the Na/LiCT activity or as determined by statistical tests.[369–375] This controversy has been unequivocally eliminated by Zerbini and coworkers, who demonstrated that the erythrocyte NHE is lacking the amiloride binding site of the transporter.[376] This finding also suggests that amiloride cannot directly compete with Na like other monovalent cations mentioned above.

Interestingly, neonatal cord erythrocytes express both amiloride-sensitive NHE1 and amiloride-insensitive Na/LiCT,[377] whereas only the latter is seen in adult RBCs.

A K (Na)/HX not related to the SLC9A group of NHEs, identified in human erythrocytes at low-ionic strength, has different properties from the NHE1 present in human RBCs and it is also present in bovine and porcine RBCs[378–380] (Section 2.9).

Na/LiCT in human RBCs is inhibited by thiol reagents such as NEM and iodoacetamide,[381,382] by plasma copper,[383] and by inhibitin, a peptide isolated from cultured leukemic promyelocytes.[384] Based on kinetic evidence, the Na-PO$_4$ cotransporter has been proposed as the molecular mechanism of human erythrocyte Na/LiCT,[107] although this was disproven by the work from Zerbini and collaborators,[376] see above. Na/LiCT in LK and HK sheep RBCs have similar kinetic properties to those described for human RBCs, except that the maximum velocity of Li fluxes was higher in LK than in HK cells.[385] These properties, together with other kinetic determination, lead to the proposal of a "ping-pong" mechanism for the system in sheep RBCs,[385] which also operates in other NHE systems.

2.5.2.3 Regulation

The human erythrocyte Na/HX is activated by insulin and by the serine-threonine phosphatases inhibitor, okadaic acid.[386] These effectors also decrease the affinity of the exchanger for Na but not for H, suggesting that insulin activates NHE by increasing the phosphorylation of the transporter.[386] Similar results have been found for Na/LiCT.[366] Protein kinase C was also found to modulate Na/HX, although its mechanism differs from that of insulin.[387] Another characteristic of the erythrocyte NHE1 is to respond to high altitude-induced chronic hypoxia by increasing its membrane density, as determined by immunoblotting.[388] In dog RBCs, cell shrinkage, intracellular Li (Li_i) and okadaic acid activate Na/HX[263,389] whereas intracellular Mg (Mg_i) inhibits[390] and the regulation of the transporter is dependent on the intracellular ionic strength.[391] Furthermore, in these cells, NHE is inhibited by Na in isotonic medium through two inhibitory sites and it is activated by cell shrinkage by decreasing the affinity of the transporter for external Na at the inhibitory sites.[392] In these cells, Li is also transported by NHE1 although its properties differ from that of the Na exchange in that Li binds to the transporter at the Na inhibitory sites.[393] In rabbit RBCs, Na/HX and Na/NaX share the same pathway.[394] In dog RBCs, Na/HX and KCC are regulated in a coordinated manner by a protein kinase-phosphatase system where shrinkage or swelling activates the Na/HX or KCC, respectively.[389] Alterations in the concentrations of cytoplasmic macromolecules induced by swelling or shrinkage are proposed as volume sensing mechanisms.[389]

Amphiuma RBCs are known as one of the first cell models where volume regulation is carried out by electroneutral transporters, i.e. K/H and Na/HX. Thermodynamically coupled to the AE1 exchange of Cl or HCO_3, cell shrinkage activates Na/HX whereas K/HX is activated by cell swelling.[73,395-397] These mechanisms are proposed to be carried out by the same transporter.[398] Furthermore, *Amphiuma* RBCs are involved in pH regulation[397,399] and because of this property, the Na/HX is involved in hypoxia-induced cell injury,[396,397] known today as ischemia/reperfusion injury. Further studies with calyculin A on the *Amphiuma tridactylum* RBCs have shown that both Na/H and K/HX are strongly activated by the phosphatase inhibitor under physiological conditions, suggesting that kinases are responsible for the activation of these exchange modes.[400] An interesting property of *Amphiuma* RBCs is the complex interplay between these two transporters or modes of transport as a function of volume. To clarify this function, a recent model has been proposed based on relaxation kinetic analysis.[401] The conclusions from this model are that for each exchanger a pair of kinase/s and phosphatase/s exists that is volume-sensitive but whose activities change reciprocally as a function of volume.[401] This would explain the reciprocal behavior of the activities of the transporters as a function of volume.

2.5.2.4 *Translational significance*

Young erythrocytes containing Hb AA, SS, and SC express a highly elevated Na/HX activity, which leads to cell volume increase.[402,403] This elevation persists after reticulocyte maturation and thus may counteract KCC-induced dehydration[403-405] (see below, Section 2.5.4). Not only has the Na/HX (Na/LiCT) been implicated in diabetes and hypertension, but insulin and blood pressure, in turn, affect the activity of this transporter.[406] The Na/HX has also been implicated in cardiac ischemia and reperfusion. Inhibition of the transporter has shown preservation of cellular integrity and function.[407] A potent inhibitor for the NHE1 isoform belonging to the family of benzyl guanidines, (2-methyl-5-(methylsulfonyl)-4-pyrroloben-zoyl) guanidine-methanesulfonate has been found to be cardioprotective when administered at the onset of acute myocardial infarction,[407] thus stressing the important role NHE1 plays in cardiac ischemia and reperfusion. Likewise, NHE1 has also been shown to be increased in erythrocytes from non-insulin-dependent diabetes mellitus patients, while incubation of these cells with tea catechins inhibits the transporter likely by changing the plasma membrane fluidity.[408]

The Na/LiCT mode of the Na/HX in erythrocytes has caused quite a turmoil in the literature due to the difficulties in showing that it did indeed belong to the typical NHE family. To a large extent this was due to its insensitivity to amiloride and derivatives that, prior, to the readily available cDNA cloning and sequencing techniques, was considered as a proof of evidence of the exchanger. Although abnormalities in Na transport in hypertensive rats existed (see references in 362), the first report of an increase in Na/LiCT in RBCs of essential but not secondary hypertensive patients was provided by our group in 1980.[362] Numerous reports followed, some confirming the finding of an abnormal Na/LiCT in essential hypertension,[329,363,365,367,406,409–418] others disproving it.[365,373,374,419–421] This lack of agreement in the field lasted for more than two decades, until in 2003 Zerbini and colleagues showed evidence, as indicated above, that the erythrocyte Na/HX was lacking the amiloride binding site of the well-characterized NHE1.[376] Figure 2.7 shows the Na/LiX has four fewer TMDS, in particular those binding amiloride, than NHE1.

This finding provided further support for the kinetic evidence that Li/Na CT was a mode of operation of the Na/H X.[412,422,423] The abnormalities in Na/Li CT found in patients with essential hypertension and diabetes, and in animal models, have been extensively investigated and the erythrocyte Na/Li CT is considered to be the best characterized intermediate phenotype.[366] Due to the simultaneous discovery of an abnormal (decreased) RBC Na-K-Cl cotransport (NKCC, see Section 2.5.3 below) in essential hypertension by Garay and colleagues,[424] it was not clear at

Properties and Membrane Transport Mechanisms of Erythrocytes

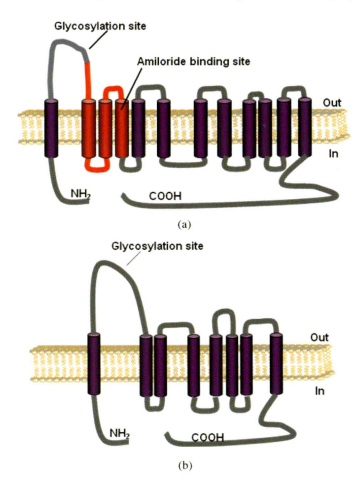

Figure 2.7 Suggested scheme for molecular changes in the NHE1 molecule (a) producing a Na/LiX mechanism (b) in human RBCs (with kind permission from Zerbini et al.[376]). Accordingly, the TMTDs in the Na/Li exchanger are reduced to 8 TMDs from the 12 in the NHE1. For details see Section 2.5.1.

the time whether the Na/LiCT and the NKCC were the same with two different operational modes, or whether these modes corresponded to two different molecules. Simultaneous studies of Na/LiCT and NKCC designed to solve this puzzle produced surprising results in that, depending on the geographical location, Na/LiCT was either normal or elevated.

In contrast, NKCC was either decreased or elevated and this variation in NKCC was found later on to be caused by both genetic diversity and dietary differences[362,363,409,412,424–429] leading to the conclusion that the two systems were separate entities and raising new questions about their heritability.[362,363,424,425]

In the earlier reports, Na/LiCT was measured as the Na-dependent or Na-stimulated Li efflux and Mg was used as Na-replacement.[362,363,410] However, divalent cations, including Mg, inhibit human RBC Na and K transport through NKCC and other ouabain- and bumetanide-insensitive pathways[430] in a dose-dependent manner. This finding raises the question whether the increase in Na/LiCT measured with Mg as a Na-replacement ion has not been over-estimated in the earlier studies. However, follow-up studies using choline, N-methyl-D-glucamine or other Na substitutes showed similar abnormal patterns in Na/LiCT and Na/HX in essential hypertensive patients and in animal models.[366,403,412,431–433]

Environmental factors such as exercise conditioning were found to decrease Na/LiCT in both normotensive and hypertensive patients with a higher decrease, the higher the basal Na/LiCT, but not in sedentary controls.[434] Similar results have been found in later studies.[435–437] These findings indicated that environmental factors may act in concert with genetic factors to influence monovalent ion transport in humans, and should be taken into account in investigations of the pathophysiological linkage between altered monovalent cation transport and essential hypertension.[412,434] Furthermore, these findings could serve as early evidence of epigenetic influence on Na/LiCT.

Besides its role in hypertension, Na/LiCT and Na/HX have been extensively studied and found altered in both insulin-dependent and insulin-independent diabetes,[366,367,383,386,406,414,416,417,438–450] in uremia, polycystic kidney disease, Bartter's syndrome,[368,451–456] and in sickle cell anemia where the Na/HX has been found highly elevated in young cells and remains elevated after reticulocyte maturation.[403] However, some studies have found either normal Na/LiCT or abnormal Na/HX but not Na/Li CT, and concluded the latter could not be used as a marker of the exchanger or simply as a marker in certain pathologies or under certain variations of a given pathology.[365,373,374,420,421,457] It appears, however, that these discrepancies can be explained, at least in part, by the conditions in which blood samples were taken such as fasting or non-fasting, due to the modulatory effects of insulin on Na/LiCT kinetics.[367,406] In addition, erythrocyte Na/Li CT was found to be strongly correlated with the level of human proximal tubule NHE3 protein and negatively related to NHE1 protein and only NHE3 and NHE1 protein levels were significant predictors of RBC Na/LiCT.[458] Thus the latter was proposed as a marker of increased NHE3 protein expression in the proximal tubule.[458]

2.5.3 *Cl-dependent cation cotransporters*

Amongst the Cl-dependent cation cotransporters, the Na-K-Cl cotransporter (NKCC), was first found and identified in the erythrocyte. First evidence, albeit with a different interpretation, proposed that human RBCs possess a second Na pump not inhibited by cardiac glycosides and dependent on external Na.[459] In 1971, simultaneous publications by Beauge and Adragna and by Sachs reported a ouabain-insensitive Na-dependent Rb influx and Na-dependent, ouabain-insensitive Na efflux, respectively, requiring ATP without hydrolysis.[49,50] Based on these findings and the emerging information on the action of loop diuretics in the kidney, in 1974 Wiley and Cooper presented evidence of a ouabain-insensitive, furosemide-sensitive coupled Na and K influx and efflux in human RBCs,[54] with similar kinetic behavior and experimental ionic conditions as previously published for these cells.[49,50] Electroneutrality was shown by Geck and Heinz in their thermodynamic study on NKCC in mouse Ehrlich ascites tumor cells. The coupling factor between the fluxes of Na and K was unity, and two between those of both monovalent cations and Cl.[56] NKCC was afterwards kinetically characterized in human RBCs[460] and Li was found to be cotransported with K by replacing Na at its binding site.[461] The idea of kinase regulation of the system arose as early as 1987[462] and before similar publications on KCC. The following sections will deal with the NKCC and KCC mechanisms but not with the NaCl cotransporter (NCC), which has not been shown yet to functionally exist in erythrocytes. A relatively recent expert review of the NKCC has been published.[463]

2.5.3.1 *Na-K-2Cl cotransport*

2.5.3.1.1 Structure

Work so far has concentrated on the two reported isoforms, NKCC1 (Table 2.2), found in most tissues, including the erythrocyte, and NKCC2, primarily and to a large extent existing only in the kidney.[464–472] This may change in the future, as we have seen with other receptors and membrane transporters, such as NCC that has recently been found outside the kidney (Ref. 320 and references therein). For excellent recent reviews on NKCC structure and function, consult.[469,472–474] NKCC1, also called BSC2 (bumetanide-sensitive cotransporter 2), occurs ubiquitously, and NKCC2 (BSC1) primarily in the apical membrane of the nephron's thick ascending limb of Henley.[475] The genes, belonging to the SLC12A family, SLC12A2 for NKCC1 map to chr 18 in mice and 5 in man, and SLC12A1 for NKCC2, to chr 2 in mice and chr 15 in man. Their open reading frames of 27 and 26 exons, respectively, encode a ca 132 kDa deglycosylated and ~170 kDa glycosylated protein with some 1212 (NKCC1) and close to 1100 (NKCC2) amino acids. Figure 2.8 is a frequently used cartoon to illustrate the features of NKCC1.

The NKCC1 isoform is presumably responsible for the coupled movements of Na, K and Cl in RBCs. NKCC1 was first cloned by Delpire and coworkers,[476,477] followed by the group of Forbush shortly thereafter,[478] and is present in RBCs. Spliced variants of NKCC1 have been recently summarized by Di Fulvio.[479] The NKCC1 polypeptide chain crosses the plasma membrane 12 times with both a long NTD and very long CTD in the cytosol and has a large hydrophilic 7–8 extracellular loop (ECL4) with two N-glycosylated polyglycan chains.[476] There is strong evidence for homodimer formation in the plasma membrane.[480] Transfected into human kidney embryonic 293 cells (HEK293), the apparent affinities (K_m, in mM) of NKCC1 were ~20 for Na, 2.7 for Rb, a K congener, and ~27 for Cl while the affinity for bumetanide was 0.16 µM.[478] Based on chimera experiments, the TMD 2 has been shown to be involved in Na and K transfer. TMDs 2, 4, 7, 11, and 12 have been implicated in bumetanide binding,[481] revealing coordinate interaction of these TMDs to confer the capability for electroneutral transport of Na, K and 2 Cl. There is abundant evidence that NKCC1 is active when phosphorylated (see Section 2.5.3.1.3). Key amino acid

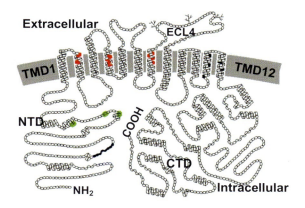

Figure 2.8 Suggestion for the membrane topology of the Na-K-2Cl cotransporter NKCC1 isoform. Shown are the 12 TMDS with the large triple-glycosylated ECL4 and the large cytoplasmic NTD (NH$_2$) and CTD (COOH). (Modified from Refs. 463, 469).

candidates for phosphorylation such as serine and threonine have been identified around residue 200 in the NTD of the molecule at T206 and T211.[482]

2.5.3.1.2 Function

The NKCC is modulated by a large number of effectors, such as cell volume, hormones, growth factors, O$_2$ tension, and Mg$_i$,[466–472,483,484] which are in many instances the same as those modulating KCC (see next section), although in a mirror image or coordinated action.[485,486] Thus, cell shrinkage, free-Mg$_i$, activate[462,484,487] whereas cell swelling,[462] acid and alkaline pH,[460] and plasma copper[383] inhibit. However, as for KCC, ATP-depletion reversibly inhibits.[434,462,484] Furthermore, the NKCC has been proposed to be outwardly directed under physiological conditions,[360] although estimations of the ratio of the product of the chemical driving forces for each of the transported species is close to unity.[462,469] Under certain experimental conditions, NKCC operates in several modes and with variable stoichiometry[488] and a diverse modality NKCC1 operation has also been shown recently.[489] For example, NKCC displayed a different behavior when the medium Mg concentration was in the physiological range 1–3 mM as compared to 75 mM Mg[490] consistent with inhibitory effect of divalent cations on Na and K transport in human RBCs.[430]

Inhibitors of the NKCC lack specificity although several drugs selectively block NKCC such as loop diuretics, bumetanide, and furosemide.[466–472,483,491,492] Analogs derived from these molecules, such as H25, have higher selectivity for NKCC than for KCC.[493] Urea also has an inhibitory effect on NKCC,[494] generally interpreted in terms of the macromolecular crowding theory.[495]

Undifferentiated mouse erythroleukemia (MEL) cells possess a highly active NKCC which is significantly reduced during the transition from proerythroblasts to reticulocytes accompanied by a decrease in cell volume.[496] In congenic mice strains possessing a common genetic background but differing only by the two alleles of the rol (resistance to osmotic lysis) gene, susceptible (rols/s) and resistant (rols/r), NKCC was threefold higher in the latter with concomitant differences in erythrocyte volume, water and cation content, and 2,3-DPG levels between the two strains.[83] A primitive fish species, the hagfish, appears to have a possible "precursor" of NKCC because, although it supports a bumetanide-sensitive Na-dependent transport of K and Cl, Na is not transported.[497] In ferret RBCs, NKCC is the primary entry pathway for K^{430} although it plays a minor role in steady-state cell volume regulation.[498]

2.5.3.1.3 Regulation

As indicated above and similarly to KCC (see Section 2.5.3.2) and in Scheme 2.1, NKCC is regulated by a cascade of kinases (SPAK) and phosphatases (PP1) to produce either the active (P-NKCCa) or the resting (NKCCr) transport mechanism, a process governed by forward ($k_{1,2}$) and backward ($k_{2,1}$) rate constants.[462,466–472,484,485,499]

Direct phosphorylation of serine and threonine residues on the transporter molecule or phosphorylation of regulatory proteins modulates the

$$\begin{array}{c} \text{PP1} \\ k_{1,2} \\ \text{P-NKCC}_A \Leftrightarrow \text{NKCC}_R + P \\ k_{2,1} \\ \text{SPAK} \end{array}$$

Scheme 2.1

transporter.[484] Calyculin A, arsenite, and deoxygenation stimulate NKCC through different signaling pathways.[484] Several mechanisms of protein-protein interactions, including NKCC complexing between monomers or with other transporters or binding to cytoskeletal proteins, regulate NKCC activity.[484] Thus calyculin A, the classic inhibitor of protein phosphatase 1 (PP1), causes activation of NKCC1. Phosphorylation declines when intracellular Cl rises, hence auto-inactivating NKCC1.[470,481]

While much work has been done to identify the kinases involved here, the most convincing data on NKCC1 phosphorylation have come from Eric Delpire's laboratory.[499–501] As shown in Fig. 2.9, the NTD of

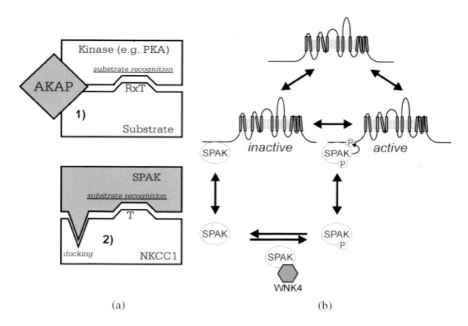

Figure 2.9 Suggested model showing the regulation of NKCC1 by the kinases SPAK and WNK4. a1) reveals a general view of a kinase such as protein kinase A, PKA, scaffolded to an A-kinase anchoring protein, AKAP, and recognizing its substrate RT. By analogy, panel a2) shows how SPAK *via* its docking site might recognize threonine (T) residues within, and thus phosphorylate the NTD of NKCC1. (b) shows the phosphorylation equilibrium of the NTD controlled by SPAK which, itself being phosphorylated by WNK4, is enabled to phosphorylate and hence activate NKCC1. (Figure and legend modified from Ref. 482).

NKCC1 has recognition or docking sites for two kinases, the stress related serine-threonine Ste-20-related proline-alanine-rich kinase (SPAK) and the oxidative stress-responsive (OSR1) kinase. Two minimal, nine residue docking motifs, (R/K)FX(V/I), occur in NKCC1 and 2, but also in KCC3,[502] one of them overlapping with a docking site for protein phosphatase 1 (PP1).[503] While SPAK directly interacts with NKCC1, it is itself under phosphorylation control by WNK4 (with no lysine kinase).[504] Delpire's group proposes that the Cl-dependent cation transporters serve as sensors for and respond to cellular stress.[505] Physical docking of SPAK, probably *via* their own CTDs[500] is required for NKCC activity under normal isotonic and hyperosmotic stress conditions.[482] Thus, the NTD of NKCC1 serves as a scaffold for both SPAK and PP1 to regulate its activity. PP1 dephosphorylates SPAK and this effect is even greater when there is protein-protein interaction between these enzymes, indicating that PP1 inhibits NKCC1 not only by dephosphorylating the transporter but also SPAK.[506] Intracellular Cl, Cl_i, regulates NKCC independently of being transported.[466,468–472,484]

2.5.3.1.4 Translational significance

Mice knock-out experiments show that NKCC1 is implicated in hearing, salivation, nociception, spermatogenesis, and volume control.[474] Translational work with human RBCs to elucidate their possible use as markers for disease is more difficult, and epidemiologically designed experiments are needed. The human erythrocyte NKCC received public health scrutiny in the 1980s due to the finding that it was abnormal in essential hypertension and in children of hypertensive patients who were still normotensive.[363,410,424]

Depending on the patient population studied, NKCC was found either normal, decreased[410,411,424,507] or elevated[363,409,418,507] with respect to normotensive controls. This apparent paradox was found to be caused by genetic traits and the composition of the diet.[426,429,508,509] In fact, the large interindividual variability in V_{max} of the transporter in human RBCs can be accounted for by genetic factors, whereas the normalization of the K_m in some hypertensive patients after reduction of Na intake points to the system as a reliable probe of environmental factors.[429] Similar results were found in erythrocytes from genetically hypertensive rats[431,510–513] although some exceptions exist.[491] Furthermore, the V_{max} of RBC NKCC from hypertensive African-American patients is reduced when compared to normotensive controls[427] whereas the K_m for the outward mode of the transporter is increased in young African-American subjects.[412,490,514] Chronic excess of humoral inhibitory factors inhibit and up-regulate RBC NKCC.[515] Defects in the gene structure of NKCC and KCC and in the structure of their regulators, in particular OSR1, SPAK, and the WNK family, are involved in pathologies such as essential hypertension.[509,516]

2.5.3.2 K-Cl cotransport

The K-Cl cotransporter (KCC) (Table 2.2) was first recognized in sheep and fish RBCs as a swelling-activated[57,58] and as a thiol-activated Cl-dependent K flux in sheep RBCs.[62] Being Cl-dependent, KCC is a Na-independent K transport system with electroneutral K and Cl movement into or out of the cell depending on the ionic chemical gradients.[517,518] Electroneutrality of KCC has been demonstrated in RBCs from different species.[518–520] Under physiological conditions, the fluxes of these ions are outwardly directed.[517,521] The fact that the protein phosphatase inhibitors okadaic acid and calyculin A inhibited KCC in rabbit RBCs established the regulation by phosphorylation/dephosphorylation equilibria,[522] helped to explain the stimulation by N-ethylmaleimide (NEM),[523] and the role of Mg and ATP in governing a matrix of kinases.[132,133] Thus, the erythrocyte's significance in the discovery of KCC again preceded the molecular developments by more than 15 years. Detailed functional and molecular accounts have been given in major reviews.[64,266,472]

2.5.3.2.1 Structure

So far, four genes have been found, the product of certainly three of these occurring in the erythrocyte[253,524,525] with almost all molecular mechanisms studied in nucleated cultured cells and in *Xenopus* oocytes.

The KCCs belong to the SLC12A solute transporter family and map to the chromosomes indicated in parenthesis: KCC1, SLC12A4 (16), KCC2, SLC12A5 (20), KCC3, SLC12A6 (15q14) and KCC4, SLC12A7 (5p15). Their 24 exons provide open reading frames for proteins with the following number of amino acids and approximate unglycosylated molecular weights (kDa) for the human isoforms: KCC1, 1085 (120), KCC2, 1115 (123), KCC3a, 1150 and KCC3b 1099 (~130), and KCC4, 1150, (120).[526–530] KCC1, sharing only 25% sequence homology with NKCC1, is widely distributed and called the "house-keeping" isoform.[526] This isoform has been cloned in several orthologs such as mice (mKCC1), mapped to chromosome 8,[531] dogs,[532] and sheep.[533] Human (K562) and mouse erythroleukemic (MEL) cells express the KCC1 isoform[534] whereas LK sheep RBCs express the KCC1 and KCC3 isoforms.[253] There is evidence for several spliced variants.[524] A cartoon displaying the characteristics of the human and mouse erythrocid KCC1 secondary protein structure is given in Fig. 2.10.

Due to a neuronal-restrictive silencing element (NRSE) in intron 1,[535] KCC2 was first considered to solely occur in the brain[527] but now is recognized to be present in at least two spliced variants: KCC2a in non-neuronal tissue, with 40 unique amino acid residues including a SPAK binding site in the NTD; and KCC2b, solely in neuronal cells.[536]

In knock-out mice, the absence of KCC2b is incompatible with life, although KCC2a may support at least rudimentary neuronal functions.[536] We recently found KCC2a in a human lens epithelial cell line (Lauf 2011, in preparation). KCC3, present in the cardiovascular,[63] renal, and central nervous systems occurs in two spliced variants, KCC3a and KCC3b whose 5′ transcriptional initiation involves two distinct exon 1 codons[537] producing proteins differing by 51 amino acid residues.[536] Consequently, the shorter KCC3b, occurring in the kidney, has its deletion with the NTD proximal to a stretch of highly conserved residues reducing their content of functionally important phosphorylation sites.[537] On the other hand, KCC3 has been recently shown to be phosphorylated at four serine and threonine residues (T991, S1023, S1029 and T1048) of its CTD by WNK1, and reduction in KCC3 protein phosphorylation in human RBCs has been demonstrated upon hypotonic swelling.[525] This finding is consistent with earlier data showing that

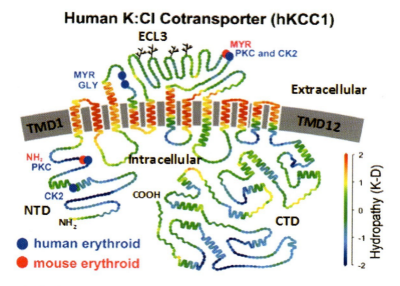

Figure 2.10 Suggested membrane topology of the human and mouse erythroid K-Cl cotransporter KCC1 isoform with 12 TMDS, the quadruple-glycosylated ECL3 between TMD5 and 6 distinguishing the KCC from the NKCC1, and the cytosolic NTD and CTD (figure modified from Ref. 534).

truncation of the KCC1 CTD leads to functional loss as assessed by NEM-stimulation[253] and that allosteric changes in the CTDs are important determinants of transport activity.[538] Finally, KCC4 occurs preferentially in the kidney.

Isoform-independent, the KCC proteins start with a relatively short NTD in the cytoplasm, cross the membrane 12 times and end with their very long CTD again in the cytoplasm. In contrast to the NKCCs, all KCCs have a large, quadruply glycosylated ECL3 between TMD 5 and 6. Chimera experiments in NKCC also suggest that in the KCC the TMD2 must participate in the hydrophilic cation "channel" transport, with two regions selectively attracting Na and Rb,[481] and by analogy to NKCC, Cl binding sites, encoded by exon 4, may be downstream from TMD2.[539] The KCCs show a relatively low affinity for loop diuretics,[540] and are inhibited by DIDS, first found in sheep RBCs.[541]

2.5.3.2.2 Function

As indicated for NKCC, the functions and regulators for KCC and NKCC are numerous and, in general, are coordinated in opposite ways. In RBCs, it was recognized early that Cl-dependent K transport participates in regulatory volume decrease or RVD.[542] Cl-dependent K flux is stimulated by cell swelling, acidification, Mg depletion, thiol modification, oxidation, and certain anions, depending on the experimental conditions,[58,64,84,133,319,485,517,543–558] urea,[559] vasodilators such as sodium nitrite and nitric oxide donors,[63,64,558,560] and peroxynitrite.[64,558,561] In contrast, cell shrinkage, ATP-depletion, elevation of cellular divalent ions, thiol alkylation, carbethoxylation, cinchona bark derivatives such as quinine and quinidine, loop diuretics and derivatives, certain anions (depending on the experimental conditions), internal protons, internal Li, DIOA [(dihydroindenyl)oxy] alkanoic acid, and cytochalasin B inhibit K-Cl cotransport in RBCs.[64,84,102,108,319,515,517,521,546–550,558,562–567] Furthermore, the inhibitory effect of loop diuretics is dependent on the presence of external cations.[540] Also, stilbene disulfonates such as H_2DIDS and DIDS are potent inhibitors of swelling-, NEM- and Mg_i-depletion-activated KCC, and their potency is modulated by K_o provided the inhibitors' binding sites are not saturated, whereas upon saturation K_o and Cl_o lose their modulatory effect.[541]

Ever since through the "NEM effect" (NEM stimulation), Cl-dependent K flux and hence KCC were discovered,[62] a search has been made for the affected SH group. An obvious target was glutathione (GSH), which following NEM treatment is irreversibly alkylated, indicating that the RBC redox system is in charge of the KCC activity.[63,102,558,560,568,569] This prediction was borne out by studies with diamide, which forms mixed disulfides with the free thiols of the cells, whether in GSH or protein-bound. Thus, as shown in Fig. 2.11, as GSH declines with increasing diamide concentrations, KCC becomes activated. This effect is reversible by either metabolically reactivating the pentose phosphate shunt or by adding reducing equivalents such as mercaptoethanol to the RBC suspensions.[548]

Similarly, KCC was irreversibly activated with the xenobiotic coupler CDNB (chloro-dinitro-benzene), and with methyl methanethiosulfonate

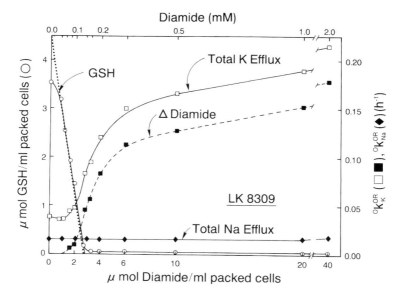

Figure 2.11 Control of K-Cl cotransport in LK sheep RBCs by the cellular redox system. Addition of increasing concentrations of diamide causes conversion of GSH into GSSG with a molar ratio of 1 diamide/2 GSH as seen in the fall of cellular GSH/diamide added on the left y-ordinate. Concomitantly, the rate constant for ouabain-resistant K efflux, $^{o}k^{OR}_{K}$, increases by fourfold while that for Na efflux, $^{o}k^{OR}_{Na}$, is unaffected. This effect was entirely Cl-dependent and reversible upon reincubation with glucose to activate the pentose phosphate shunt and hence the regeneration of GSH (from Ref. 548).

(MMTS). The effect of this, surprisingly, could not be fully reversed indicating conformational changes within the KCC and its regulatory enzymes.[549] Using ^{3}H-NEM, ~2 × 10^4 diamide-reactive SH groups/cell were labelled in both LK and HK sheep RBCs, the former having a significantly higher KCC activity than the latter.[570] Nevertheless, the firm conclusion was reached that the redox status of RBC controls the activity of KCC through a limited number of thiols. An increase in temperature revealed deocclusion of inhibitory thiols in sheep erythrocytes.[571] Another conclusion is that thiol modification of KCC, dependent on the type of thiol, reveals thiol heterogeneity and differs in its response to anisosmotic media suggesting that a "volume-sensing" component is part of the system.[550,569,572]

An interesting model of KCC activity as a function of cell density is found in high K (HK) and high glutathione (HG) dog erythrocytes. Here light cells are the mature, older and larger RBCs whereas the dense cells are the younger and smaller RBCs, yet KCC is about ten-fold higher in the latter.[257] The paradox between KCC activity and cell density can be explained by the inverse correlation with the GSH concentration, which is higher in the lighter and lower in the older cells.[259]

Horse RBCs express K-Cl cotransport with characteristics similar to those found in other species, i.e. volume-, thiol- and pH-dependencies, inhibition by DIOA,[573] calyculin A and okadaic acid, and phosphate. Sensitivity to partial oxygen pressure (pO_2)[574–576] is independent of the free intracellular Mg (Mg_i)[575] and may account for the inter-individual variability observed in equine RBCs,[574,576] and likely in RBCs from other species.

Thermodynamics

The ratio of K efflux over K influx under physiological conditions with 5 mM $[K]_o$ in the medium always exceeds unity, whether K-Cl cotransport is stimulated by swelling or chemical interventions.[57,62] This is due to the product of the chemical driving forces $[K]*[Cl]$ being larger within the cell than outside. A rigorous test for this paradigm was to study bidirectional K fluxes in Mg-depleted sheep RBCs (Mg depletion interferes with the phosphorylation by kinases) pH-clamped by pretreatment with DIDS to change simultaneously both intra-cellular Cl and extra-cellular K. If only the products of the chemical gradients, not in combination with V_m, govern the directionality of the K-Cl cotransport, then the flux reversal point (FRP), i.e. the $[K]_o$ at which the inward K flux, ^iM_K, over the outward K flux, ^oM_K, flux becomes unity should be a linear function with unity slope of the ratio of $[K]_i$, $[Cl]_I$, and $[Cl]_o$ as follows:

$$\text{FRP}\ \{([K]_o\ \text{at}\ ^iM_K/^oM_K = 1)\} = \{[K]_i*[Cl]_i\}/[Cl]_o$$

Indeed, this linearity was found, which is of course commensurate with the Donnan ratio of K and Cl ions across the plasma membrane.[518] Furthermore, Cl-dependent K fluxes were not affected by changes in V_m introduced by the Na ionophore hemisodium.[112]

Kinetics

The kinetic constants for RBCs of sheep, man, and rabbit are published.[517] In comparison to various transporters discussed in this chapter and the NKCC, the apparent affinities for K are low and even lower for Cl.[517] In hindsight, with the knowledge that RBCs, like smooth muscle cells[64] have the KCC1, 3, and 4 isoforms,[253,524] these ionic affinity constants are of relative value, since they are most likely determined by oligomerization. In contrast, individual KCC1, 3 and 4 expression in *Xenopus* oocytes provides kinetic values[472] indicating that the sequence for K_m values is KCC3 < KCC4 < KCC1, which is interesting with respect to the kinetic property of an individual KCC isoform but of little practical value in mammalian cells with several isoforms. Thus based on the published values[517] one can only speculate that in RBCs, perhaps KCC3 dominates over KCC1 and 4.[78] This yet unclear situation also tempers the interpretations of the kinetic characteristics of Cl-dependent K fluxes defined by us some 20 years ago, well before we knew the molecular mechanism. In LK sheep RBCs with the internal K composition altered by the nystatin method,[97] K influx and efflux were studied under zero-K trans conditions under which the translocation of the loaded carrier was rate limiting. It was found that, at the external surface of the membrane, Cl bound first followed by K, and both were released and bound randomly at the cytoplasmic side.[577] This model differs from that given for NKCC where, externally, the cation binds first, followed by the anion, both being released inside in the same order[578] and where trans-K stimulates K movements from the cis-side.[489]

2.5.3.2.3 Regulation

Regulation of KCC activity is diverse and, mostly based on erythrocyte work supplanted by additional insight from molecular biological studies, can be partitioned into metabolic, ionic and "other" mechanisms. KCC has been found to be involved in the maintenance of Cl_i levels away from electrochemical equilibrium and in the K buffering capability during neuronal function.[527,579] As indicated above, one of the major characteristics of KCC is to regulate cellular volume after osmotic challenge[64,108,517,521,546,558,572,580–582] and to respond with activation under oxidative conditions.[64,517,543,547,548,556,558] These primeval and critical cellular functions require activation of a complex network of regulatory mechanisms as soon as the system is subject to minute deviations of its normal steady state.

Metabolic regulation
A cascade of kinases and phosphatases affecting protein conformers *via* different ATP-driven phosphorylation levels has been proposed as the main regulatory machinery controlling KCC.[63,64,108,485,558,568,583–588] Based on inspection of the activation kinetics after acute swelling in hyposmotic solutions and subsequent shrinkage in hyperosmotic solution, as well as the time-dependence of the NEM stimulation, Jennings proposed a bimolecular reaction mechanism controlling by phosphorylation and dephosphorylation the K-Cl cotransport mechanism as adapted in Scheme 2.2.[583,588] On the left side is the resting (R) phosphorylated and on the right side the activated (A) dephosphorylated KCC mechanism. The parameters k_{12} and k_{21} are the forward and backward bimolecular rate constants, respectively, which are controlled by a calyculin-sensitive PP1 and serine-threonine phosphokinase S/T-K, most likely SPAK *via* the NTD,[489] and WNK1 *via* the CTD,[525] with the length of the arrows indicating the prevailing equilibrium in isotonic media. This model, originally based on kinetic data obtained in rabbit RBCs,[538,588] has principally stood the test of time and promoted search for the phosphatases and kinases involved. Accordingly, activation by either swelling and thiol modification then would reduce $k_{2,1}$, so that $k_{1,2}$ becomes the dominant rate leading to activation of KCC by dephosphorylation.

Comparing this scheme with the findings on NKCC regulation by phosphorylation/dephosphorylation in Section 2.5.3.1.3 (Scheme 2.1), it is obvious that both KCC and NKCC are antithetically regulated. While

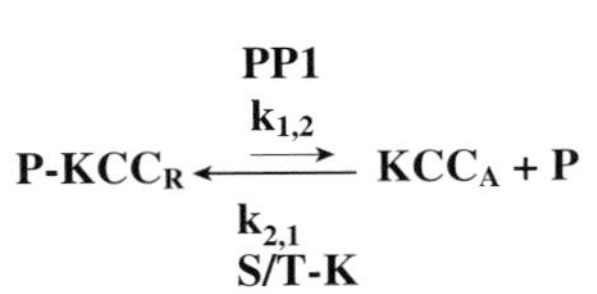

$$\text{P-KCC}_R \underset{\underset{\text{S/T-K}}{k_{2,1}}}{\overset{\overset{\text{PP1}}{k_{1,2}}}{\rightleftarrows}} \text{KCC}_A + \text{P}$$

Scheme 2.2

upon phosphorylation NKCC is activated, KCC is at rest, and as KCC is dephosphorylated and hence activated, NKCC becomes inactive. Commensurate with this conclusion of a coordinated action, MgATP is needed to maintain the KCC at rest in sheep RBCs.[133] Mg depletion by the A23187 method[132,547] led to a several fold activation of KCC, independent of whether cells were swollen or shrunken, i.e. independent of cell volume. Furthermore, the effects of Mg and MgATP were independent of each other.

In spite of being practically silent under physiological conditions,[545] the KCC has surprised investigators given the different regulatory mechanisms controlling its activity. At the same time, such signaling pathways have obscured interpretation of data obtained when more than one modulator was used to uncover interactions between "sites", activation processes, equilibrium states, or biophysical parameters, such as volume set point, pH set point, and others. A detailed dissection and identification of the different enzymes involved has proved to be challenging, in part due to the lack of a specific inhibitor for this transporter. However, some "selective" inhibitors have been instrumental such as H74,[493] DIOA[64,108,558] (although recently shown to inhibit IK channels *via* Cl channels in lens epithelial cells),[589] and others (see above). Recently, Delpire and colleagues have identified inhibitors for the neuronal KCC2 isoform.[589]

The enzymes known to be involved in the regulation of this transport system have been studied not only in the erythrocyte but also in other cell types. Exploration of the kinase(s)/phosphatase(s) involved has made use of inhibitors or activators of these proteins. In sheep RBCs, staurosporine, a general protein kinase inhibitor, NEM and other thiol reagents, and glutathione removal, activate KCC[64,108,558,568,590] whereas calyculin A, a protein phosphatase inhibitor, inhibits KCC.[63,64,108,523,553,558] Both serine/threonine protein phosphatase-1 (PP1) and -2 A (PP2A) localized in human RBC membranes activate KCC and this activation correlates with the activity of these enzymes.[591–593] Genistein, an inhibitor of tyrosine kinases, inhibits activation of KCC by Mg_i-depletion, staurosporine and NEM.[63,64,108,523,558] In mammalian RBCs, ML-7, a selective myosine light kinase inhibitor, activates KCC in a volume-dependent manner, suggesting inhibition of a putative volume-sensitive kinase.[594] The Src-family kinases Fgr and Hck negatively regulate KCC in mouse erythrocytes.[595,596] However, studies are still being made of the mechanism(s) although recent information indicates that the src kinase pathway inhibits the activity of PP1α and in turn KCC activity.[596] Volume-clamping of sheep RBCs pointed to participation of kinases/phosphatases in KCC regulation (see above) by pH, Mg, and ATP.[133] Protein kinase C has also been found to modulate swelling-activated KCC through the phosphatidylinositol pathway[64,108,558,567] and, apparently, protein kinase G and the cGMP pathway as well,[63] although further studies to document existence of this pathway in erythrocytes are still lacking. PKC also inhibits KCC in frog erythrocytes, which is mostly regulated by tyrosine kinases and phosphatases.[597,598] In dog RBCs, the coordinated regulation of KCC and Na/HX has been proposed to occur through a kinase/phosphatase cascade.[263] In HK dog RBCs, nitrite activates KCC and increases the K_m values of the transporter.[599] In this species, KCC1 was found to be dominant although no difference was found in KCC protein expression between HK and LK RBCs, pointing to a difference in regulation and susceptibility to oxidation between the two RBC types.[599]

As with NKCC1 and 2, the stress-related serine-threonine kinases SPAK and OSR1 interact with KCC3, but not with KCC1 and KCC4[502] suggesting that KCC3 is the isoform responding to stress.[64,108,558] WNK kinases also regulate the KCCs. Both WNK3 and WNK4 inhibit these transporters under different experimental conditions and in certain pathologies.[504] Peroxynitrate stimulates KCC in a dose-dependent manner, independently of the src kinase pathway, suggesting a direct oxidative effect by peroxynitrate on KCC.[561]

Ionic regulation
As discussed before, KCC as NKCC responds to manipulations of a variety of monovalent and divalent ions, the latter with and without combination of ATP. In dog RBCs, KCC and Na/HX are regulated in a coordinated fashion whereby shrinkage, Mg_i and Li_i activate Na/HX and inhibit KCC, whereas swelling inhibits the exchanger but activates KCC.[263,390] The regulation of these transporters is affected by the intracellular ionic strength, perhaps affecting macromolecular crowding.[391,495]

Intracellular Mg (Mg_i) and ATP are important modulators of KCC in SRBCs.[64,133,319,558] The effect of ATP depletion in these cells is a function of medium osmolality and thus of cell volume, whereas the inhibitory effect of Mg_i or activation of KCC in Mg_i-depleted cells is independent of medium osmolality.[132] Furthermore, Mg_i or its complex with ATP do not interfere with binding of Cl or K to the transporter.[132] The complex effect of pH, Mg, and ATP in volume-clamped sheep RBC KCC supports the conclusion of a regulation by phosphorylation/dephosphorylation cascades.[133]

In LK SRBCs treated with DIDS (to clamp the intracellular and extracellular pH to be varied either simultaneously or independently), NEM was shown to interfere with the internal but not with the external pH-sensitive site.[600] Using the pH-clamp technique,[552] it was found that internal protons modulate KCC activity in LK sheep RBCs stripped of their Mg. Between pH_i 9 and 6 and at constant pH_o, K-Cl cotransport decreased in a hyperbolic fashion from which a K_d of 6.5–7 was calculated. This suggests that histidine (HIS) residues are involved in the KCC transport mechanism. A Hill coefficient of 1 suggests no cooperativity and perhaps one specific HIS group.[102] These early data have been eagerly confirmed in *Xenopus* oocytes transfected with KCC1-4: rbKCC1 and −2 showed a similar K flux activation between pH 6.5 and 7.5; rt KCC2 with a shift to pH 7.5–8; KCC3 a bell-shaped pH activation peaking at pH 7, and KCC4 an inactivation between pH 7 and 8.[601] Again, the RBC work was paving the way for molecular biology.

Physiologically relevant is that HCO_3 ions inhibit KCC in sheep RBCs with a Hill coefficient of 2, suggesting more than one inhibitory site.[602] Inhibition of KCC by HCO_3 was corroborated in human HbSS RBCs exposed to solutions augmented with autologous plasma.[603] Foreign anions were shown to modulate KCC under different conditions, and these effects have been interpreted as modulations of the volume set point of the transporter, with differential behavior between HK and LK cells. Interestingly, exposure of LK sheep red cells at 37°C to non-Cl anions prior to Cl caused a reversible activation of Cl-dependent K transport with the following sequence: SCN>I>NO3>Br.[604] This unusual chaotropic anion effect was explained in terms of change within the lipid bilayer that are sensed by the K-Cl cotransport mechanism, a finding certainly of interest in terms of elucidating the volume sensor of these coupled anion-cation cotransporters. The phenomenon may be also related to shielding of intracellular charges and related to the polyamine effects reported by Sachs 1994 in human RBC ghosts: soluble polycations and cationic amphiphiles inhibit KCC independently of the intracellular K concentration and through negative charges at the inner membrane surface.[605]

Other effectors

Stimulation and inhibition of K-Cl cotransport varies between intact and erythrocyte ghosts,[606] and according to the species.[517,583] For instance, in human, sheep, and pig, basal K-Cl cotransport decreases with cellular maturation.[493,517,545,572,607,608] In young human and pig RBCs, KCC appears to be active and becomes silent although responsive in mature erythrocytes.[493,521,608–610] Undifferentiated MEL cells also possess KCC activity that decreases ten-fold during their transition to young reticulocytes.[496] However, the physiological inactivation appears to be prevented in low-K sheep RBCs.[517,521]

Expression of mKCC1 in *Xenopus* oocytes is modulated and regulated by effectors and signaling pathways as described in the erythroid KCC.[531] Also in *Xenopus* oocytes and HEK293 cells, the KCCs form oligomers when expressed in the membrane,[611] and, with implications for RBCs, KCC4 and NKCC1 hetero-oligomerized as suggested from immunoprecipitation experiments when co-expressed in *Xenopus* oocytes.[612]

Numerous other effectors modulate KCC activity in RBCs. In sheep the activity is higher in LK than in high K (HK) erythrocytes under all conditions tested. As pointed out in Section 2.4.2.3, sheep RBCs are either of LK or HK nature. The apparently significantly higher KCC activity in LK cells than in HK cells can be down-regulated by an immunological reaction involving the L_l antigen and the anti-L_l antibody.[602] In common with the nature of the L_p antigen, that of the L_l antigen is unknown. This modulation is not affected by interfering with kinase inhibitors.[253] In adult human RBCs, high hydrostatic pressure activates KCC and this stimulation involves modification of the phosphorylation/dephosphorylation cascade.[613] Furthermore, oxygenation-deoxygenation transitions of Hb appear to be involved in the re-organization of the RBC cytoskeleton and lead to signaling pathway activation/deactivation in an O_2-dependent manner.[129] As for NKCC and NHE, elevation of temperature affects KCC. Mild warming (37–41°C) activates KCC in guinea pig RBCs.[614] Preincubation of rat RBCs at 49°C irreversibly arrests shrinkage-activated Na/HX and NKCC and swelling-activated KCC with concomitant denaturation of spectrin, commensurate with involvement of the cytoskeleton in erythrocyte volume-regulation and different mechanisms of regulation for these transporters.[615] Prior to the cloning of KCC,[526,533] inhibition by cinchona bark derivatives uncovered heterogeneity of the transporter in SRBCs,[566] probably caused by isoform- or signaling pathway-specific effects.

2.5.3.2.4 Translational significance

Abnormalities in the KCCs gene structures, function or regulation have been implicated in the manifestation of hematological diseases and pathological conditions. In human hemoglobinopathies, including sickle cell anemia (SCA), K-Cl cotransport has been shown to cause cellular dehydration and volume decrease[403,405,490,517,558,616–625] and even more elevated in young RBCs.[403–405,617–620] A mathematical model to explain reticulocyte volume, pH, and ion content under physiological and pathological conditions led to a novel mechanism of the derivation of irreversibly sickled cells directly from reticulocytes.[231] In this model, KCC was shown to play an important role in the irreversibility of reticulocyte dehydration.[231] Numerous studies have followed earlier findings in erythrocytes from human and animal models of different species, including new therapeutic avenues *in vitro* and *in vivo* addressing the mechanisms of RBC dehydration, ion transport, and associated phenomena such as Hb polymerization and RBC adhesion to vascular cells. These studies are reported or reviewed elsewhere.[405,406,522,617–621,624–635] Furthermore, KCC was shown to be involved in K loss and dehydration of RBCs stored under blood bank conditions (4°C) and afterwards incubated in autologous plasma at 37°C due to activation by acidification and swelling of the erythrocytes at 4°C.[636]

In RBCs from patients with mutations in the KCC3 gene, as in hereditary motor and sensory neuropathy with agnesis of corpus callosum (HMSN-ACC), K-Cl cotransport is affected differently by thiol alkylation and Mg_i removal compared with normal KCC3 human RBCs, emphasizing the importance of this isoform in overall KCC function and regulation.[78] Furthermore, KCC3 knock-out ($KCC3^{-/-}$) mice are severely hypertensive when compared to wild type.[64] Also, KCC3 has recently been found to be involved in nerve conduction, which is impaired in KCC3 knock-out mice,[637] further supporting the validity of this model for HMSN-ACC patients. Disruption of both the KCC1 and KCC3 genes in mice indicates that RBC KCC activity is primarily mediated by KCC3 in this model, whereas disruption of both KCC1 and KCC3 in transgenic sickle cell (SAD) mice rescues the RBC dehydration phenotype, except for the densest fraction.[634] Human cervical cancer cells have an increased mRNA expression of KCC1, KCC3, and KCC4 transcripts and the system plays an important role in cell volume-regulation in these cells.[638] However, while the KCC1 and KCC4 isoforms are involved in regulatory volume decrease, KCC3 is refractory to osmotic swelling.[639] Instead, KCC3 is more selectively inhibited by DIOA than the other isoforms.[639] Mutations in the human cervical cancer cell KCC1 isoform indicate that this isoform participates in growth and invasion.[640] KCC3 is involved in breast cancer cell proliferation; it is abundantly expressed in cervical cancer[641] and plays an important role in epithelial-mesenchymal transition, an indicator of malignancy.[642,643] Alterations in KCC4 were found to cause deafness.[644] This isoform also appears to play a major role in the spreading of breast, cervical, and ovarian cancer.[642,643]

2.6 The Ca-Activated (Gardos) or $K_{Ca}3.1$/IK/KCNN4/sK4 Potassium Channel

2.6.1 *History*

In 1956 Gardos and Staub reported that treatment of human RBCs with iodoacetamide, a SH–alkylating agent and metabolic poison used to search for evidence of active Na/K transport, caused intracellular ATP to fall and a loss of K and cell volume.[11–13] At that time, the constant field equation was the dominating theory explaining ion movements in general permeability terms rather than channels, so Gardos himself did not associate this phenomenon with the term "Gardos channel". However, he realized that Ca overload of the erythrocyte caused the effect since the presence of external Ca chelators attenuated the K loss.[645,646] Ever since this discovery, the search was on for the molecular basis of this "Gardos effect". Indeed, a maximum value of Ca uptake of 3–7 micromoles/liter cells was found to suffice for the increased K permeability effect.[66] With the molecular mechanism unraveled, George Gardos was honored by the naming of the $K_{Ca}3.1$, KCNN4, small conductance 4 (SK4) or intermediate conductance K channel (IK) "the Gardos channel". This name is used, if not in all systems, where it was found later, but certainly in RBCs,[9] and in the cardiovascular system.[647] Based on this classic work in the red cell, the now-defined "Gardos" effect or IK channel activation takes a paramount position in cellular physiology and pathology.

2.6.2 *Structure*

The structural identity of the $K_{Ca}3.1$, IK, sK4, or KCNN4 channel (Table 2.2) gene maps to chromosome 19q13.2[648–651] and its open reading frame encodes a ~>50 kDa protein with 425 amino acids spanning the plasma membrane six times, with both a short NTD and long CTD in the cytoplasm (see Fig. 2.12).

The IK channel has now been seen in the pancreas[648]; salivary gland, placenta, and trachea[651]; vascular smooth muscle[652]; lymphocytes and hematopoietic system[649]; and is volume-activated in human lens epithelial cells.[653] Using RT-PCR on reticulocyte RNA extracts and Western blots in mature RBC membranes, hSK4 (KCNN4) — but no other SK channel — was shown beyond doubt to be present in erythrocytes and thus identified as the "Gardos" channel.[9] As an inward rectifier and with a unit conductance of 30–50 pS,[648,652] KCNN4/IK/SK4 moderately hyperpolarizes the membrane potential, inactivates voltage-gated Ca channels, and opens voltage-independent channels (transient receptor potential channels, Trps) in endothelial cells.[647] Its ionic selectivity is K = Rb>Cs>>Na, Li. Erythrocytes from KCNN4 wild type (wt) mice shrink in the presence of ionomycin and Ca while those from KCNN4 null mice fail to do so.[654] The MOF in RBCs from KCNN4 knock-out mice, exposed to A23187 and Ca, is 0.47% and from the KCNN4 controls 0.28% relative tonicity consistent with the protective effect of the presence of the IK channel.[654] Selective inhibitors should show effects similar to deletion of KCNN4. The calmodulin-binding motif in

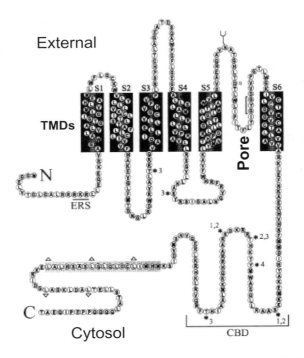

Figure 2.12 Suggested membrane topology of the KCa3.1, IK, SK4, KCNN4, or Gardos K channel. The 425 long polypeptide crosses the membrane in six TMDs (S1–S6), with the pore between TMD 5 and 6, and has a cytosolic NTD (N) and CTD (C). ERS: endoplasmic reticulum retention sequence; CBD: the putative calmodulin-binding domain; and the numbers with asterisks 1–4 denote phosphorylation consensus sites for PKA, PKG, PKC, and TK, respectively (figure modified from Ref. 652).

the CTD is important for the regulation and activation of IK, although it is not the K-conducting pore.[655] There is no voltage-sensitive sequence, consistent with the fact that all SK channels are only Ca-calmodulin-activated.[647]

2.6.3 *Function*

Inhibitors

Selective IK channel inhibitors are the anti-mycotic histidine-derivative chlotrimazole (CTZ), the triarylmethane TRAM-34 and the scorpion venom charybdotoxin (CTX). For more details consult.[647] At pH 8 and low ionic strength ~120 and 126 ^{125}IChTX, binding sites per human and rabbit RBC were found with a K_d of about 94 and 37 pM, corresponding to an IC_{50} of IK channel inhibition of 21 and 25 pM, respectively.[656] Given the fact that in human RBCs, Ca-activated K efflux measures in the order of >100 mmol/L cells × h (unpublished data), we estimate an IK or "Gardos" channel turnover of ~2×10^5 K ions/sec for human RBCs. There is evidence for proton modulation of the IK channel.[657] The "Gardos" channel plays a major role in the dehydration of HbS RBCs and hence contributes significantly to the polymerization of deoxy-Hb, a process which is a 10^{20+} power function of the available cell water. Thus the search for *in vivo* compatible inhibitors is crucial for therapeutic success, and CTZ has long been used in clinical trials to reduce red cell dehydration.[406,626,658–660] The chemical composition of CTZ is 1-[(2-chlorophenyl)(diphenyl)methyl]-1H-imidazole, and it has been shown that the histidine moiety is not responsible for the IK inhibition[661] and there is evidence that it binds/acts from the inside. ICA-17043 is a more recent drug that inhibits IK channels in mice with higher potency than CTZ.[662] Other attempts to interfere with the dehydration of RBCs, especially of HbS cells, made use of dimethyl adipimidate acting through P_{sickle},[663] or inhibitors of Cl conductance.[621]

Calcium entry

Whereas it is known that a rise in cytosolic Ca activates the hyperpolarizing IK channel, the big question still under study is by which mechanism Ca enters the cell. A tentative mechanistic explanation of how we need to understand the activation of the IK channel is given in a superb illustration from Lew's laboratory[663] in Fig. 2.13. In HbS cells, stochastic Ca permeabilization determines dehydration *via* the Gardos channel.[664] In normal human RBCs the Ca permeability is randomly activated.[665] Current data suggest Ca must enter *via* various proposed mechanisms involving nonselective voltage-dependent ion channels that are still to be defined,[666] possibly through P_{sickle} or closely related pathways,[667] and involving a coupled feedback between the Gardos channel and the Ca pump.[668] A nifedipine-sensitive Ca influx induced by the Ca-pump inhibitor vanadate suggested a Ca channel.[669] Nitrendipene has been tested as potential inhibitor of the Gardos channel[670] blocking the Ca-entry step.[671] Most recently, it has been shown that carbon monoxide-induced deoxygenation of

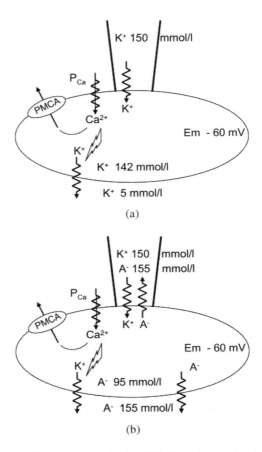

Figure 2.13 Diagram illustrating the Gardos (IK) K channel activation. (a) Erythrocyte-attached high K pipette in series with cytosol and extra-cellular space. Entrance of Ca ions through various possible mechanisms discussed in Section 2.6 and 2.9, overwhelming the PMCA, causes in (b) opening of IK channels, cytosolic entrance of K from the pipette, and membrane hyperpolarization with simultaneous anion (A$^-$) loss through conductive pathways (with permission and adapted from Ref. 137).

HbS inhibits the HbS polymerization-induced change in Ca entry needed to precede the Gardos channel activation.[672] A Ca-ATPase channel has also been proposed as Ca entry similar to the B-type Ca channels in cardiac myocytes.[673] Oxidation of the Ca pump by xenobiotic coupling reagents such as 1-Cl-2, 4, dinitrobenzene (CDNB) seems to be another IK channel stimulant[674,675] acting by reducing GSH and, indirectly, its anti-oxidative effect on the Ca pump.

2.6.4 Regulation

The physiological role of the IK channel is still being explored.[676] Prostaglandin E2 released from stimulated platelets activates the Gardos effect in human red cells, and through IK promotes K, Cl, and water efflux leading to cell shrinkage and contributing to RBC aggregation during thrombus formation. Experimentally, PGE2 action is stimulated by NO^3 ions more readily passing band 3.[677,678] Homocysteine binding to rat erythrocyte N-methyl-D-aspartate (NMDA) receptors[679] induces Ca entrance and thus activates IK channels and cell shrinkage possibly contributing to clot formation.[680]

Recently, Florian Lang and collaborators have proposed a novel hypothesis that erythrocytes may utilize the IK or Gardos channels to undergo apoptosis, termed eryptosis, during which the associated membrane phosphatidylderine (PS) asymmetry breakdown is recognized as a signal for their removal by the reticulo-endothelial system.[681,682] Corroborating are the findings that *Escherichia coli* α-hemolysin appears to exert a similar effect, such as PS externalization and Gardos channel activation.[683] Given that the anion conductance may be rate-limiting for the IK channel, it is not surprising that the ionomycin + calcium-induced K loss, annexin V-positive staining of PS externalization, and subsequent eryptosis were attenuated by the Cl channel inhibitors 5-nitro-2-(3-phenylpropylamino)benzoic acid (NPPB) and niflumic acid.[684] The relationship of this Cl channel to the acid-sensitive outwardly rectifying (ASOR) Cl channel that is both inhibited by DIDS and NPPB but not niflumic acid[685] awaits further clarification.

It is likely that any stress on the erythrocyte such as osmotic, oxidative, or energy depletion and a variety of causative agents including drugs and divalent heavy metals[686] causes opening of non-selective Ca-permeable cation channels with the end result of clearance of the eryptotic cells.[682] Heavy metals such as lead may exert multiple effects, not only on Ca entry but also replacing Ca in the activation of IK and are causative in lead-induced anemias (see Chapter 7). Loss of K *via* the Gardos channel, with Cl and obligatory water, has been reported to protectively attenuate the tendency toward hemolysis (and probably eryptosis) of erythrocytes with cytoskeletal defects in hereditary spherocytosis.[687]

Normal RBCs have about 238 endothelin-1(ET-1) receptors to which ET-1 binds with a K_d of 128 nM.[688] ET-1 stimulates a CTZ-sensitive K influx in both HbA and HbS red cells that is also prevented by ETB receptor antagonists.[688] Also, in HbS RBCs of transgenic sickle (SAD) mice, elevated ET-1 seems to modulate the Gardos channels since ET-1 receptor antagonists decrease red cell density, i.e. increase red cell hydration.[689,690] That the ET-1 receptor protects RBCs from eryptosis by a Ca-entry-elicited mechanism leading to IK channel activation, was nicely shown in ET-1 receptor etb$^{-/-}$ knock-out mice which exhibited a greater phosphatidylserine externalization and a faster clearance from circulation.[691]

Other species

That adult, but not fetal, sheep RBCs can be ATP-depleted by deoxy-D-glucose in Ca-containing buffers without loss of K,[79,319] is due to the absence of Gardos channel activity.[318] Adult sheep RBCs also have a very low PMCA[317] and cattle RBCs exhibit an extremely low PMCA as well.[692] Thus, in analogy to sheep RBCs, cattle RBCs are probably also devoid of significant IK channel activity. In spherocytic RBCs of murine band 4.1 knock-out mice, the Gardos channel is proposed to protect against hemolysis due to membrane instability since treatment with CTZ increased hemolysis.[687]

2.7 Cl/HCO$_3$ Anion Exchange/Conductance or "Band 3", "Capnophorin", AE1, and SLC4A1

2.7.1 *History and significance*

In 1959 Daniel C. Tosteson[46] published a paper on rapid halide equilibration in cattle RBCs and stated in his summary:

> The ratios of cell to medium concentrations of the various halides at equilibrium were measured by isotope dilution and found to be 0.58, 0.63, 0.50 and 1.06 for ^{38}Cl, ^{82}Br, ^{18}F and ^{131}I, respectively, in bovine red cells. Similar ratios were observed in human cells. The rate constants for ^{38}Cl

outflux from bovine red cells washed and suspended in bromide-phosphate and iodide-phosphate media were 1.8 and 0.6 sec-1 respectively, both much lower than the value of 3.1 observed in chloride phosphate medium.

This finding established that some Cl was "exchanging" across the red cell plasma membrane, the remainder by diffusion, and thus set the stage for the future discovery of the Cl/HCO_3 anion exchanger, AE1, first coined "band 3" on electrophoretic basis.[20]

Depending on the level of activity, the pulmonary capillary transit time for human red cells is between 0.3 and 0.7 sec. Thus the erythrocyte requires an anion transporter that permits fast exchange of Cl with HCO_3 arising from the carbonic anhydrase-controlled hydration and dehydration CO_2 equilibrium:

$$HCO^-_3 + H^+.$$

As shown in Fig. 2.14, metabolically derived CO_2 diffuses through the now-recognized specific transport proteins (Section 2.8) from the tissue into the red cell that, in contrast to the surrounding plasma, contains carbonic anhydrase which accelerates by 10^4-fold the formation of HCO_3 to be exchanged for Cl. In the lung, differences in partial pressures drive CO_2 into the alveoli whilst HCO_3 enters the RBC in exchange for Cl. This cyclic process is tightly coupled to and requires the proton-buffering capacity of oxy-Hb (Haldane effect), which in turn defines the trans-membrane distribution of Cl *via* the conductive property of band 3 (see below) and thus V_m. Absence of band 3 protein, as in a cattle mutant (see below), interferes with tissue-derived upload of CO_2 into the blood, increasing its partial pressure difference. In order to achieve this gas transfer, band 3 protein or AE1 constitutes some 30% of the total RBC membrane protein, which, as estimated by binding of negatively charged DIDS, first used as radioactive H_2DIDS, translates into ca 10^6 sites/cell.[103,104] Molecular Cl/HCO_3 turnover estimates are $10^6/(AE1 \times sec)$ at low temperature. Based on this and the numbers of AE1/cell and RBC/L, respectively, one can only marvel at the capacity of AE1 to exchange some 10^5 mol of $HCO_3/(L$ of cells $\times$ h) and hence indirectly CO_2, justifying its name "capnophorin" from the Greek καπνοσ φερειν (smoke carrying).[693,694]

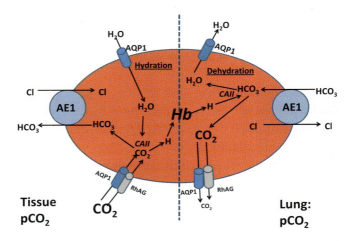

Figure 2.14 The significance of erythrocyte AE1 in the carbonic anhydrase-II (CAII) catalyzed CO_2 hydration and dehydration reactions (Jacob–Parpart–Stewart cycle and Hamburger shift, see also Section 2.3.2.). *Left half*: In tissue with high pCO_2, CO_2 enters the cell *via* AQP1 and RhAG (see Section 2.8). In the presence of CAII, CO_2 is hydrated to H_2CO_3 dissociating instantaneously into H, buffered by hemoglobin (Hb), and HCO_3, exchanging with external Cl and creating the external alkali reserve. *Right half*: In lung with low pCO_2, the process is reversed and in exchange for cellular Cl, HCO_3 enters the erythrocytes reacting with H from Hb to form H_2CO_3. Upon CAII-catalyzed dissociation of the latter, CO_2 leaves *via* AQP1 and RhAG, and water exits through AQP1. In Fig. 2.18 it will be shown that CAII is intimately associated with AE1 in its function to catalyze hydration to and dehydration from H_2CO_3. With the CAII inhibitor azetazolamide (Diamox) or a natural or genetically engineered absence of AE1 (see Sections 2.7.5.1–2.7.5.3) the hydration/dehydration rates for CO_2 and hence the HCO_3/Cl are sharply reduced.

Even given this phenomenal exchange capacity, a biophysical analysis of the rate-limiting steps suggests that the alveolar transit time may not be sufficient for the required CO_2 release and HCO_3 uptake, and must become gas delivery rate limiting under exercise or anemic conditions.[693]

2.7.2 Structure

The anion exchanger (Table 2.2) is now assigned as solute carrier 4A1 SLC4A1[695] to chromosome 17 in man,[696] 11 in mouse,[697] and 10 in the

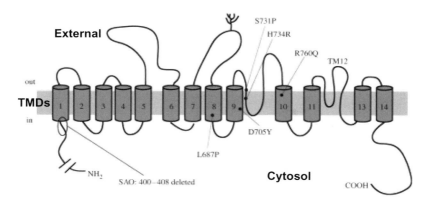

Figure 2.15 Cartoon with AE1 membrane topology. The eight amino acid deletions at the NTD in SAO, and five important single point mutations conferring cation selectivity are indicated on or near TMDs 8, 9, and 10 (figure modified from Ref. 755).

rat.[698] The SLC4A1 gene is co-expressed with other erythroid genes during mammalian and avian erythropoiesis[699] and hence its protein is present early in RBC development. As suggested in Fig. 2.15, with about 95 kDa molecular weight and 911 amino acids, the human red cell AE1 has up to 14 transmembrane domains (TMDs) and is N-glycosylated at Asn642 by a complex high mannose-oligosaccharide in the extracellular loop (ECL) 4.[700–702]

Electronmicroscopically, AE1 exists mainly in dimers due to inter-helice interactions between the TMDs of the CTD.[703,704] Each monomer has two cytoplasmic domains connected by a hinge. The 52 kDa CTD conducts the Cl/HCO_3 exchange,[700] and is important for traffic of AE1 to the plasma membrane.[705] The CTD self-rotates within the lipid bilayer due to its hinge connection with the 43 kDa NTD which, in turn, connects the entire AE1 *via* ankyrin (band 2.1) to the cytoskeleton provided by $\alpha\beta$ spectrin heterodimers.[700] The association of AE1 with ankyrin requires Cys201 of one monomer to interact with Cys317 of the other, and derivatization of these and additional 2 Cys residues abolishes ankyrin binding.[706] An 11 amino acid-long hairpin loop of the NTD interacts with ankyrin.[707] Adducin, by stabilizing the interaction between actin and spectrin, may exert long-distance effects through band 3 on the transport properties of the red cell.[708] Based on crystallographic maps at 20 Å resolution, an apparent pore — a rectangular core around a central depression — is formed by two subdomains of each monomeric 52 kDa CTD.[709]

2.7.3 Function

The AE1-mediated Cl exchange flux shows a somewhat lopsided bell-shaped pH-dependence rising from pH 7.2–7.8 to a pH maximum when lowering the temperature from 38°C to 0°C.[710] The $K_{0.5}$ for Cl decreases from 65 to 28 mM at the two temperatures, indicating an increase of the apparent Cl affinity at lower temperature. However, Cl transport increases by 200-fold between 0°C and 38°C.[710,711] Based on the pH dependence of AE1-mediated Cl fluxes in human RBCs between pH 5.7 and 9.2, Robert B. Gunn postulated in his titrated carrier model that protons titrate the carrier into a less functional form.[15] The pk_a of this titrated group of ~5.2 suggests a glutamic acid residue involved in the titration from the monovalent to a divalent anion exchanger.[712] Based on bromide equilibrium exchange studies, the anion exchange has been shown to be a ping-pong mechanism with reciprocating sites of asymmetric anion affinities.[713,714] More recently, Jennings confirmed that the catalytic cycle of AE1 consists of one pair of anions per monomer band 3 protein exchanging in a ping-pong mode.[715]

The nature of the anion binding site was in part explored with the aid of the negatively charged DIDS whose binding involves sites that interact with anions, in contrast to other inhibitors such as fluvenamic acid acting at different sites[716] or NAP(N-(4-azido-2-nitrophenyl)-2-aminoethylsulfonate)-taurine acting at modifier sites.[717,718] With a K_d of 3×10^{-8} M at equal internal and external Cl concentrations, DIDS is the most powerful inhibitor of AE1.[719] DIDS binds in a kinetically first fast phase *via* its sulfonyl groups to Lys539, and in a second slow time- and temperature-dependent irreversible phase to Lys851, a difficult site to access.[720,721] The DIDS binding site may be located in a cleft at the outer anion hemi-channel[722] inhibiting transport by allosteric effects rather than direct block of the presumable pore.[723] In murine RBCs, Lys558 and Lys869 have been implicated in binding of H_2DIDS.[724] Expression in oocytes of mouse AE1 mutated at Arg509 and 748 created protein misfolding and transport inactivation.[725] In Southeast Asian ovalocytosis, SAO, deletion of 9 amino acids at the NTD (400–408) near TMD1 causes hydrophilic residues lower than 400 to become a TMD1, thus affecting downstream the DIDS-binding site and glycosylation of ECL2.[726,727]

Further insight into the functioning of AE1 comes from specific *in situ* site modifications or from expression studies using mutants in *Xenopus* oocytes. Based on the transport of SO_4, a second external Cl site was proposed at glutamate 681 present in all AE1 across species.[728] Carboxyl side chain modification of E681, by treatment with Woodward's reagent and Schiff's borohydrate reduction to a primary alcohol, inhibited Cl exchange and accelerated SO_4 exchange and conductance[728] consistent with a proton-titratable site[15] and cotransport of SO_4 and H *via* AE1.[728] Thus, normal influx/efflux coupling does not require electroneutrality, because the negative carboxyl group on E681 travels together with one Cl across the membrane electric field, with two positive charges in exchange for the divalent negative SO_4.[729] To accomplish this, energy transfer experiments suggest coordinated large-scale conformational changes in AE1 during the transport cycle.[730] As shown in Fig. 2.16, the free energy

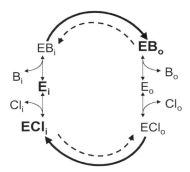

Figure 2.16 A qualitative interpretation of kinetic data revealing asymmetry of the Cl/HCO_3 (B) exchanger (AE1). Bold letters E_i>>E_o define the distribution of the empty carrier; ECl_i>>ECl_o and EB_o>>EB_i define the carrier loaded with Cl and HCO_3 (B). The major reason for the exchange asymmetry is that the exchange of B_i and Cl_o uses the faster rates as defined by the clockwise movements of the **solid** arrows, while the counter-clockwise Cl_i and B_o exchange occurs at slower rates, indicated by dashed arrows. Both processes are suggested to lead at any time to a preponderance of E_i >>E_o, i.e. functional carrier asymmetry (figure with permission and legend modified from Ref. 732).

is much lower for the inward than for the outward-facing loaded and unloaded AE1 conformations, which can be altered by replacing Cl in the external medium with HCO_3.[716,731,732] The cause of this seeming asymmetry of the Cl/HCO_3 exchanger remains unexplained. Since the asymmetry is largely temperature-independent the possibility of phosphorylation at various sites[733] may be excluded. There are differences in other species, such as trout, which lack cotransport of protons with SO_4.[734]

There is sufficient evidence that the conductive anion pathway utilizes the AE1 protein. The total RBC Cl tracer exchange "P_{Cl}" is about $>10^{-6}$ cm/sec, i.e. $>10^4$ times faster than Pcat (10^{-10}cm/sec),[97] whilst the net KCl flux in gramicidin-permeabilized RBCs is 0.7×10^{-8} cm/sec.[735] In the presence of sufficient DIDS concentrations, "P_{Cl}" also falls to $\sim 10^{-8}$ cm/sec due to inhibition of the electrically silent Cl exchange. This residual P_{Cl} is still 100-fold larger than P_{cat} and thus not rate-limiting for normal cation movements. In contrast to the competitive action of HCO_3 at the Cl site, halides not only modulate Cl-self exchange non-competitively *via* modifier sites[736] but are also exchanged for Cl with the following anion series in declining rate order: Br>>SCN>Cl = I>SO_4.[52] This rate limitation of the anion conductance explains the well-known phenomenon that replacing Cl — for example with SCN, or with NO_3, another highly permeable anion[737] — causes larger K fluxes through the [A23187 + Ca]-activated Gardos channel (Section 2.6). To understand the nature of the conductive pathway, DNDS, a less potent inhibitor than DIDS, was used in developing the concept of anion slippage or tunneling for Cl ions passing occasionally through the empty carrier.[106] However, continued increase of valinomycin-induced K flux at Cl concentrations saturating AE1 has been seen at variance with this concept.[738]

2.7.4 *Regulation*

Band 3's association with glycolytic enzymes with major functional consequences for glycolytic rate changes and Hb has long been recognized, especially by Philip Low and his group. Band 3 facilitates elevated glycolytic flux and lower pentose phosphate shunt activity in mammalian RBCs, by oxygenation/deoxygenation-dependent binding and debinding of glycolytic enzymes.[739] This is achieved *via* the NTD[740] directly interacting with the enzymes glyceraldehyde-3-phosphate dehydrogenase, aldolase, phosphofructokinase, and pyruvate kinase, which in the deoxygenated cell are within the cytosol and complex with band 3 in the oxygenated cell.[739] Apparently, band 3's NTD can be phosphorylated by the p72syk Y-kinase.[741,742] Phosphorylation by Y kinases at Y8 and Y21 of the NTD reversibly releases the glycolytic enzymes tying the process together with Hb oxygenation.[733] Simultaneously, band 3's N-terminus becomes available to enter the central cavity of the deoxygenated Hb.[743] As 2,3 DPG also binds to protein 4.1, the band 3-ankyrin interaction is modified upon oxygenation.[744]

Carbonic anhydrase isoform II, CAII, is tethered to the acidic residues D887ADD of the NTD of AE1. In the presence of H_2O and CO_2, the band 3- enzyme complex, known as a metabolon, directs the freshly catalyzed HCO_3 to the CTD residing anion "pore".[745] The CAII inhibitor acetazolamide and inactive CAII inhibit this process in AE1-transfected HEK293 cells.[746] Hence, band 3 becomes a nucleation site[747] at the membrane for a variety of metabolic functions, and must affect short and long range interactions with other proteins (see further below).

As discussed in Section 2.10, immunoprecipitation studies reveal that AE1 apparently forms macrocomplexes with a variety of integral and peripheral membrane proteins such as the Rh antigens, the Rh-associated glycoproteins, RhAG, CD 47, GPB, and the Landsteiner–Wiener antigen (LW).

2.7.5 Translational implications

2.7.5.1 AE1 mutations

More recent observations of the existence of a considerable number of human mutations in SLC4A1[748] (summarized recently by Ref. 749) and their pathophysiology (see Chapter 7 on hereditary spherocytosis) led to the discovery of dual modes of operation of AE1 with transitions to cation channels dependent on the mutation. Although it is estimated that the turnover of cations through mutated AE1 is about 1/10,000, the large number of functional AE1 proteins in the membrane seems to guarantee that $[Na]_i$ rises sufficiently in high Na plasma, which is followed by Cl and water entrance and hence cell swelling and certainly increased Na/K pump activity. Blood group En(a-) red cells lacking glycoprotein A (GPA) or GPA mutants with truncated C terminus exhibit reduced DIDS-sensitive transport, due to V_m changes for monovalent and K_m changes for divalent anions, respectively, although the number of AE1/cell is unchanged.[750] These apparent conformational changes in AE1 are prevented during the normal interaction of its proposed TMDs 9-12, and a region somewhere between residues 59 toward the C-terminus of GPA. In red cells infected with malaria parasites (see Chapter 8), two molecularly still undefined anion conductances occur, one of which is stretch-activated, exhibiting Ohmic behavior[751] and apparently similar to a channel activated by these parasites in chicken red blood cells.[752]

In the Southeast Asian ovalocytosis (SAO) or elliptocytosis, deletion of eight amino acid residues (400–408) at the NTD causes an entrapment of AE1 in the spectrin mesh[744] since the first TMD of AE1 is critical for membrane insertion of the entire protein.[753] Spontaneous hemolysis, first discovered in SAO cells when blood samples were shipped, as usual, at temperature below 10°C[754] has now been explained as due to a 10–100-fold elevated non-selective ion (Li, Na, K, Rb, Ca, and even taurine) leak *via* the abnormal AE1 that is non-saturated with the cation concentration[750,754] (see Section 2.9). Research is focusing on TMD 9 and 10 carrying the possible mutations causing functional changes.[755] The finding is also reminiscent of the earlier work by Stewart of a paradoxical low temperature-increased cation permeability in human red cells.[756] Other single point mutations in AE1 have similar effects, producing non-selective ion leaks, all of which are attenuated by AE1 inhibitors such as DIDS, SITS, and dipyridamole, ruling out involvement of Ca-gated K channels. Because of the low-temperature cation flux-associated cell swelling, the term cryohydrocytosis has been coined.[757] SAO probands exhibit distal renal tubular acidosis due to the presence of a similarly misfolded AE1 not properly incorporated into the luminal plasma membrane (for further details consult reviews Ref. 754,755).

2.7.5.2 *AE1 and rare blood groups*

Antithetical mutations in AE1 of Glu 658 which interacts with R61 of GPA give rise to the formation of the blood group antigen Wright

(Wra and Wrb).[757] An AE1 variant in the H_2DIDS-binding site is correlated with an extracellular conformational change and the Diego blood group antigen,[758] whereas the Waldner blood group antigen is associated with a single amino acid mutation in AE1, val557met.[759]

2.7.5.3 Absence of AE1

One would think that the absence of AE1 would be incompatible with life, as the red cell's Hamburger shift of HCO_3 with Cl becomes rate limiting for CO_2 exchange from the tissue. Indeed, cases of the homozygous form of SAO have been suspected to be incompatible with life, but cattle with a complete AE1 deficiency and lack of ankyrin, spectrin, actin, and band 4.2 survive, although the animals tend to be hemolytic, especially when stressed.[760] Physiologically, these animals have low serum total bicarbonate and a mild acidosis. The tendency to hemolysis may perhaps be due to the fact that the conductive anion permeability in these $AE1^{-/-}$ cattle RBCs is rate-limiting for the Ca-activated IK channel (Section 2.6.3.) to protect against cell swelling, and that other Cl channels, recently reported to compensate, do not suffice.[685] However, after various pro-eryptotic[686] challenges, $AE1^{-/-}$ mice RBCs show a greater Ca-entry-associated phospholipid scrambling and annexin V binding than their wt counterparts.[761] In the absence of AE1, one would expect an impairment of IK channel-mediated protection against hemolysis.[762] It remains to be seen by analogy whether cattle RBCs, like sheep RBCs,[79,318] lack the Ca-induced Gardos effect and thus also a naturally built-in protection against hemolysis.

If DIDS does not bind and reduce the anion permeability, and anti-AE1 antibody does not detect a 100 kDa protein, does it mean that AE1 is absent? Lamprey red cells which have a low HCO_3 exchange permeability (10^{-9} cm/sec) are DIDS-insensitive,[763] do not possess a mouse anti-trout AE1-immuno-reactive 97 kDa protein, but instead a ~200 kDa molecular weight protein,[764] suggesting a lamprey-specific AE1 is either missing[765] or constitutes an aggregate with different biophysical properties and physiological functions. Although carbonic anhydrase is present in these cells,[764] the low HCO_3 permeability limits an effective Jacob Stewart cycle in RBCs of this evolutionarily primitive species and hence gas exchange in the gills.[766,767]

2.8 Channels for H_2O, CO_2, and NH_3

2.8.1 *Aquaporins*

As discussed in Section 2.3.1, Hg reduced water P_f *close to P_d values signaling the presence of protein-mediated water* pores. In 1988, Peter Agre, in his quest to define the nature of the Rh substance (see Section 2.8.3), purified a 28 kDa protein, which on closer look was not the Rh antigen but a water channel, now called aquaporin (AQP)1.[3] Of the more than 13 AQPs discovered during the last 20 years, the red cell has, depending on the species, AQP 1, 3, and 9, of which only AQP1 transports water whereas the other two transport glycerol.[768,769] Therefore, of the some 13 genes known today, only the products for AQP 1, 3, and 9 will be discussed.

The 269 amino acids of AQP1, initially called Chip28 (channel-forming integral protein, molecular weight 28 kDa),[770] are encoded by an 807 bp open reading frame of four exons of a single gene located on human chromosome 7p14[771] and chr 6 in the mouse.[772] AQP1 has 6 α-helical TMDS, with the CTD and NTD in the cytosol.[770] Oocytes injected with CHIP28 mRNA showed a P_f increased by 50-fold which was reduced by Hg as predicted.[773] Out of four cysteines, Hg targets cysteine 189 and reduces P_f, as does its mutation to serine.[774] Since then, AQPs have been shown to play a major role in transcellular water movement throughout the animal and plant kingdom. Reconstitution of AQP1 into proteoliposomes yielded a water permeability of ~4 × 10⁹/(channel subunit × sec).[775] There are four individually active monomers in a functional tetramer as shown by 2-dimensional crystals[776] and the tetramer is required for stable membrane insertion.[777] Due to gene duplication, AQP1 consists of two halves arranged in an "hourglass" structure with the NPA (asparagines-proline-alanine) motifs in each half oriented 180° to each other, with the second half close to Cys 189 where Hg inhibits P_f (778) (see Fig. 2.17). The aqueous pore length of each monomer was calculated from the P_f/P_d ratio to be 36 Å.[779] In a 3.8 Å resolution model, the aqueous pore is lined with hydrophobic residues with a constriction to a diameter of 3 Å, indicating that water but not protons pass the pore.[780]

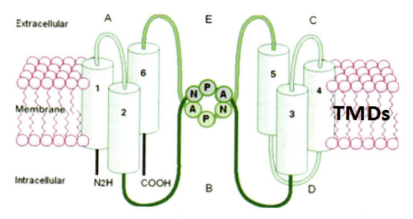

Figure 2.17 The "hourglass" membrane topology of an aquaporin monomer. The polypeptide chain transits the membrane in six TMDs interconnected by extra- and intracellular loops (a–e) with both the NTD and the CTD remaining in the cytoplasm. The two halves of the molecule consisting of TMD 1, 2, and 6, and TMD 3, 4, and 5 coalesce to form the water pore with their NPA motifs. Aquaporin exists as a homo-tetramer in the membrane. Modified from Chen et al. 2005.[917]

Recently, the paradigm that AQP only transports water was challenged by the finding that AQP1 in planar lipid bilayers showed an ionic conductance in the presence of cGMP. However, the ratio of water transporting to ion-conducting units was >10^6.[781] In contrast, AQP6 does not function as a water channel but can be induced by Hg to carry Na, and the NO_3/Cl ion ratio is ~10,[782,783] raising the interesting possibility of environmentally induced non-selective ion leaks through AQPs.

Antibodies against Colton (Co) blood groups bind to an epitope on the ECL1 of AQP1's first half, with alanine and valine at residue 45 for Co(a) and Co(b).[784] However, due to the presence of a variety of AQPs in other organs, there was no clinical phenotype in the absence of AQP1 and low P_f in red cells of Co(a–,b–) individuals.[785] Human and rat RBC AQP3,[768] as measured by stopped flow light scattering, also transports glycerol in Colton (a–b–) cells with a permeability of 10^{-6} cm/sec and thus is independent of the water flux through AQP1. In mouse red cells, AQP9 is the major pathway for glycerol influx and feeds malarial parasites, reducing the survival of normal mice in comparison to $AQP9_{null}$ mice.[769] Based on emerging regulatory aspects of non-red cell AQPs (c.f. calmodulin binding of aquaporin 6),[786] further discoveries of signaling pathways involving AQPs including AQP1, 3, and 9 in the red cell can be expected.

2.8.2 Non-AQP-mediated non-electrolyte solute transporters

Urea permeates human RBCs with a permeability of $\sim 10^{-4}$ cm/sec, i.e. close to that of water of 2.4×10^{-3} cm/sec.[99,787] The fact that PCMBS and phloretin reduced the urea movements to a ground permeability, suggested early the presence of drug-inhibited and carrier-mediated urea transport.[787] The finding that blood cells negative for the Kidd blood group (Jk(a–b–)) showed diminished urea (but not Cl) transport suggested early that a specific urea transport protein existed[788] that had no relationship to AE1. These data were confirmed by immunological detection of the urea transporter B (UTB) in human and rat erythrocytes.[789,790] A more recent update on the kinetics of glucose transport through glut 1 was given, suggesting earlier complicated transport models are redundant.[791] For further information on non-AQP-mediated glucose transport, the reader may consult Naftalin's recent review.[792]

2.8.3 Rhesus (Rh) antigen, Rh-associated glycoproteins (RhAG), and gas transport

The exciting story of the discovery of the Rh antigen and subsequent concept development of the Rh antigen complex and its importance in gas transport is unique in red cell membrane research. Landsteiner and Wiener[6] thought they had discovered the Rh antigen, but the human antibody agglutinating Rhesus monkey RBCs was actually against an epitope later named the Landsteiner–Wiener (LW) antigen in their honor. In the pursuit of the cause of severe erythroblastosis fetalis or hemolytic disorder of the newborn (HDN), Levin and Stetson identified the Rhesus D antigen as a case of intra-Rhesus group agglutination.[8] Some 20 years later, anti-D immunoglobulin was made by the Connaught Laboratories in Toronto to protect Rhesus D negative mothers against immunization with the RhD-positive fetal red cells carrying the father's product of the RhD allele. In search for the nature of the RhD antigen, it was detected in

Hb-free ghosts but not after extraction of the ghosts with organic solvents such as butanol and reductive cleavage, suggesting lipid dependence and intramolecular thiols are required for its integrity.[23,793] Rh_{null} erythrocytes lacking the RhD, Cc, Ee antigens (see below) exhibit elevated Na/K pump and K leak properties commensurate with being stomatocytes and a tendency to hemolyze,[187,794] supporting an early contention that membrane antigen/antibody reactions may modify membrane function and *vice versa*.[32]

2.8.3.1 *Rhesus antigens*

Two Rh genes, derived by duplication on chromosome 1p34–36, encode by ten exons antidromally 5′ to 3′ and 3′ to 5′, respectively, the RhD and RhCcEe proteins.[29,795] Deletion of the 5′–3′ RhD gene results in the RhD-negative product. There are some 10^5 RhD antigens/cell[10,796] or less, depending on the antibody chosen.[797] The non-glycosylated and cysteine-palmitoylated RhD and RhCc/Ee proteins each have 417 amino acids with ~32 kDa molecular weight, and 12 TMDs which differ by 35 amino acids, whereas Cc and Ee differ by 5 amino acids. As shown in Fig. 2.18, in normal RhD positive or negative RBCs, two Rh peptides are combined into the 170 kDa Rh core complex with 2 RhAG peptides *via* helix/helix interactions utilizing the GXXXG motif containing three additional accessory proteins, glycoprotein B, the LW antigen or ICAM, and CD47 or IAP.[798] Recent modeling with the renal RhCG homologues suggests that erythrocyte RhAG, RhD, and RhCE proteins may assemble stochastically into a trimeric Rh membrane complex.[799] During RBC differentiation and maturation, the RhAG proteins are necessary for delivering the Rh antigens to the plasma membrane.[798,800] Thus, RhAG gene deletion produces defective cells that have neither Rh nor RhAG, the so called "regulatory" Rh_{null} type,[801] accounting for 80% of all Rh_{null} cases.[798,800] The amorphous Rh_{null} state (20% of all Rh_{null}) derives from the absence of the Rh polypeptides and a >80% reduction of the RhAG.[802] For the Rh gene evolution and further Rh subtypes and variants deriving from point mutations and gene rearrangements consult available excellent reviews.[29,798,800,803]

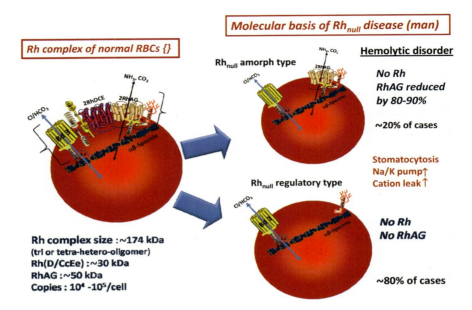

Figure 2.18 On the left side the normal "Rhesus Membranosome Complex" consisting of 2 RhDCcEe polypeptides, 2 RhAG proteins mediating NH_3 and CO_2 transport, each one LW or ICAM, CD47 or IAP, glycoprotein B (GPB), and an AE1 or band 3 homodimer. On the right side (top) the deletion of the RhDCcEe causes the "amorphous" Rh_{null} type in which the RhAG is reduced (20% of all Rh_{null} cases). On the right side (bottom) the deletion of RhAG causes failure to regulate the transport of the RhDCcEe products to the membrane surface, producing the "regulatory" Rh_{null} type (80% of all Rh_{null} cases). Absence of either Rh or RhAG peptides, or both, causes the hemolytic disorder with stomatocytosis and altered membrane permeabilities (see Section 2.8.3. for further details).

2.8.3.2 Rh-associated glycoproteins (RhAG in man, Rhag in others)

The significance of the RhAG proteins (Table 2.2) lies in their capability to transport CO_2 and NH_3, and thus they are recognized as major players in pH regulation and NH_3 detoxification. The red cell RhAG protein in man (Fig. 2.19) contains 409 amino acids, with a molecular weight of ~50 kDa due to glycosylation on the first extracellular loop of 12 TMDs.[29] The gene locus encoding RhAG lies on chromosome 6p12–21, and there are 2×10^5 molecules/red cell.[10] The allelic non-erythroid RhBG and RhCG proteins occur in liver, brain, and kidney, and transport ammonium, with

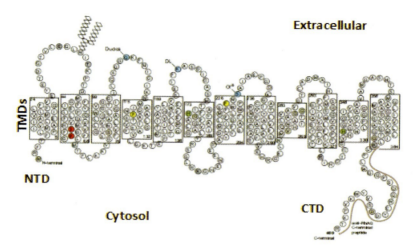

Figure 2.19 Suggested transmembrane topology of the RhAG protein. The RhAG polypeptide spans the membrane with 12 TMDs and the NTD and CTD are located in the cytosol. Two phenylalanine residues, # 120 in the 4th and # 225 in the 7th TMD form the so-called "phe-gate" at the presumable opening of the transmembrane pore for NH_3/NH_4 transport (with permission from Ref. 814).

RhCB and RhCG mapping to chromosomes 1q21.3[804] and 15q25,[805] respectively. As noted above, the RhAG protein is essential for directing the Rh peptides to the plasma membranes, which does not occur in the so-called Rh_{null} regulatory type.[798] Rh_{null} type mice with inactivated Rh and Rhag genes have been created using insertional targeting.[806]

RhAG belongs to an ancient protein family related to the ammonium transporters Amt/MEP channel proteins for ammonium and methyl ammonium in microbes, fungi, and green algae[807] and shares 1/3 of its identity with NeRh50 from *Nitrosomonas europaea*.[808] Proteins of these primitive organisms form trimers which contain a hydrophobic pore that at similar size also occurs in RhAG. Early, it was reported that *Xenopus* oocytes expressing RhAG protein showed a membrane potential-independent uptake of tracer methyl-ammonium coupled to proton counter-transport, which was also suggested based on pH experiments.[809,810] A similar conclusion of the presence of NH_4/H exchange was reached from studies of RhAG expressed in a *Saccharomyces cerevisiae* mutant suggesting net flux of NH_3.[811] Using stop flow experiments to measure NH_3 transport, a P_{NH3} of 4.5 and 3.9×10^{-5} cm/sec was found in RhD-positive and negative ghosts, respectively, close to that of intact red cells.[10,812] In contrast, ghosts from Rh_{null} RBCs had a reduced P_{NH3} of 2.5×10^{-5} cm/sec that was commensurate with the RhAG expression level. This suggests that the RhAG but not the Rh proteins contained the NH_3 channels, while the remainder of the gas was passing the lipid bilayer.[10] From these data and the number of RhAG polypeptides per cell, a turnover number of 2×10^{6}/(sec $\times$ Rh complex) was calculated.[10] The pore is gated by two extracellular phenylalanine residues (the "phe-gate") on TMDs 4 and 7. Two histidine residues at the pore center, however, may be charge barriers for monovalent cation flow, and are important for the ongoing argument whether NH_3 or NH_4 traverses the pore of RhAG. Indeed, a monovalent cation leak measured by Li fluxes across a RhAG mutant was recently suggested in a case of overhydrated stomatocytosis.[813] Expressed in *Xenopus* oocytes, the identified Ile61Arg and Phe65Ser substitutions of this mutant RhAG enlarged the transport cavity to accommodate monovalent cations with a hydrated radius of between 3.3 (K and NH_4) and 3.6 (Na) Å, respectively. Also, the wild type RhAG was capable of a small cation-non selective leak, i.e. NH_4 transport,[814] the magnitude of which is a function of pH, or of the K_d of NH_4. Consequently, incubating oocytes expressing these variant RhAGs in high Na media lead to a large inversion of the K/Na ratio, to levels seen in LK erythrocytes.[814]

Recently it was shown that RhAG is also permeable to the gas CO_2. Indeed, the P_{CO_2} in rat RBC with >90% inhibited carbonic anhydrase is ~7×10^{-3} cm/sec,[815] lower than that estimated for lipid bilayers under maximum velocities at high $[HCO_3 + CO_3]/CO_2$,[816] thus suggesting channel-mediated CO_2 transfer. Using ^{18}O, half of the CO_2 permeability of human red cells occurs through RhAG and is DIDS-sensitive, while the other half, also DIDS-sensitive, is AQP1-mediated (see Section 2.8.1).[817] Rate estimates are ~10^6 CO_2 per RhAG/(protein $\times$ sec) and about 2×10^5 CO_2/(AQP1 $\times$ sec). At pH 7.4, NH_4Cl inhibits CO_2 transport with an IC_{50} for NH_3 of 0.3 mM which means that both gases compete for the RhAG pore.[817] In Rh_{null} red cells, P_{CO2} is reduced to 0.05 cm/sec and DIDS reduces this further to 0.015 cm/sec by inhibiting AQP1. The implication for Rh_{null} patients lacking the RhAG polypeptides is that under exercise conditions, the gas exchange through the remaining AQP1 may be barely commensurate with the shortened lung capillary red cell transit time.[817] Although the DIDS inhibition has been argued not to be due to AE1 inhibition, it is still possible that DIDS, by inhibiting AE1, may affect both the nearby AQP1 as well as the RhAG within the Rhesus macro-complex; this will be addressed in Section 2.10.

2.9 "Non-Selective" Ion Channels (NSIC): Anions, Cations, and Amino Acids

Non-selective ion channels (NSIC) do not discriminate between charges and within limits and sizes, and there is less demarcation between the various forms of NSIC described in the literature. As a matter of fact, each of the systems listed below shows some transitional behaviors also present in any other listed. The following "systems" can be considered "non-selective" ion channels.

2.9.1 *Mutated AE1 ion conductance*[750,755]

Based on the analysis of 11 human pedigrees, a series of single amino acid mutations was found within the TMDs (9–10) of AE1 converting its anion to a cation conductance.

2.9.2 *RhAG protein mediated ion transport (see Section 2.8)*

2.9.3 *Oxidation induced "Cl-dependent" cation conductance*[818]

This Cl-dependence needs to be understood not in terms of the typical Cl-dependent Na-K or K cotransport but rather as a Cl-dependent inactivation of a Na channel. The whole-cell conductance is 238 pS for this NSIC, and it is stimulated by external Cl replacement with gluconate, Br> I = SCN and sorbitol, and hence V_m-dependent.[818] With a partially amiloride-inhibited perm-selectivity of Cs> K> Na = Li and permeability to Ca, this NSIC is activated by oxidation with t-butylhydroperoxide. Oxidation-induced elevation of this NSIC may contribute to certain hemolytic disorders (glucose-6-phosphate-dehydrogenase deficiency) since normal human red cells exposed to t-butyl hydroperoxide hemolyze in physiological media (colloid osmotic hemolysis), with Cl replaced by either NO_3 or SCN, suggesting the generation of "non-specific holes" in the membrane.

2.9.4 *Acid pH sensitive and hyperosmotically activated outward rectifying (ASOR) anion channel*[685]

The ASOR channel's reversal potential is –40 mV and loses its rectification at reversal potentials close to 0 mV in the absence of external Cl, i.e. under hyperpolarizing conditions. This channel is also different from that in Section 2.9.3 because DIDS and NPPB inhibit but niflumic acid and amiloride do not. The interrelations between the channel described here and Section 2.9.3 need further study.

2.9.5 *Low-ionic strength stimulated cation fluxes (LISCF)*

LISCF are basal Na and K conductances arising in red cells suspended in low-ionic strength media[819] and, by virtue of their partial inhibition by DIDS and amiloride, have similarity to the NSIC in Section 2.9.3. Some of these LIS-activated fluxes exhibit a unity stoichiometry with protons and hence are considered basal Na/H and K/H exchange mechanisms, but are, however, different from the NHE system inhibited by amiloride[380] (see Section 2.5.2). These fluxes are inhibited by charged compounds such as quinacrine[379] and they are activated at 40MPa (mega pascals) pressures.[819]

2.9.6 *"P_{sickle}"*

"P_{sickle}" in HbS-containing RBCs, but not in other hemoglobinopathies, is partially inhibited by DIDS like the NSIC, in addition by dipyridamole and stochastically activated under deoxygenation.[820] While P_{sickle} has been shown to mediate Na and Ca transport, there is also novel evidence for deoxygenation-elicited non-electrolyte transfer in low-ionic strength conditions.[821] The existence and role of transient receptor channels (TRPs) needs to be further explored.

2.9.7 Taurine and ion transport

Since the early work of Fugelli,[822] teleost red cells are known to have a hypotonically activated efflux of K and of ninhydrin-positive compounds, most likely taurine, a predominant solute in cells from these species. It is now clear that AE1 from nucleated RBCs from various fish, but not from mammals, exercises multiple transport functions when challenged hyposmotically. Apart from the participation of Cl-dependent K efflux and hence KCC,[58] swelling induces an NEM- and staurosporine-inhibited Cl-independent efflux of K and taurine sharing the same transport pathway.[823] The same anion channel is inhibited by DIDS in flounder red cells.[824] Hydrostatic pressure also inhibits taurine fluxes and stimulates LISCF.[825] As opposed to cation and taurine movements, there is evidence in erythroleukemia cells for hyposmotically induced taurine transport through anion channels inhibited by chloride channel blockers.[826] In skate RBCs, taurine efflux is accelerated after hyposmotic rather than isosmotic swelling.[827] This process requires PI3 kinase since wortmannin and LY294002 inhibited hyposmotic taurine efflux. Under hyposmotic stress, SkAE1, the skate's RBC anion exchanger, is moved to the plasma membrane by exocytosis,[828] which coincides with its phosphorylation by syk and activates taurine efflux.[829] A low osmolarity-induced tetramerization of skAE1 may provide the pore through which taurine passes in response to osmotic stress.[69] Trout AE1 expressed in hyposmotically swollen oocytes is capable of transporting taurine and other uncharged solutes.[830] Additional details in (Ref. 831).

2.9.8 Complement (C') lesion-induced colloid osmotic hemolysis

Of all systems described here, the C'-elicited membrane lesion is molecularly best described, whereas its ion discriminatory ability is the lowest. When the nine C' components are inserted into the plasma membrane of the red cell, a membrane attack complex[832] is formed. This was first thought to be stable with fixed dimensions, a 150 Å cylinder perpendicular to the membrane displaying a 100 Å internal diameter,[833] supporting the then widely accepted doughnut hypothesis of the complement membrane attack complex.[834] This hypothesis was not compatible with osmotic studies with desialidated sheep red cells in which the diffusion through this presumable doughnut hole was found to be at least an order of magnitude smaller than predicted from the dimension of the hole.[835] Subsequent work in human ghosts revealed that, upon insertion of the ninth C′ component, a lesion was formed that again showed sucrose diffusion restriction by 1–2 orders of magnitude. This indicates a fluctuating rather than a fixed membrane attack complex through which cations and anions, like through gramicidin channels, flow in an indiscriminatory manner, causing massive shifts of water into the cell and rupture by hemolysis (immune hemolysis).[836,837] Others have shown that insertion of the C5b-8 components suffices to increase Na and K permeability and to reduce the K/Na ratio.[838] During the complement lesion-mediated solute entry, that of Ca has been shown to cause activation of the Gardos effect and thus counteracts the colloid osmotic swelling.[839]

2.10 Integrated View: Macromolecular Complexes, Transport Metabolons

Evidence has accumulated over the past two decades indicating that, although ion channels and exchangers can be studied individually, they appear not to function as separate entities in the plasma membrane. Rather, entire groups of structural and functional proteins like to cluster *via* hydrophobic effects, short range van der Waals bonds and dipole interactions into major macromolecular complexes, and perhaps, membrane rafts within the lipid bilayer of the red cell membrane. Such a view would have been unthinkable in the days of the Davson–Danielli–Robertson phospholipid bilayer model with proteins attached as beta sheets. A few examples already established and others that await experimental resolution are listed below.

2.10.1 *The AE1-centered membrane macromolecular complex*

The dynamics of the formation of such macrocomplexes became evident, especially with the discovery of the "primeval" metabolon between AE1 and carbonic anhydrase isoform II (CaII) by Reithmeyer's group,[746] of AE1 mutants and of deletions and associated heterogeneities of minor blood groups by Tanner's group[840] and the unraveling of the multi-component nature of the Rhesus antigen complex.[798] There is evidence for protein-protein interactions between ankyrin and the C-termini of the Rh and RhAG polypeptides, thus connecting the Rh complex *via* ankyrin to the cytoskeleton without involvement of AE1.[29,841] On the other hand, coimmuno-precipitation experiments show that in human, and especially in mouse AE1$^{-/-}$ red cells, the Rhesus and the RhAG polypeptides, the IAP or CD47, ICAM or LW, and band 4.2 proteins are markedly reduced or absent (mouse) suggesting that AE1 plays a crucial role in forming the Rhesus antigen macromolecular complex.[842] According to Fig. 2.20, band 3's strong association with the spectrin cytoskeleton through ankyrin is utilized to attach the Rh macromolecular complex *via* band 4.2, and thus form a gas and anion transporter super-metabolon of about ~1 MDa, containing 4 AE1 proteins (ca 400 kDa), 4 GPA + B polypeptides (ca 140 kDa), 2 Rh proteins (~60 kDa), 2 RhAG proteins (~100 kDa), CD 47 (~50 kDa), LW (~42 kDa), carbonic anhydrase II (30 kDa), and ankyrin (~110 kDa).

This is consistent with the conclusion that protein band 4.2 regulates the membrane CD47 protein levels and affects membrane attachment of RhD, RhAG and band 3 proteins.[843] When the erythrocyte passes a lung capillary during a transit time of 0.3–1 sec, it may attach to the endothelial lining *via* the CD47 protein, while it delivers CO_2 through the complex-associated RhAG gas channels. Thus, CO_2 derived *via* carbonic anhydrase from a combined inward exchange of HCO_3/Cl by AE1 is accompanied by water flux across AQP1, although this peptide does not belong to the band 3-anchored macromolecular complex. Nevertheless, AQP1 seems to travel with GPA from the cytosol to the plasma membrane.[842] As the CAII is not rate limiting in the formation of CO_2 from HCO_3 and H_2O, the advantage of such a close association is to further cut the delivery time of the gas. In Rh_{null} cells, the complex role of the gas and anion transporter super-metabolon delineated here is perturbed, the mechanical stability of the red cell compromised and the susceptibility to phagocytosis increased due to reduced CD47 levels.[29]

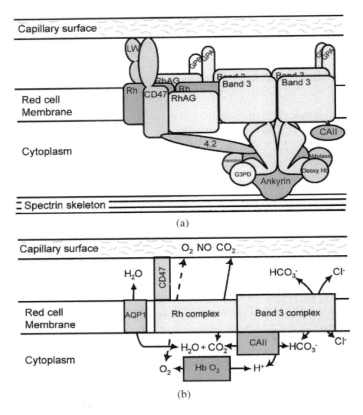

Figure 2.20 An intriguing proposal of a functional >>1 mDa "macrometabolon" of the band 3(AE1)-metabolon and the Rhesus complex. (a) The [band 3(AE1)-CAII]-metabolon is in the center and contains 2 homodimers of band 3, with 2 GPA molecules attached, and CAII, G3PD (GAPDH), aldolase, deoxy-Hb, and ankyrin connecting to the cytoskeleton, whereas the attached Rhesus complex entails the components described in Fig. 2.18 (2 RhDCcEe peptides, 2 RhAG peptides, CD47 or IAP, LW or ICAM and GPB). (b) During circulation in the lung, the band 3 metabolon facilitates *via* CAII the dehydration of H_2CO_3 to CO_2 (see Fig. 2.14), escaping the erythrocyte through the RhAG proteins, and H_2O through a close by Aquaporin 1. While HCO_3 enters the red cell in exchange for Cl, Hb assumes the relaxed oxygenated conformation and associates with the AE1 metabolon. During arrival of the red cell in the tissue capillary, the [band 3 + CAII]-metabolon catalyzes hydration of CO_2 to H_2CO_3 entering with the high P_{CO_2} gradient. Upon dissociation, the Bohr protons are buffered by Hb which, in its now tense deoxy-conformation, dissociates from the [band 3 + CAII]-metabolon, and HCO_3 leaves the red cell in exchange for Cl. CD 47 of the Rh complex may retard the red cell's transit by adhesion to the capillary membrane and thus lengthen the transit time (Figure with permission from Ref. 842).

Properties and Membrane Transport Mechanisms of Erythrocytes

Additional examples for putative macromolecular complexes in red blood cells of various species for further exploration and verification are listed below.

2.10.2 *An AE1-KCC "functionom"?*

In LK sheep red cells, DIDS inhibits K-Cl cotransport *via* KCC1, 3, and 4 isoforms[253,541] suggesting short range interactions between AE1 and KCC, a testable hypothesis.

2.10.3 *A band 4.2-NKCC-KCC-NHE-IK channel "functionom"?*

The red cells of band 4.2 null mice have a lower K/Na ratio and increased NKCC, KCC, Na/H exchange and IK channel activities,[844] commensurate with increased transport activities in the combined deficiencies of spectrin, ankyrin, and band 3 and 4.2 protein in human RBCs.[845]

2.10.4 *A protein 4.1, p55, and glycoprotein C "functionom"?*

A ternary complex formation has been reported in which the multiple binding sites of protein 4.1R have been narrowed down to a 30 kDa domain, encoded by its exons 10 and 8, which interfaces with p55 and GPC, respectively. The unitary subcomponent interaction of this complex is dynamically and reversibly modulated by Ca-calmodulin: the affinity between the partners of this complex inversely decreases or increases with rising or falling intracellular Ca levels.[846]

2.10.5 *The AE1-centered hemoglobin-2,3 DPG metabolon*

See Section 2.7 for details.

2.10.6 *The Na/K pump-supporting ATP pool and glycolytic pathway*

Since the early 1970s, a membrane-located ATP pool has been suspected to directly feed the pyruvate kinase-generated metabolic energy into the

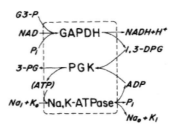

Figure 2.21 The concept that both pyruvate glycerate kinase (PGK) and the Na/K pump have access to a membrane ATP pool (yet to be located; indicated by broken rectangle), may also apply to the PMCA[217] (figure with permission to publish from Ref. 216).

Na/K pump, being simultaneously available for both the enzyme and the Na/K pump as shown in Fig. 2.21. The arrangement is thought to provide both energy supply for the pump and also to act as a putative transduction mechanism for the cell to sense its energy stores.[189–191] The exact location of this membrane ATP pool that serves both the Na/K pump and the Ca pump, however, is unknown.[217]

2.10.7 *The Na/K pump-KCC modifying Lp-Ll blood group antigen "functionom"*

See Section 2.4.3. Detailed aspects of this system have been published.[847]

2.11 Functional Surface Receptors

2.11.1 *Insulin*

Human RBCs express active insulin receptors.[848] These receptors constitute specific binding sites for insulin, 2000/cell and 14 sites/μm^2, with insulin-binding characteristics as found in other cell types.[848]

2.11.2 *Leptin*

A spliced variant of the leptin receptor has been found in both human and murine hematopoietic stem cell populations,[849] and red cells from obese humans possess a lower Na/HX activity and lower leptin binding than those of normal subjects.[850]

2.11.3 β Adrenoceptors

These are G-protein coupled receptors characterized by seven transmembrane domains that bind epinephrine and norepinephrine. Early evidence of beta adrenergic receptors expression in RBCs was found in birds,[851] amphibians,[852] and fish. A large body of information was obtained between the late 1970s and 1990s. Turkey RBCs have been extensively studied as a model for beta adrenergic receptors and their interactions with agonists, antagonists, and other effectors.[853–874] These receptors have also been found in pigeon erythrocytes[875,876] and other species. In frog RBCs, purine nucleotides were found to decrease the affinity of the receptors for their agonists,[854,877] the receptors becoming desensitized by their antagonists,[878] and there exist interactions between the receptors and the cytoskeleton.[879] In carp erythrocytes, norepinephrine and other catecholamines induce cell swelling by activation of the NaHX, and this activation is higher under hypoxic than in normoxic conditions and with concomitant accumulation of camp.[880] These receptors are also expressed in trout RBCs, controlling βNHE during the stress response,[881–884] and in immature rabbit bone marrow erythroblasts[885,886]; rat reticulocytes[887–891]; mouse erythroleukemia cells,[892] rat fetal erythroid cells,[893] and in human,[894–897] and pig RBCs.[898]

2.11.4 NMDA and homocysteine

Recently, functional NMDA receptors have been found in rat reticulocytes and peripheral blood.[679] The affinity of these receptors is higher for homocysteine and homocysteinic acid than for NMDA,[679] raising the question of a potential contribution of elevated homocysteine in the pathology of human cardiovascular disease.[680]

2.11.5 Other receptors

Erythrocytes and/or their precursors express numerous agonist receptors, some of which will be only mentioned in this section. These are the glucocorticoid and erythropoietin receptors in human erythroid cells[899–901]; the ligand-gated cation channel and P2X,[7] a member of the P2X receptor family[902-904]; G protein, a potential target for anti-malarial chemotherapy[905]; P1 or A1 adenosine receptors in pig RBCS[898]; P2Y purinergic receptor in turkey RBCs,[906-910] phosphoinositidase C-linked purinergic receptors in chicken erythrocytes,[911] P2Y4-like purinoceptor in the erythrocytes of the lizard *Ameiva ameiva*[912] and salamander *Necturus*,[913] and ET(B) receptors that bind endothelin in mouse RBCs and regulate the Gardos channel.[689]

2.12 Conclusion

This chapter has attempted to show the vast contribution erythrocyte research provided over the past 80 years to the advancement of science, in membrane composition, protein and lipid constituents and structure, primary and secondary "active" and passive membrane transport, and their molecular mechanisms in volume, ion, and water homeostasis. The authors of this chapter have indirectly witnessed and directly participated over decades in the excitement over the big discoveries from pumps to exchangers and cotransporters, and wish to pass on this torch of enthusiasm to those who will follow us in exploring fundamental biophysical aspects of living cells using the erythrocyte as a model.

Acknowledgements

The authors' research at Duke, Harvard, and Wright State Universities, described in this chapter, was financially supported for decades at different times by different institutions such as the National Institute of Health and the American Heart Association; the Max Planck Society during sabbatical leaves; and by the Fogarty Foundation, as well as by other State and private agencies. Although the days of funding for the red blood cell field appear to have passed for many investigators, we want to express our gratitude for this support we enjoyed over more than 45 years. This financial support enabled us to establish ourselves through our work and meet our colleagues at national and international meetings where we presented our research. We want to thank the many undergraduate and graduate students, postdoctoral fellows, visiting faculty and technical personnel for their contributions, in words and deeds, to our accomplishments. These undertakings could not have happened without them. Finally, there were and there are our families, who understood us and our decades-long quest for a glimpse into the workings of an erythrocyte. We are especially grateful to Dr Adrian Lauf, our personal engineer, for his unfailing support of our computer networks at home and at work at any time in need. Lastly, we want to thank Ms. Tanja Loch for the final formatting of this manuscript.

Dedication

This chapter is dedicated to the memory of our common mentor Daniel C Tosteson (1925–2009), Chair of Physiology & Pharmacology at Duke University (PKL) and Dean at Harvard Medical School (NCA), and to the many friends and colleagues over the past 50 years, who are not with us anymore and whose work we attempted to cite faithfully.

Bibliography

1. Mueller P, Rudin DO, Ti Tien H, Wescott WC. Reconstitution of cell membrane structure *in vitro* and its transformation into an excitable system. *Nature*. 1962;194(4832):979–980.
2. MacKinnon R. Potassium channels and the atomic basis of selective ion conduction (Nobel Lecture). *Angew Chemie Intern Edit*. 2004;43(33):4265–4277.
3. Agre P. Nobel lecture. Aquaporin water channels. *Biosci Rep. [Nobel Lecture]*. 2003;24(3):127–163.
4. Morth JP, Pedersen BP, Toustrup-Jensen MS, Sorensen TLM, Petersen J, Andersen JP et al. Crystal structure of the sodium-potassium pump. *Nature*. 2007;450(7172):1043–1049.
5. Gadsby DC, Takeuchi A, Artigas P, Reyes N. Review. Peering into an ATPase ion pump with single-channel recordings. *Philos Trans R Soc Lond B Biol Sci*. 2009;364(1514):229–238.
6. Landsteiner K, Wiener AS. An agglutinable factor in human blood recognized by immune sera for rhesus blood. *Proc Soc Exp Medicine & Biol*. 1940;43:223.
7. Landsteiner K, Wiener AS. Studies on an agglutinogen (Rh) in human blood reacting with anti-rhesus sera and with human isoantibodies. *J Exp Med*. 1941;74(4):309–320.

8. Levine P, Stetson RE. An unusual case of intra-group agglutination. *J Am Med Assoc.* 1939;113(2):126–127.
9. Hoffman JF, Joiner W, Nehrke K, Potapova O, Foye K, Wickrema A. The hSK4 (KCNN4) isoform is the Ca^{2+}-activated K^+ channel (Gardos channel) in human red blood cells. *Proc Natl Acad Sci USA.* 2003; 100(12):7366–7371.
10. Ripoche P, Bertrand O, Gane P, Birkenmeier C, Colin Y, Cartron JP. Human Rhesus-associated glycoprotein mediates facilitated transport of NH_3 into red blood cells. *Proc Natl Acad Sci USA.* 2004;101(49):17222–17227.
11. Gardos G. The permeability of human erythrocytes to potassium. *Acta Physiologica Hungarica.* 1956;10(2–4):185–189.
12. Gardos G. Potassium permeability of human erythrocytes. *Acta Physiologica Hungarica.* 1957;11(Suppl):31–32.
13. Gardos G, Straub FB. The role of adenosine-triphosphoric acid (ATP) in the potassium permeability of human erythrocytes. *Acta Physiologica Hungarica.* 1957;12(1–3):1–8.
14. Lambert A, Lowe AG. Chloride/bicarbonate exchange in human erythrocytes. *J Physiol.* 1978;275:51–63.
15. Gunn RB, Dalmark M, Tosteson DC, Wieth JO. Characteristics of chloride transport in human red blood cells. *J Gen Physiol.* 1973;61(2):185–206.
16. Kleinzeller A. Ernest Overton's contribution to the cell membrane concept: a centennial appreciation. *News Physiol Sci.* 1997;12(1):49–53.
17. Deuticke B. Anion permeability of the red blood cell. *Naturwissenschaften.* 1970;57(4):172–179.
18. Robertson JD. Membrane structure. *J Cell Biol.* 1981;91(3):189s–204s.
19. Robertson JD. The ultrastructure of cell membranes and their derivatives. *Biochem Soc Symp.* 1959;16:3–43.
20. Fairbanks G, Steck TL, Wallach DF. Electrophoretic analysis of the major polypeptides of the human erythrocyte membrane. *Biochem.* 1971;10(13):2606–2617.
21. Hillier J, Hoffman JF. On the ultrastructure of the plasma membrane as determined by the electron microscope. *J Cell Comp Physiol.* 1953; 42(2):203–247.
22. Hoffman JF, Hillier J, Wolman IJ, Parpart AK. New high density particles in certain normal and abnormal erythrocytes. *J Cell Comp Physiol.* 1956; 47(2):245–259.

23. Poulik MD, Lauf PK. Heterogeneity of water-soluble structural components of human red cell membrane. *Nature.* 1965;208(5013):874–876.
24. Singer SJ, Nicolson GL. The fluid mosaic model of the structure of cell membranes. *Science.* 1972;175(23):720–731.
25. Tanford C. *The hydrophobic effect: formation of micelles and biological membranes.* New York: John Wiley & Sons, Inc.; 1973.
26. Tanford C. The hydrophobic effect and the organization of living matter. *Science.* 1978;200(4345):1012–1018.
27. Morgan WT, Watkins WM. Genetic and biochemical aspects of human blood-group A-,B-H-,Lea and Leab specificity. *Br Med Bull.* 1969;25(1):30–34.
28. Springer GF, Desai PR. Common precursors of human blood group MN specificities. *Biochem Biophys Res Commun.* 1974;61(2):470–475.
29. Van Kim CL, Colin Y, Cartron J-P. Rh proteins: key structural and functional components of the red cell membrane. *Blood Reviews.* 2006;20(2):93–110.
30. Rasmusen BA, Hall JG. Association between potassium concentration and serological type of sheep red blood cells. *Science.* 1966;151(717):1551–1552.
31. Ellory JC, Tucker EM. Stimulation of the potassium transport system in low potassium type sheep red cells by a specific antigen antibody reaction. *Nature.* 1969;222(5192):477–478.
32. Lauf PK. Antigen-antibody reactions and cation transport in biomembranes: immunophysiological aspects. *Biochim Biophys Acta — Reviews Biomembranes.* 1975;415(2):173–229.
33. Lohr GW, Waller HD, Karges O, Schlegel B, Muller AA. Biochemistry of aging of human erythrocytes. *Klinische Wochenschrift.* 1958;36(21): 1008–1013.
34. Skou JC. The influence of some cations on an adenosine triphosphatase from peripheral nerves. *Biochim Biophys Acta.* 1957;23:394–401.
35. Skou JC. Further investigations on a Mg^{++} + Na^+-activated adenosintriphosphatase, possibly related to the active, linked transport of Na^+ and K^+ across the nerve membrane. *Biochim Biophys Acta.* 1960;42:6–23.
36. Tosteson DC, Hoffman JF. Regulation of cell volume by active cation transport in high and low potassium sheep red cells. *J Gen Physiol.* 1960; 44(1):169–194.
37. Post RL, Sen AK, Rosenthal AS. A Phosphorylated intermediate in adenosine triphosphate-dependent sodium and potassium transport across kidney membranes. *J Biol Chem.* 1965;240:1437–1445.

38. Albers RW. Biochemical aspects of active transport. *Annu Rev Biochem.* 1967;36(1):727–756.
39. Schatzmann HJ. Cardiac glycosides as inhibitors of active potassium and sodium transport by erythrocyte membrane. *Helvetica Physiologica et Pharmacologica Acta.* 1953;11(4):346–354.
40. Dodge JT, Mitchell C, Hanahan DJ. The preparation and chemical characteristics of hemoglobin-free ghosts of human erythrocytes. *Arch Biochem Biophys.* 1963;100:119–130.
41. Ling G. *A Physical Theory of The Living State: The Association-Induction Hypothesis.* New York: Blaisdell; 1962.
42. Pollack G. *Cells, Gels and The Engines of Life: A New, Unifying Approach to Cell Function.* Seattle, WA: Ebner & Sons; 2001.
43. Lingrel JB. The physiological significance of the cardiotonic steroid/ouabain-binding site of the Na,K-ATPase. *Annu Rev Physiol.* 2010;72(1):395–412.
44. Bernhardt I, Weiss E. Passive membrane permeability for ions and the membrane potential. In: Bernhardt I, Ellory CJ, editors. *Red Cell Membrane Transport in Health and Disease*: Berlin, Heidelberg, New York, Hong Kong, London, Milan, Paris, Tokyo: Springer; 2003, pp. 82–109.
45. Schwoch G, Passow H. Preparation and properties of human erythrocyte ghosts. *Mol Cell Biochem.* 1973;2(2):197–218.
46. Tosteson DC. Halide transport in red blood cells. *Acta Physiologica Scandinavica* 1959;46(1):19–41.
47. Passow H. Molecular aspects of band 3 protein-mediated anion transport across the red blood cell membrane. *Rev Physiol Biochem Pharmacol.* 1986;103:61–203.
48. Steck TL, Fairbanks G, Wallach DFH. Disposition of the major proteins in the isolated erythrocyte membrane. *Proteolytic dissection. Biochemistry.* 1971;10(13):2617–2624.
49. Beauge LA, Adragna N. The kinetics of ouabain inhibition and the partition of rubidium influx in human red blood cells. *J Gen Physiol.* 1971;57(5):576–592.
50. Sachs JR. Ouabain-insensitive sodium movements in the human red blood cell. *J Gen Physiol.* 1971;57(3):259–282.
51. Beauge LA, Adragna N. pH dependence of rubidium influx in human red blood cells. *Biochim Biophys Acta.* 1974;352(3):441–447.

52. Dalmark M, Wieth JO. Temperature dependence of chloride, bromide, iodide, thiocyanate and salicylate transport in human red cells. *J Physiol.* 1972;224(3):583–610.
53. Kregenow FM. The response of duck erythrocytes to hypertonic media. Further evidence for a volume-controlling mechanism. *J Gen Physiol.* 1971;58(4):396–412.
54. Wiley JS, Cooper RA. A furosemide-sensitive cotransport of sodium plus potassium in the human red cell. *J Clin Invest.* 1974;53(3):745–755.
55. Schmidt WF, 3rd, McManus TJ. Ouabain-insensitive salt and water movements in duck red cells. I. Kinetics of cation transport under hypertonic conditions. *J Gen Physiol.* 1977;70(1):59–79.
56. Geck P, Pietrzyk C, Burckhardt BC, Pfeiffer B, Heinz E. Electrically silent cotransport on Na^+, K^+ and Cl^- in Ehrlich cells. *Biochim Biophys Acta.* 1980;600(2):432–447.
57. Dunham PB, Ellory JC. Passive potassium transport in low potassium sheep red cells: dependence upon cell volume and chloride. *J Physiol.* 1981;318:511–530.
58. Lauf PK. Evidence for chloride dependent potassium and water transport induced by hyposmotic stress in erythrocytes of the marine teleost, *Opsanus tau*. *J Comp Physiol B: Biochem Syst & Env Physiol.* 1982;146(1):9–16.
59. Garcia-Romeu F, Cossins AR, Motais R. Cell volume regulation by trout erythrocytes: characteristics of the transport systems activated by hypotonic swelling. *J Physiol.* 1991;440:547–567.
60. Borgese F, Motais R, Garcia-Romeu F. Regulation of Cl-dependent K transport by oxy-deoxyhemoglobin transitions in trout red cells. *Biochim Biophys Acta.* 1991;1066(2):252–256.
61. Motais R, Guizouarn H, Garcia-Romeu F. Red cell volume regulation: the pivotal role of ionic strength in controlling swelling-dependent transport systems. *Biochim Biophys Acta.* 1991;1075(2):169–180.
62. Lauf PK, Theg BE. A chloride dependent K^+ flux induced by N-ethylmaleimide in genetically low K^+ sheep and goat erythrocytes. *Biochem Biophys Res Commun.* 1980;92(4):1422–1428.
63. Adragna NC, White RE, Orlov SN, Lauf PK. K-Cl cotransport in vascular smooth muscle and erythrocytes: possible implication in vasodilation. *Am J Physiol Cell Physiol.* 2000;278(2):C381–390.

64. Adragna NC, Fulvio M, Lauf PK. Regulation of K-Cl cotransport: from function to genes. *J Membr Biol.* 2004;201(3):109–137.
65. Alvarez-Leefmans FJ, Delpire E. *Physiology and Pathology of Chloride Transporters and Channels in the Nervous System: From Molecules to Diseases.* Amsterdam; Boston: Elsevier/Academic Press; 2009.
66. Kregenow FM, Hoffman JF. Some kinetic and metabolic characteristics of calcium-induced potassium transport in human red cells. *J Gen Physiol.* 1972;60(4):406–429.
67. Ponder E. *Hemolysis and Related Phenomena.* New York: Grune & Stratton; 1948.
68. Altman PL, Katz DD. *Blood and Other Body Fluids/Analysis and Compilation.* Dittmer DS, editor. Washington, DC: Federation of American Societies for Experimental Biology; 1961.
69. Perlman D, Goldstein L. The anion exchanger as an osmolyte channel in the skate erythrocyte. *Neurochem Res.* 2004;29(1):9–15.
70. Bourne PK, Cossins AR. On the instability of K^+ influx in erythrocytes of the rainbow trout, Salmo gairdneri, and the role of catecholamine hormones in maintaining *in vivo* influx activity. *J Exp Biol.* 1982;101:93–104.
71. Nikinmaa M. Membrane transport and control of hemoglobin-oxygen affinity in nucleated erythrocytes. *Physiol Rev.* 1992;72(2):301–321.
72. Hartman FA, Lessler MA. Erythrocyte measurements in fishes, amphibia and reptiles. *Biol Bull.* 1964;126(1):83–88.
73. Cala P. Volume regulation by flounder red blood cells in anisotonic media. *J GenPhysiol.* 1977;69(5):537–552.
74. Hoffman JF, Laris PC. Determination of membrane potentials in human and Amphiuma red blood cells by means of fluorescent probe. *J Physiol.* 1974;239(3):519–552.
75. Stoner LC, Kregenow FM. A single-cell technique for the measurement of membrane potential, membrane conductance, and the efflux of rapidly penetrating solutes in Amphiuma erythrocytes. *J Gen Physiol.* 1980;76(4):455–478.
76. Aldrich KJ, Saunders DK, Sievert LM, Sievert G. Comparison of erythrocyte osmotic fragility among amphibians, reptiles, birds and mammals. *Trans Kansas Acad Sci.* 2006;109:149–158.
77. Ryan DH. Examination of the Blood. In: Beutler E, Lichtman MA, Coller BS, Kipps TJK, Selingsohn U, editors. *Williams Hematology.* 6th ed.: McGraw-Hill Professional; 2000, pp. 9–16.

78. Lauf PK, Adragna NC, Dupre N, Bouchard JP, Rouleau GA. K-Cl cotransport in red blood cells from patients with KCC3 isoform mutants. *Biochem Cell Biol.* 2006;84(6):1034–1044.
79. Lauf PK, Dessent MP. Effect of metabolic state on immune-hemolysis of L-positive low potassium (LK) sheep red blood cells by ISO-Immune anti-L serum and rabbit serum complement. *Immun Invest.* 1973;2(2):193–212.
80. Evans JV. Relationships between red blood cell potassium concentrations, medial corpuscular fragility and haemoglobin type in merino and southdown sheep. *Nature.* 1961;192(4802):567–568.
81. See JRH, Adragna NC, Lauf, PK. Osmotic resistance of low and high potassium sheep red blood cells: comparison with the "ROL" gene model. *Blood* 1996;88:54a.
82. Armsby CC, Stuart-Tilley AK, Alper SL, Brugnara C. Resistance to osmotic lysis in BXD-31 mouse erythrocytes: association with upregulated K-Cl cotransport. *Am J Physiol Cell Physiol.* 1996;270(3):C866–877.
83. Fernandes PR, Dewey MJ. Genetic control of erythrocyte volume regulation: effect of a single gene (rol) on cation metabolism. *Am J Physiol.* 1994;267(1 Pt 1):C211–C219.
84. Fujise H, Lauf PK. Swelling, NEM, and A23187 activate Cl^--dependent K^+ transport in high-K^+ sheep red cells. *Am J Physiol Cell Physiol.* 1987;252(2):C197–204.
85. Peters LL, Lambert AJ, Zhang W, Churchill GA, Brugnara C, Platt OS. Quantitative trait loci for baseline erythroid traits. *Mamm Genome.* 2006;17(4):298–309.
86. Hoffman JF. On the mechanism and measurement of shape transformations of constant volume of human red blood cells. *Blood Cells.* 1987;12(3):565–588.
87. Deuticke B. Membrane lipids and proteins as a basis of red cell shape and its alteration. In: Bernhardt I, Ellory CJ, editors. *Red Cell Membrane Transport in Health and Disease.* Berlin, Heidelberg, New York, Hong Kong, London, Milan, Paris, Tokyo. Springer; 2003, pp. 2–760.
88. Fischer TM. Human red cell shape and the mechanical characteristics of the mdembrane. In: Bernhardt I, Ellory CJ editors. *Red Cell Membrane Transport in Health and Disease.* Berlin, Heidelberg, New York, Hong Kong, London, Milan, Paris, Tokyo: Springer; 2003, pp. 62–82.
89. Haest CWM. Distribution and movement of membrane lipids. In: Bernhardt I, Ellory CJ, editors. *Red Cell Membrane Transport in Health and Disease.*

Berlin, Heidelberg, New York, Hong Kong, London, Milan, Paris, Tokyo: Springer; 2003, pp. 1–17.
90. Luna E, Hitt A. Cytoskeleton–plasma membrane interactions. *Science*. 1992;258(5084):955–964.
91. Snyers L, Umlauf E, Prohaska R. Oligomeric nature of the integral membrane protein stomatin. *J Biol Chem*. 1998;273(27):17221–17226.
92. Wang Y, Morrow JS. Identification and characterization of human SLP-2, a novel homologue of stomatin (Band 7.2b) present in erythrocytes and other tissues. *J Biol Chem*. 2000;275(11):8062–8071.
93. Gallagher P, Forget BG. The red cell membrane. In: Beutler E, Lichtman MA, Coller BS, Kipps TJ, Selingsohn U, editors. *Williams Hematology*. 6th ed: McGraw-Hill Professional; 2000, pp. 93–102.
94. Yawata Y. *Cell Membrane: The Red Blood Cell as a Model*. Wiley VCH; 2003.
95. Chien S. Red cell deformability and its relevance to blood flow. *Annu Rev Physiol*. 1987;49:177–192.
96. Lee P, Kirk RG, Hoffman JF. Interrelations among Na and K content, cell volume, and buoyant density in human red blood cell populations. *J Membr Biol*. 1984;79(2):119–126.
97. Cass A, Dalmark M. Equilibrium dialysis of ions in nystatin-treated red cells. *Nat New Biol*. 1973;244(132):47–49.
98. Freedman JC, Hoffman JF. Ionic and osmotic equilibria of human red blood cells treated with nystatin. *J Gen Physiol*. 1979;74(2):157–185.
99. Brahm J. Diffusional water permeability of human erythrocytes and their ghosts. *J Gen Physiol*. 1982;79(5):791–819.
100. Moura TF, Macey RI, Chien DY, Karan D, Santos H. Thermodynamics of all-or-none water channel closure in red cells. *J Membr Biol*. 1984;81(2):105–111.
101. Macey RI, Farmer RE. Inhibition of water and solute permeability in human red cells. *Biochim Biophys Acta*. 1970;211(1):104–106.
102. Lauf PK, Adragna NC. Functional evidence for a pH sensor of erythrocyte K-Cl cotransport through inhibition by internal protons and diethylpyrocarbonate. *Cell Physiol Biochem*. 1998;8(1–2):46–60.
103. Shami Y, Rothstein A, Knauf PA. Identification of the Cl-transport site of human red blood cells by a kinetic analysis of the inhibitory effects of a chemical probe. *Biochim Biophys Acta — Biomembranes*. 1978;508(2):357–363.

104. Cabantchik ZI, Knauf PA, Rothstein A. The anion transport system of the red blood cell. The role of membrane protein evaluated by the use of "probes". *Biochim Biophys Acta — Reviews on Biomembranes.* 1978; 515(3):239–302.
105. Lepke S, Fasold H, Pring M, Passow H. A study of the relationship between inhibition of anion exchange and binding to the red blood cell membrane of 4,4'-diisothiocyano stilbene-2,2'-disulfonic acid (DIDS) and its dihydro derivative (H_2DIDS). *J Membr Biol.* 1976;29(1):147–177.
106. Fröhlich O, Leibson C, Gunn RB. Chloride net efflux from intact erythrocytes under slippage conditions. Evidence for a positive charge on the anion binding/transport site. *J Gen Physiol.* 1983;81(1):127–152.
107. Elmariah S, Gunn RB. Kinetic evidence that the Na-PO_4 cotransporter is the molecular mechanism for Na/Li exchange in human red blood cells. *Am J Physiol Cell Physiol.* 2003;285(2):C446–456.
108. Lauf P, Adragna NC. K-Cl cotransport: properties and molecular mechanisms. *Cell Physiol Biochem.* 2000;10:341–354.
109. Beutler MDE. Composition of the erythrocyte. In: Beutler E, Lichtman MA, Coller BS, Kipps TJ, Selingsohn U, editors. *Williams Hematology.* 6th ed.: McGraw Hill Professional; 2000, pp. 289–294.
110. Sims PJ, Waggoner AS, Wang C-H, Hoffman JF. Mechanism by which cyanine dyes measure membrane potential in red blood cells and phosphatidylcholine vesicles. *Biochemistry.* 1974;13(16):3315–3330.
111. Tosteson DC, Cook P, Andreoli T, Tieffenberg M. The effect of valinomycin on potassium and sodium permeability of HK and LK sheep red cells. *J Gen Physiol.* 1967;50(11):2513–2525.
112. Kaji DM. Hemisodium, a novel selective Na ionophore. Effect on normal human erythrocytes. *J Gen Physiol.* 1992;99(2):199–216.
113. Evans JV. Electrolyte concentrations in red blood cells of British breeds of sheep. *Nature.* 1954;174(4437):931–932.
114. Evans JV, King JWB. Genetic control of sodium and potassium concentrations in the red blood cells of sheep. *Nature.* 1955;176(4473):171–172.
115. Ellory JC, Carleton S. (Na^+-/K^+)-activated ATPase in cattle erythrocytes. *Biochim Biophys Acta — Biomembranes.* 1974;363(3):397–403.
116. Sachs JR, Ellory JC, Kropp DL, Dunham PB, Hoffman JF. Antibody-induced alterations in the kinetic characteristics of the Na:K pump in goat red blood cells. *J Gen Physiol.* 1974;63(4):389–414.

117. Tucker EM. Genetic variation in the sheep red blood cell. *Biol Rev Camb Philos Soc.* 1971;46(3):341–386.
118. Lauf PK, Rasmusen BA, Hoffman PG, Dunham PB, Cook P, Parmelee ML et al. Stimulation of active potassium transport in LK sheep red cells by blood group-L-antiserum. *J Membr Biol.* 1970;3(1):1–13.
119. Lauf PK, Stiehl BJ, Joiner CH. Active and passive cation transport and L antigen heterogeneity in low potassium sheep red cells: evidence against the concept of leak-pump interconversion. *J Gen Physiol.* 1977;70(2):221–242.
120. Fujise H, Nakayama T, Iwase N, Tsubota T, Komatsu T. Comparison of cation transport and regulatory volume decrease between red blood cells from the japanese black bear and the dog. *Comp Clin Pathol.* 2003;12(1):33–39.
121. Willis JS, Nelson RA, Gordon C, Vilaro P, Zhao ZH. Membrane transport of sodium ions in erythrocytes of the American black bear, Ursus americanus. *Comp Biochem Physiol A Comp Physiol.* 1990;96(1):91–96.
122. Willis JS, Nelson RA, Livingston B, Marjanovic M. Membrane transport of potassium ions in erythrocytes of the American black bear, Ursus americanus. *Comp Biochem Physiol A Comp Physiol.* 1990;96(1):97–105.
123. Garfinkel L, Garfinkel D. Magnesium regulation of the glycolytic pathway and the enzymes involved. *Magnesium.* 1985;4(2–3):60–72.
124. Flatman PW, Lew VL. Magnesium buffering in intact human red blood cells measured using the ionophore A23187. *J Physiol.* 1980;305(1):13–30.
125. Fujise H, Cruz P, Reo NV, Lauf PK. Relationship between total magnesium concentration and free intracellular magnesium in sheep red blood cells. *Biochim Biophys Acta — Molecular Cell Research.* 1991;1094(1):51–54.
126. Flatman PW. Magnesium transport. In: Bernhardt I, Ellory CJ, editors. *Red Cell Membrane Transport in Health and Disease.* Berlin, Heidelberg, New York, Hong Kong, London, Milan, Paris, Tokyo. Springer; 2003, pp. 408–434.
127. Lüdi H, Schatzmann HJ. Some properties of a system for sodium-dependent outward movement of magnesium from metabolizing human red blood cells. *J Physiol.* 1987;390:367–382.
128. Schatzmann HJ. Asymmetry of the magnesium sodium exchange across the human red cell membrane. *Biochim Biophys Acta.* 1993;1148(1):15–18.
129. Barvitenko NN, Adragna NC, Weber RE. Erythrocyte signal transduction pathways, their oxygenation dependence and functional significance. *Cell Physiol Biochem.* 2005;15(1–4):1–18.

130. Ortiz OE, Lew VL, Bookchin RM. Deoxygenation permeabilizes sickle cell anaemia red cells to magnesium and reverses its gradient in the dense cells. *J Physiol.* 1990;427(1):211–226.
131. Olukoga AO, Adewoye HO, Erasmus RT, Adedoyin MA. Erythrocyte and plasma magnesium in sickle-cell anaemia. *East Afr Med J.* 1990; 67(5):348–354.
132. Delpire E, Lauf PK. Magnesium and ATP dependence of K-Cl co-transport in low K^+ sheep red blood cells. *J Physiol.* 1991;441:219–231.
133. Ortiz-Carranza O, Adragna NC, Lauf PK. Modulation of K-Cl cotransport in volume-clamped low-K sheep erythrocytes by pH, magnesium, and ATP. *Am J Physiol Cell Physiol.* 1996;271(4):C1049–1058.
134. Joiner CH, Jiang M, Fathallah H, Giraud F, Franco RS. Deoxygenation of sickle red blood cells stimulates KCl cotransport without affecting Na^+/H^+ exchange. *Am J Physiol Cell Physiol.* 1998;274(6):C1466–1475.
135. Romero PJ, Romero EA. The role of calcium metabolism in human red blood cell ageing: a proposal. *Blood Cells Mol Dis.* 1999;25(1):9–19.
136. Kabaso D, Shlomovitz R, Auth T, Lew VL, Gov NS. Curling and local shape changes of red blood cell membranes driven by cytoskeletal reorganization. *Biophys J.* 2010;99(3):808–816.
137. Dyrda A, Cytlak U, Ciuraszkiewicz A, Lipinska A, Cueff A, Bouyer G et al. Local membrane deformations activate Ca^{2+}-dependent K^+ and anionic currents in intact human red blood cells. *PLoS One.* 2010;5(2):e9447.
138. Tiffert T, Lew VL. Cytoplasmic calcium buffers in intact human red cells. *J Physiol.* 1997;500 (Pt 1):139–154.
139. Brown AM, Lew VL. The effect of intracellular calcium on the sodium pump of human red cells. *J Physiol.* 1983;343(1):455–493.
140. Cueff A, Seear R, Dyrda A, Bouyer G, Egée S, Esposito A et al. Effects of elevated intracellular calcium on the osmotic fragility of human red blood cells. *Cell Calcium.* 2010;47(1):29–36.
141. Tiffert T, Bookchin RM, Lew VL. Calcium homeostasis in normal and abnormal human red cells. In: Bernhardt I, Ellory CJ, editors. *Red Cell Membrane Transport in Health and Disease.* Berlin, Heidelberg, New York, Hong Kong, London, Milan, Paris, Tokyo. Springer; 2003, pp. 373–405.
142. Horn NM, Thomas AL, Oakley F. Trace metal transport. In: Bernhardt I, Ellory CJ, editors. *Red Cell Membrane Transport in Health and Disease.*

Berlin, Heidelberg, New York, Hong Kong, London, Milan, Paris, Tokyo. Springer; 2003, pp. 436–450.
143. Hokin LE, Hokin MR. Evidence for phosphatidic acid as the sodium carrier. *Nature*. 1959;184(4692):1068–1069.
144. Glynn IM, Slayman CW, Eichberg J, Dawson RM. The adenosine-triphosphatase system responsible for cation transport in electric organ: exclusion of phospholipids as intermediates. *Biochem J*. 1965;94:692–699.
145. Tanford C. Equilibrium state of ATP-driven ion pumps in relation to physiological ion concentration gradients. *J Gen Physiol*. 1981;77(2):223–229.
146. Reynolds JA, Johnson EA, Tanford C. Incorporation of membrane potential into theoretical analysis of electrogenic ion pumps. *Proc Natl Acad Sci USA*. 1985;82(20):6869–6873.
147. Garrahan PJ, Rega AF. Cation loading of red blood cells. *J Physiol*. 1967;193(2):459–466.
148. Hoffman JF. Physiological characteristics of human red blood cell ghosts. *J Gen Physiol*. 1958;42(1):9–28.
149. Bodemann H, Passow H. Factors controlling the resealing of the membrane of human erythrocyte ghosts after hypotonic hemolysis. *J Membr Biol*. 1972;8(1):1–26.
150. Steck TL, Weinstein RS, Straus JH, Wallach DF. Inside-out red cell membrane vesicles: preparation and purification. *Science*. 1970;168(3928):255–257.
151. Kant JA, Steck TL. Cation-impermeable inside-out and right-side-out vesicles from human erythrocyte membranes. *Nat New Biol*. 1972 Nov 1;240(96):26–28.
152. Xu ZC, Dunham PB, Munzer JS, Silvius JR, Blostein R. Rat kidney Na-K pumps incorporated into low-K^+ sheep red blood cell membranes are stimulated by anti-Lp antibody. *Am J Physiol Cell Physiol*. 1992;263(5):C1007–1014.
153. Munzer JS, Silvius JR, Blostein R. Delivery of ion pumps from exogenous membrane-rich sources into mammalian red blood cells. *J Biol Chem*. 1992;267(8):5202–5210.
154. Skou JC, Esmann M. The Na,K-ATPase. *J Bioenerg Biomembr*. 1992; 24(3):249–261.
155. Shull GE, Lane LK, Lingrel JB. Amino-acid sequence of the beta-subunit of the ($Na^+ + K^+$)ATPase deduced from a cDNA. *Nature*. 1986;321(6068):429–431.

156. Schoner W, Scheiner-Bobis G. Endogenous and exogenous cardiac glycosides and their mechanisms of action. *Am J Cardiovasc Drugs.* 2007; 7(3):173–189.
157. Forbush B, 3rd, Kaplan JH, Hoffman JF. Characterization of a new photoaffinity derivative of ouabain: labeling of the large polypeptide and of a proteolipid component of the Na, K-ATPase. *Biochemistry.* 1978; 17(17):3667–3676.
158. Yang-Feng TL, Schneider JW, Lindgren V, Shull MM, Benz EJ, Lingrel JB et al. Chromosomal localization of human Na^+,K^+-ATPase [alpha]- and [beta]-subunit genes. *Genomics.* 1988;2(2):128–138.
159. Chehab FF, Kan YW, Law ML, Hartz J, Kao FT, Blostein R. Human placental Na^+,K^+-ATPase alpha subunit: cDNA cloning, tissue expression, DNA polymorphism, and chromosomal localization. *Proc Natl Acad Sci USA.* 1987;84(22):7901–7905.
160. Specht SC, Sweadner KJ. Two different Na,K-ATPases in the optic nerve: cells of origin and axonal transport. *Proc Natl Acad Sci USA.* 1984; 81(4):1234–1238.
161. Dhir R, Nishioka Y, Blostein R. Na,K-ATPase isoform expression in sheep red blood cell precursors. *Biochim Biophys Acta.* 1990;1026(2):141–146.
162. Hoffman JF, Wickrema A, Potapova O, Milanick M, Yingst DR. Na pump isoforms in human erythroid progenitor cells and mature erythrocytes. *Proc Natl Acad Sci USA.* 2002;99(22):14572–14577.
163. Woo AL, James PF, Lingrel JB. Characterization of the fourth α isoform of the Na,K-ATPase. *J Membr Biol.* 1999;169(1):39–44.
164. Pu HX, Cluzeaud F, Goldshleger R, Karlish SJ, Farman N, Blostein R. Functional role and immunocytochemical localization of the gamma a and gamma b forms of the Na,K-ATPase gamma subunit. *J Biol Chem.* 2001;276(23):20370–20378.
165. Shull GE, Schwartz A, Lingrel JB. Amino-acid sequence of the catalytic subunit of the ($Na^+ + K^+$) ATPase deduced from a complementary DNA. *Nature.* 1985;316(6030):691–695.
166. Kawakami K, Ohta T, Nojima H, Nagano K. Primary structure of the alpha-subunit of human Na,K-ATPase deduced from cDNA sequence. *J Biochem.* 1986;100(2):389–397.
167. Shamraj OI, Lingrel JB. A putative fourth Na^+,K^+-ATPase alpha-subunit gene is expressed in testis. *Proc Natl Acad Sci USA.* 1994;91(26):12952–12956.

168. Sweadner KJ, Wetzel RK, Arystarkhova E. Genomic organization of the human FXYD2 gene encoding the gamma subunit of the Na,K-ATPase. *Biochem Biophys Res Commun.* 2000;279(1):196–201.
169. Arystarkhova E, Wetzel RK, Asinovski NK, Sweadner KJ. The gamma subunit modulates Na$^+$ and K$^+$ affinity of the renal Na,K-ATPase. *J Biol Chem.* 1999;274(47):33183–33185.
170. Post RL, Kume S. Evidence for an aspartyl phosphate residue at the active site of sodium and potassium ion transport adenosine triphosphatase. *J Biol Chem.* 1973;248(20):6993–7000.
171. Kuntzweiler TA, Wallick ET, Johnson CL, Lingrel JB. Amino acid replacement of Asp369 in the sheep alpha 1 isoform eliminates ATP and phosphate stimulation of [3H]ouabain binding to the Na$^+$, K$^+$-ATPase without altering the cation binding properties of the enzyme. *J Biol Chem.* 1995;270(27):16206–16212.
172. Jewell-Motz EA, Lingrel JB. Site-directed mutagenesis of the Na,K-ATPase: consequences of substitutions of negatively-charged amino acids localized in the transmembrane domains. *Biochemistry.* 1993;32(49):13523–13530.
173. Blostein R, Wilczynska A, Karlish SJ, Arguello JM, Lingrel JB. Evidence that Ser775 in the alpha subunit of the Na,K-ATPase is a residue in the cation binding pocket. *J Biol Chem.* 1997;272(40):24987–24993.
174. Schultheis PJ, Wallick ET, Lingrel JB. Kinetic analysis of ouabain binding to native and mutated forms of Na,K-ATPase and identification of a new region involved in cardiac glycoside interactions. *J Biol Chem.* 1993;268(30):22686–22694.
175. Feng J, Lingrel JB. Analysis of amino acid residues in the H5-H6 transmembrane and extracellular domains of Na,K-ATPase alpha subunit identifies threonine 797 as a determinant of ouabain sensitivity. *Biochemistry.* 1994;33(14):4218–42124.
176. Skou JC, Butler KW, Hansen O. The effect of magnesium, ATP, Pi, and sodium on the inhibition of the (Na$^+$+K$^+$)-activated enzyme system by g-strophanthin. *Biochim Biophys Acta — Biomembranes.* 1971;241(2):443–461.
177. Hansen O, Skou JC. A study on the influence of the concentration of Mg^{2+}, P$_i$, K$^+$, Na$^+$, and tris on (Mg^{2+}+P$_i$)-supported g-strophanthin binding to (Na$^+$+K$^+$)-activated ATPase from ox brain. *Biochim Biophys Acta — Biomembranes.* 1973;311(1):51–66.

178. Skou JC. Effect of ATP on the intermediary steps of the reaction of the (Na$^+$+K$^+$)-dependent enzyme system. I. Studied by the use of N-ethylmaleimide inhibition as a tool. *Biochim Biophys Acta — Biomembranes.* 1974;339(2):234–245.
179. Esmann M, Skou JC. The effect of K$^+$ on the equilibrium between the E2 and the K$^+$-occluded E2 conformation of the (Na$^+$+K$^+$)-ATPase. *Biochim Biophys Acta — Protein Structure and Molecular Enzymology.* 1983;748(3):413–417.
180. Esmann M, Skou JC. Occlusion of Na$^+$ by the Na, K-ATPase in the presence of oligomycin. *Biochem Biophys Res Commun.* 1985;127(3):857–863.
181. Skou JC, Esmann M. Effect of magnesium ions on the high-affinity binding of eosin to the (Na$^+$+K$^+$)-ATPase. *Biochim Biophys Acta.* 1983;727(1):101–107.
182. Cornelius F, Skou JC. Na$^+$-Na$^+$ exchange mediated by (Na$^+$+K$^+$)-ATPase reconstituted into liposomes. Evaluation of pump stoichiometry and response to ATP and ADP. *Biochim Biophys Acta — Biomembranes.* 1985;818(2):211–221.
183. Cornelius F, Skou JC. The sided action of Na$^+$ and of K$^+$ on reconstituted shark (Na$^+$+K$^+$)-ATPase engaged in Na$^+$-Na$^+$ exchange accompanied by ATP hydrolysis. I. The ATP activation curve. *Biochim Biophys Acta — Biomembranes.* 1987;904(2):353–364.
184. Cornelius F, Skou JC. The sided action of Na$^+$ on reconstituted shark Na$^+$/K$^+$-ATPase engaged in Na$^+$-Na$^+$ exchange accompanied by ATP hydrolysis. II. Transmembrane allosteric effects on Na$^+$ affinity. *Biochim Biophys Acta — Biomembranes.* 1988;944(2):223–232.
185. Cornelius F, Skou JC. The effect of cytoplasmic K$^+$ on the activity of the Na$^+$/K$^+$-ATPase. *Biochim Biophys Acta — Biomembranes.* 1991;1067(2):227–234.
186. Hoffman JF. The interaction between tritiated ouabain and the Na-K Pump in red blood cells. *J Gen Physiol.* 1969;54(1):343–353.
187. Lauf P, Joiner C. Increased potassium transport and ouabain binding in human Rhnull red blood cells. *Blood.* 1976;48(3):457–468.
188. Joiner CH, Lauf PK. The correlation between ouabain binding and potassium pump inhibition in human and sheep erythrocytes. *J Physiol.* 1978;283:155–175.

189. Hoffman JF, Proverbio F. Membrane ATP and the functional organization of the red cell Na:K pump. *Ann New York Acad Sci.* 1974;242(1):459–460.
190. Proverbio F, Hoffman JF. Membrane compartmentalized ATP and its preferential use by the Na,K-ATPase of human red cell ghosts. *J Gen Physiol.* 1977;69(5):605–632.
191. Hoffman JF. ATP compartmentation in human erythrocytes. *Curr Op Hematol.* 1997;4(2):112–115.
192. Sachs JR. Kinetics of the inhibition of the Na-K pump by external sodium. *J Physiol.* 1977;264(2):449–470.
193. Sachs JR. Competitive effects of some cations on active potassium transport in the human red blood cell. *J Clin Invest.* 1967;46(9):1433–1441.
194. Sachs JR, Welt LG. The concentration dependence of active potassium transport in the human red blood cell. *J Clin Invest.* 1967;46(1):65–76.
195. Sachs JR. Interaction of external K, Na, and cardioactive steroids with the Na-K pump of the human red blood cell. *J Gen Physiol.* 1974;63(2):123–143.
196. Hoffman PG, Tosteson DC. Active sodium and potassium transport in high potassium and low potassium sheep red cells. *J Gen Physiol.* 1971;58(4):438–466.
197. Garay RP, Garrahan PJ. The interaction of sodium and potassium with the sodium pump in red cells. *J Physiol.* 1973;231(2):297–325.
198. Rodland KD, Dunham PB. Kinetics of lithium efflux through the (Na,K)-pump of human erythrocytes. *Biochim Biophys Acta.* 1980;602(2):376–388.
199. Wuddel I, Apell HJ. Electrogenicity of the sodium transport pathway in the Na,K-ATPase probed by charge-pulse experiments. *Biophys J.* 1995;69(3):909–921.
200. Repke KR, Schon R. Flip-flop model of (NaK)-ATPase function. *Acta Biologica et Medica Germanica* 1973;31(4):(Suppl):K19–30.
201. Glynn IM. A hundred years of sodium pumping. *Annu Rev Physiol.* 2002;64(1):1–18.
202. Sachs JR. Kinetic evaluation of the Na-K pump reaction mechanism. *J Physiol.* 1977;273(2):489–514.
203. Sachs J. Na^+/K^+ Pump. In: Bernhardt I, Ellory CJ, editors. *Red Cell Membrane Transport in Health and Disease.* Berlin, Heidelberg, New York, Hong Kong, London, Milan, Paris, Tokyo. Springer; 2003, pp. 111–133.
204. Garrahan PJ, Glynn IM. The stoicheiometry of the sodium pump. *J Physiol.* 1967;192(1):217–235.

205. Karlish SJ, Yates DW, Glynn IM. Conformational transitions between Na^+-bound and K^+-bound forms of (Na^++K^+)-ATPase, studied with formycin nucleotides. *Biochim Biophys Acta.* 1978;525(1):252–264.
206. Garrahan PJ, Glynn IM. The sensitivity of the sodium pump to external sodium. *J Physiol.* 1967;192(1):175–188.
207. Garrahan PJ, Glynn IM. The behaviour of the sodium pump in red cells in the absence of external potassium. *J Physiol.* 1967;192(1):159–174.
208. Glynn IM, Karlish SJ. ATP hydrolysis associated with an uncoupled efflux of Na through the Na pump. *J Physiol.* 1975;250(1):33P–34P.
209. Glynn IM, Karlish SJ. ATP hydrolysis associated with an uncoupled sodium flux through the sodium pump: evidence for allosteric effects of intracellular ATP and extracellular sodium. *J Physiol.* 1976;256(2):465–496.
210. Cavieres JD, Glynn IM. Sodium-sodium exchange through the sodium pump: the roles of ATP and ADP. *J Physiol.* 1979;297(0):637–645.
211. Dissing S, Hoffman JF. Anion-coupled Na efflux mediated by the human red blood cell Na/K pump. *J Gen Physiol.* 1990;96(1):167–193.
212. Marín R, Hoffman JF. Phosphate from the phosphointermediate (EP) of the human red blood cell Na/K pump is coeffluxed with Na, in the absence of external K. *J Gen Physiol.* 1994;104(1):1–32.
213. Garrahan PJ, Glynn IM. The incorporation of inorganic phosphate into adenosine triphosphate by reversal of the sodium pump. *J Physiol.* 1967;192(1):237–256.
214. Glynn IM, Lew VL. Synthesis of adenosine triphosphate at the expense of downhill cation movements in intact human red cells. *J Physiol.* 1970;207(2):393–402.
215. Glynn IM, Lew VL, Luthi U. Reversal of the potassium entry mechanism in red cells, with and without reversal of the entire pump cycle. *J Physiol.* 1970;207(2):371–391.
216. Mercer RW, Dunham PB. Membrane-bound ATP fuels the Na/K pump. Studies on membrane-bound glycolytic enzymes on inside-out vesicles from human red cell membranes. *J Gen Physiol.* 1981;78(5):547–568.
217. Hoffman JF, Dodson A, Proverbio F. On the functional use of the membrane compartmentalized pool of ATP by the Na^+ and Ca^{++} pumps in human red blood cell ghosts. *J Gen Physiol.* 2009;134(4):351–361.
218. Dunham PB, Senyk O. Lithium efflux through the Na/K pump in human erythrocytes. *Proc Natl Acad Sci USA.* 1977;74(7):3099–3103.

219. Blostein R. Na$^+$ATPase of the mammalian erythrocyte membrane. Reversibility of phosphorylation at 0 degrees. *J Biol Chem.* 1975;250(15):6118–6124.
220. Blostein R. Side-specific effects of sodium on (Na,K)-ATPase. Studies with inside-out red cell membrane vesicles. *J Biol Chem.* 1979;254(14):6673–6677.
221. Blostein R. Proton-activated rubidium transport catalyzed by the sodium pump. *J Biol Chem.* 1985;260(2):829–833.
222. Glynn IM, Hara Y, Richards DE. The occlusion of sodium ions within the mammalian sodium-potassium pump: its role in sodium transport. *J Physiol.* 1984;351:531–547.
223. Glynn IM, Howland JL, Richards DE. Evidence for the ordered release of rubidium ions occluded within the Na,K-ATPase of mammalian kidney. *J Physiol.* 1985;368:453–469.
224. Glynn IM, Hara Y, Richards DE, Steinberg M. Comparison of rates of cation release and of conformational change in dog kidney Na, K-ATPase. *J Physiol.* 1987;383:477–485.
225. Garrahan PJ, Rega AF. Potassium activated phosphatase from human red blood cells. The effects of p-nitrophenylphosphate on carbon fluxes. *J Physiol.* 1972;223(2):595–617.
226. Drapeau P, Blostein R. Interactions of K$^+$ with (Na,K)-ATPase orientation of K$^+$-phosphatase sites studied with inside-out red cell membrane vesicles. *J Biol Chem.* 1980;255(16):7827–7834.
227. Blostein R. Measurement of Na$^+$ and K$^+$ transport and Na$^+$,K$^+$-ATPase activity in inside-out vesicles from mammalian erythrocytes. In: Fleischer S, Fleischer B, Abelson J, Simon M. *Methods in Enzymology.* Academic press; 1988, pp. 171–178.
228. Polvani C, Blostein R. Protons as substitutes for sodium and potassium in the sodium pump reaction. *J Biol Chem.* 1988;263(32):16757–16763.
229. Skou JC. The dynamics of the cell membrane coupling of the reaction of the Na, K-ATPase with ATP to the reaction with the cations. *Tokai J Exp Clin Med.* 1982;7(Suppl):1–6.
230. Post RL, Albright CD, Dayani K. Resolution of pump and leak components of sodium and potassium ion transport in human erythrocytes. *J Gen Physiol.* 1967;50(5):1201–1220.
231. Lew VL, Freeman CJ, Ortiz OE, Bookchin RM. A mathematical model of the volume, pH, and ion content regulation in reticulocytes. Application to the pathophysiology of sickle cell dehydration. *J Clin Invest.* 1991;87(1):100–112.

232. Dunham PB, Hoffman JF. Active cation transport and ouabain binding in high potassium and low potassium red blood cells of sheep. *J Gen Physiol.* 1971;58(1):94–116.

233. Lauf PK, Tosteson DC. The M-antigen in HK and LK sheep red cell membranes. *J Membr Biol.* 1969;1(1):177–193.

234. Ellory JC, Glynn IM, Lew VL, Tucker EM. Effects of an antibody and of potassium ions on the apparent affinity for sodium of the sodium pump in low-potassium (LK) goat red cells. *J Physiol.* 1971;217(1):61P–62P.

235. Lauf PK, Sun WW. Binding characteristics of M and L isoantibodies to high and low potassium sheep red cells. *J Membr Biol.* 1976;28(1):351–372.

236. Smalley CE, Tucker EM, Dunham PB, Ellory JC. Interaction of L antibody with low potassium-type sheep red cells: resolution of two separate functional antibodies. *J Membr Biol.* 1982;64(3):167–174.

237. Snyder JJ, Rasmusen BA, Lauf PK. The nature of the antibody in iso-immune anti-L sera affecting active potassium transport in LK sheep red cells. *J Immunol.* 1971;107(3):772–781.

238. Bratcher RL, Kanik-Ennulat CL, Logue PJ, Dunham PB. Stimulation of the Na/K pump in LK sheep erythrocytes by immunoglobulin fragments. *Immunol Commun.* 1983;12(6):565–571.

239. Sachs JR, Dunham PB, Kropp DL, Ellory JC, Hoffman JF. Interaction of HK and LK goat red blood cells with ouabain. *J Gen Physiol.* 1974;64(5):536–550.

240. Wiedmer T, Lauf PK. Properties of the M antigen solubilized from genetically high potassium sheep red cells. *Mol Membr Biol.* 1981;4(1):31–47.

241. Smalley CE, Tucker EM, Clive Ellory J, Young JD. The solubilization of the L and M antigens from sheep red cell membranes. *Biochim Biophys Acta — Biomembranes.* 1983;733(2):283–285.

242. Pittman SJ, Ellory JC, Tucker EM, Newbold CI. Identification of a 25 kDa polypeptide associated with the L antigen in low potassium-type sheep red cells. *Biochim Biophys Acta.* 1990;1022(3):408–410.

243. Zouzoulas A, Therien AG, Scanzano R, Deber CM, Blostein R. Modulation of Na,K-ATPase by the gamma subunit: studies with transfected cells and transmembrane mimetic peptides. *J Biol Chem.* 2003; 278(42):40437–40441.

244. Zouzoulas A, Blostein R. Regions of the catalytic alpha subunit of Na,K-ATPase important for functional interactions with FXYD 2. *J Biol Chem.* 2006;281(13):8539–8544.

245. Pu HX, Scanzano R, Blostein R. Distinct regulatory effects of the Na,K-ATPase gamma subunit. *J Biol Chem.* 2002;277(23):20270–20276.
246. Segall L, Daly SE, Blostein R. Mechanistic basis for kinetic differences between the rat alpha 1, alpha 2, and alpha 3 isoforms of the Na,K-ATPase. *J Biol Chem.* 2001;276(34):31535–31541.
247. Lee P, Woo A, Tosteson DC. Cytodifferentiation and membrane transport properties in LK sheep red cells. *J Gen Physiol.* 1966;50(2):379–390.
248. Lauf PK, Valet G. Cation transport in different volume populations of genetically low K^+ lamb red cells. *J Cell Physiol.* 1980;104(3):283–293.
249. Tucker EM, Smalley CE, Ellory JC, Dunham PB. The transition from HK to LK phenotype in the red cells of newborn genetically LK lambs. *J Gen Physiol.* 1982;79(5):893–915.
250. Dunham PB, Blostein R. Active potassium transport in reticulocytes of high-K^+ and low-K^+ sheep. *Biochim Biophys Acta.* 1976; 455(3): 749–758.
251. Xu ZC, Dunham PB, Dyer B, Blostein R. Decline in number of Na-K pumps on low-K^+ sheep reticulocytes during maturation is modulated by L_p antigen. *Am J Physiol.* 1994;266(5 Pt 1):C1173–1181.
252. Dunham PB, Farquharson BE, Bratcher RL. Stimulation of Na^+-K^+ pump in sheep red blood cells by heteroimmune anti-sheep red cell antibodies. *Am J Physiol.* 1984;247(1 Pt 1):C120–123.
253. Lauf PK, Zhang J, Delpire E, Fyffe REW, Mount DB, Adragna NC. K-Cl co-transport: immunocytochemical and functional evidence for more than one KCC isoform in high K and low K sheep erythrocytes. *Comp Biochem Physiol — Part A: Mol & Integr Physiol.* 2001;130(3):499–509.
254. Campbell EH, Ellory JC, Gibson JS. Effects of protein kinase and phosphatase inhibitors and anti-L antisera on K^+ transport in LK sheep red cells. *Bioelectrochemistry.* 2000;52(2):151–159.
255. Farquharson BE, Dunham PB. Intracellular potassium promotes antibody binding to an antigen associated with the Na/K pump of sheep erythrocytes. *Biochem Biophys Res Commun.* 1986;134(2):982–988.
256. Parker JC. Dog red blood cells. Adjustment of density *in vivo*. *J Gen Physiol.* 1973;61(2):146–157.
257. Fujise H, Abe K, Kamimura M, Ochiai H. K^+-Cl^- cotransport and volume regulation in the light and the dense fraction of high-K^+ dog red blood cells. *Am J Physiol Reg Integ Comp Physiol.* 1997;273(3):R991–998.

258. Fujise H, Higa K, Nakayama T, Wada K, Ochiai H, Tanabe Y. Incidence of dogs possessing red blood cells with high K in Japan and East Asia. *J Vet Med Sci.* 1997;59(6):495–497.

259. Fujise H, Hishiyama N, Ochiai H. Heredity of red blood cells with high K and low glutathione (HK/LG) and high K and high glutathione (HK/HG) in a family of Japanese Shiba Dogs. *Exp Anim.* 1997;46(1):41–46.

260. Maede Y, Amano Y, Nishida A, Murase T, Sasaki A, Inaba M. Hereditary high-potassium erythrocytes with high Na, K-ATPase activity in Japanese shiba dogs. *Res Vet Sci.* 1991;50(1):123–125.

261. Parker JC. Volume-responsive sodium movements in dog red blood cells. *Am J Physiol.* 1983;244(5):C324–330.

262. Parker JC. Sodium and calcium movements in dog red blood cells. *J Gen Physiol.* 1978;71(1):1–17.

263. Parker JC, Colclasure GC, McManus TJ. Coordinated regulation of shrinkage-induced Na/H exchange and swelling-induced [K-Cl] cotransport in dog red cells. Further evidence from activation kinetics and phosphatase inhibition. *J Gen Physiol.* 1991;98(5):869–880.

264. Parker JC, Glosson PS. Interactions of sodium-proton exchange mechanism in dog red blood cells with N-phenylmaleimide. *Am J Physiol.* 1987;253(1 Pt 1):C60–65.

265. Parker JC. Volume-activated transport systems in dog red blood cells. *Comp Biochem Physiol A Comp Physiol.* 1988;90(4):539–542.

266. Gibson J, S. Comparative Physiology of Red Cell Membrane Transport. In: Bernhardt I, Ellory CJ, editors. *Red Cell Membrane Transport in Health and Disease.* Berlin, Heidelberg, New York, Hong Kong, London, Milan, Paris, Tokyo: Springer; 2003, pp. 722–748.

267. Gadsby DC, Nakao M, Bahinski A. Voltage dependence of transient and steady-state Na/K pump currents in myocytes. *Mol Cell Biochem.* 1989;89(2):141–146.

268. Bahinski A, Nakao M, Gadsby DC. Potassium translocation by the Na^+/K^+ pump is voltage insensitive. *Proc Natl Acad Sci USA.* 1988;85(10):3412–3416.

269. Yaragatupalli S, Olivera JF, Gatto C, Artigas P. Altered Na^+ transport after an intracellular alpha-subunit deletion reveals strict external sequential release of Na^+ from the Na/K pump. *Proc Natl Acad Sci USA.* 2009;106(36):15507–15512.

270. Nakao M, Gadsby DC. Voltage dependence of Na translocation by the Na/K pump. *Nature*. 1986;323(6089):628–630.
271. Habermann E. Palytoxin acts through Na^+,K^+-ATPase. *Toxicon*. 1989;27(11):1171–1187.
272. Halperin JA, Brugnara C, Van Ha T, Tosteson DC. Voltage-activated cation permeability in high-potassium but not low-potassium red blood cells. *Am J Physiol*. 1990;258(6 Pt 1):C1169–1172.
273. Tosteson MT, Halperin JA, Kishi Y, Tosteson DC. Palytoxin induces an increase in the cation conductance of red cells. *J Gen Physiol*. 1991;98(5):969–985.
274. Tosteson MT, Thomas J, Arnadottir J, Tosteson DC. Effects of palytoxin on cation occlusion and phosphorylation of the (Na^+,K^+)-ATPase. *J Membr Biol*. 2003;192(3):181–189.
275. Artigas P, Gadsby DC. Na^+/K^+-pump ligands modulate gating of palytoxin-induced ion channels. *Proc Natl Acad Sci USA*. 2003;100(2):501–505.
276. Takeuchi A, Reyes N, Artigas P, Gadsby DC. The ion pathway through the opened Na^+,K^+-ATPase pump. *Nature*. 2008;456(7220):413–416.
277. Rakowski RF, Artigas P, Palma F, Holmgren M, De Weer P, Gadsby DC. Sodium flux ratio in Na/K pump-channels opened by palytoxin. *J Gen Physiol*. 2007;130(1):41–54.
278. Artigas P, Gadsby DC. Ouabain affinity determining residues lie close to the Na/K pump ion pathway. *Proc Natl Acad Sci USA*. 2006;103(33):12613–12618.
279. Rossini GP, Bigiani A. Palytoxin action on the Na^+,K^+-ATPase and the disruption of ion equilibria in biological systems. *Toxicon*. 2011;57:424–439.
280. Schatzmann HJ. ATP-dependent Ca^{2+}-extrusion from human red cells. *Experientia*. 1966;22(6):364–365.
281. Knauf PA, Proverbio F, Hoffman JF. Electrophoretic separation of different phophosproteins associated with Ca-ATPase and Na,K-ATPase in human red cell ghosts. *J Gen Physiol*. 1974;63(3):324–336.
282. Shull GE, Greeb J. Molecular cloning of two isoforms of the plasma membrane Ca^{2+}-transporting ATPase from rat brain. Structural and functional domains exhibit similarity to Na^+,K^+- and other cation transport ATPases. *J Biol Chem*. 1988;263(18):8646–8657.

283. Strehler EE, Zacharias DA. Role of alternative splicing in generating isoform diversity among plasma membrane calcium pumps. *Physiol Rev.* 2001;81(1):21–50.
284. Burk SE, Lytton J, MacLennan DH, Shull GE. cDNA cloning, functional expression, and mRNA tissue distribution of a third organellar Ca^{2+} pump. *J Biol Chem.* 1989;264(31):18561–18568.
285. Greeb J, Shull GE. Molecular cloning of a third isoform of the calmodulin-sensitive plasma membrane Ca^{2+}-transporting ATPase that is expressed predominantly in brain and skeletal muscle. *J Biol Chem.* 1989; 264(31):18569–18576.
286. Keeton TP, Burk SE, Shull GE. Alternative splicing of exons encoding the calmodulin-binding domains and C termini of plasma membrane Ca^{2+}-ATPase isoforms 1, 2, 3, and 4. *J Biol Chem.* 1993;268(4):2740–2748.
287. Rossi JP, Schatzmann HJ. Is the red cell calcium pump electrogenic? *J Physiol.* 1982;327:1–15.
288. Romero PJ, Ortiz CE. Electrogenic behavior of the human red cell Ca^{2+} pump revealed by disulfonic stilbenes. *J Membr Biol.* 1988; 101(3):237–246.
289. Niggli V, Sigel E, Carafoli E. The purified Ca^{2+} pump of human erythrocyte membranes catalyzes an electroneutral Ca^{2+}-H^+ exchange in reconstituted liposomal systems. *J Biol Chem.* 1982;257(5):2350–2356.
290. Hao L, Rigaud JL, Inesi G. Ca^{2+}/H^+ countertransport and electrogenicity in proteoliposomes containing erythrocyte plasma membrane Ca-ATPase and exogenous lipids. *J Biol Chem.* 1994;269(19):14268–14275.
291. Brini M, Carafoli E. Calcium signalling: a historical account, recent developments and future perspectives. *Cell Mol Life Sci.* 2000;57(3): 354–370.
292. Niggli V, Adunyah ES, Penniston JT, Carafoli E. Purified (Ca^{2+}-Mg^{2+})-ATPase of the erythrocyte membrane. Reconstitution and effect of calmodulin and phospholipids. *J Biol Chem.* 1981;256(1):395–401.
293. Zimmerman ML, Daleke DL. Regulation of a candidate aminophospholipid-transporting ATPase by lipid. *Biochemistry.* 1993;32(45):12257–12263.
294. Daleke DL, Lyles JV. Identification and purification of aminophospholipid flippases. *Biochim Biophys Acta.* 2000;1486(1):108–127.
295. Daleke DL. Regulation of transbilayer plasma membrane phospholipid asymmetry. *J Lipid Res.* 2003;44(2):233–242.

296. Daleke DL. Phospholipid flippases. *J Biol Chem.* 2007;282(2):821–825.
297. Smriti, Nemergut EC, Daleke DL. ATP-dependent transport of phosphatidylserine analogues in human erythrocytes. *Biochemistry.* 2007; 46(8):2249–2259.
298. Daleke DL. Regulation of phospholipid asymmetry in the erythrocyte membrane. *Curr Op Hematol.* 2008;15(3):191–195.
299. Dagher G, Lew VL. Maximal calcium extrusion capacity and stoichiometry of the human red cell calcium pump. *J Physiol.* 1988;407:569–586.
300. Schatzmann H, J. Dependence on calcium concentration and stoichiometry of the calcium pump in human red cells. *J Physiol.* 1973;235:559–569.
301. Garcia ML, Strehler EE. Plasma membrane calcium ATPases as critical regulators of calcium homeostasis during neuronal cell function. *Front Biosci.* 1999;4:D869–D882.
302. Rega AF, Garrahan PJ. Calcium ion-dependent phosphorylation of human erythrocyte membranes. *J Membr Biol.* 1975;22(3–4):313–327.
303. Barrabin H, Garrahan PJ, Rega AF. Vanadate inhibition of the Ca^{2+}-ATPase from human red cell membranes. *Biochim Biophys Acta.* 1980;600(3):796–804.
304. Varecka L, Carafoli E. Vanadate-induced movements of Ca^{2+} and K^+ in human red blood cells. *J Biol Chem.* 1982;257(13):7414–7421.
305. Caride AJ, Rega AF, Garrahan PJ. Effects of p-nitrophenylphosphate on Ca^{2+} transport in inside-out vesicles from human red-cell membranes. *Biochim Biophys Acta.* 1983;734(2):363–367.
306. Garrahan PJ, Rega AF. Activation of partial reactions of the Ca^{2+}-ATPase from human red cells by $Mg2^+$ and ATP. *Biochim Biophys Acta.* 1978;513(1):59–65.
307. Caride AJ, Rega AF, Garrahan PJ. The reaction of Mg^{2+} with the Ca^{2+}-ATPase from human red cell membranes and its modification by Ca^{2+}. *Biochim Biophys Acta.* 1986;863(2):165–177.
308. Adamo HP, Rega AF, Garrahan PJ. Magnesium-ions accelerate the formation of the phosphoenzyme of the (Ca^{2+} + Mg^{2+})-activated ATPase from plasma membranes by acting on the phosphorylation reaction. *Biochem Biophys Res Commun.* 1990;169(2):700–705.
309. Kratje RB, Garrahan PJ, Rega AF. Two modes of inhibition of the Ca^{2+} pump in red cells by Ca^{2+}. *Biochim Biophys Acta.* 1985;816(2):365–378.

310. Adamo HP, Rega AF, Garrahan PJ. The E2 in equilibrium E1 transition of the Ca^{2+}-ATPase from plasma membranes studied by phosphorylation. *J Biol Chem*. 1990;265(7):3789–3792.
311. Herscher CJ, Rega AF, Garrahan PJ. The dephosphorylation reaction of the Ca^{2+}-ATPase from plasma membranes. *J Biol Chem*. 1994;269(14):10400–10406.
312. Rossi JP, Garrahan PJ, Rega AF. Reversal of the calcium pump in human red cells. *J Membr Biol*. 1978;44(1):37–46.
313. Chiesi M, Zurini M, Carafoli E. ATP synthesis catalyzed by the purified erythrocyte Ca-ATPase in the absence of calcium gradients. *Biochemistry*. 1984;23(12):2595–2600.
314. Ronner P, Gazzotti P, Carafoli E. A lipid requirement for the $(Ca^{2+} + Mg^{2+})$-activated ATPase of erythrocyte membranes. *Arch Biochem Biophys*. 1977;179(2):578–583.
315. Rossi JP, Garrahan PJ, Rega AF. The activation of phosphatase activity of the Ca^{2+}-ATPase from human red cell membranes by calmodulin, ATP and partial proteolysis. *Biochim Biophys Acta*. 1986;858(1):21–30.
316. Wüthrich A, Schatzmann HJ, Romero P. Net ATP synthesis by running the red cell calcium pump backwards. *Experientia*. 1979;35(12):1589–1590.
317. Eaton JW, Berger E, Nelson D, White JG, Rundquist O. Intracellular calcium: lack of effect on ovine red cells. *Proc Soc Exp Biol Med*. 1978;157(3):506–510.
318. Brown AM, Ellory JC, Young JD, Lew VL. A calcium-activated potassium channel present in foetal red cells of the sheep but absent from reticulocytes and mature red cells. *Biochim Biophys Acta — Biomembranes*. 1978;511(2):163–175.
319. Lauf PK. Thiol-dependent passive K^+-Cl^- transport in sheep red blood cells. V. Dependence on metabolism. *Am J Physiol Cell Physiol*. 1983;245(5):C445–C448.
320. Lauf PK, Chimote AA, Adragna NC. Lithium fluxes indicate presence of Na-Cl cotransport (NCC) in human lens epithelial cells. *Cell Physiol Biochem*. 2008;21(5–6):335–346.
321. Brett CL, Donowitz M, Rao R. Evolutionary origins of eukaryotic sodium/proton exchangers. *Am J Physiol Cell Physiol*. 2005;288(2):C223–C239.

322. Kinsella JL, Aronson PS. Amiloride inhibition of the Na^+-H^+ exchanger in renal microvillus membrane vesicles. *Am J Physiol Renal Physiol.* 1981;241(4):F374–F379.
323. Aronson PS, Suhm MA, Nee J. Interaction of external H^+ with the Na^+-H^+ exchanger in renal microvillus membrane vesicles. *J Biol Chem.* 1983;258(11):6767–6771.
324. Mahnensmith R, Aronson P. The plasma membrane sodium-hydrogen exchanger and its role in physiological and pathophysiological processes. *Circ Res.* 1985;56(6):773–788.
325. Knickelbein R, Aronson PS, Atherton W, Dobbins JW. Sodium and chloride transport across rabbit ileal brush border. I. Evidence for Na-H exchange. *Am J Physiol Gastrointest Liver Physiol.* 1983;245(4):G504–G510.
326. Sardet C, Franchi A, Pouysségur J. Molecular cloning, primary structure, and expression of the human growth factor-activatable Na^+/H^+ antiporter. *Cell* 1989;56(2):271–280.
327. Szpirer C, Szpirer J, Rivière M, Levan G, Orlowski J. Chromosomal assignment of four genes encoding Na/H exchanger isoforms in human and rat. *Mamm Genome.* 1994;5(3):153–159.
328. Malakooti J, Dahdal RY, Dudeja PK, Layden TJ, Ramaswamy K. The human Na^+/H^+ exchanger NHE2 gene: genomic organization and promoter characterization. *Am J Physiol Gastrointest Liver Physiol.* 2001;280(4):G763–G773.
329. Kemp G, Young H, Fliegel L. Structure and function of the human Na^+/H^+ exchanger isoform 1. *Channels (Austin).* 2008;2(5):329–336.
330. Orlowski J, Grinstein S. Na^+/H^+ Exchangers of Mammalian Cells. *J Biol Chem.* 1997;272(36):22373–22376.
331. Wakabayashi S, Ikeda T, Iwamoto T, Pouyssegur J, Shigekawa M. Calmodulin-binding autoinhibitory domain controls "pH-sensing" in the Na^+/H^+ exchanger NHE1 through sequence-specific interaction. *Biochemistry.* 1997;36(42):12854–12861.
332. Sardet C, Fafournoux P, Pouyssegur J. Alpha-thrombin, epidermal growth factor, and okadaic acid activate the Na^+/H^+ exchanger, NHE-1, by phosphorylating a set of common sites. *J Biol Chem.* 1991;266(29):19166–19171.
333. Borgese F, Malapert M, Fievet B, Pouyssegur J, Motais R. The cytoplasmic domain of the Na^+/H^+ exchangers (NHEs) dictates the nature of the hormonal response: behavior of a chimeric human NHE1/trout beta NHE antiporter. *Proc Natl Acad Sci USA.* 1994;91(12):5431–5435.

334. Fliegel L, Walsh MP, Singh D, Wong C, Barr A. Phosphorylation of the C-terminal domain of the Na^+/H^+ exchanger by Ca^{2+}/calmodulin-dependent protein kinase II. *Biochem J.* 1992;282(Pt 1):139–145.

335. Bertrand B, Wakabayashi S, Ikeda T, Pouyssegur J, Shigekawa M. The Na^+/H^+ exchanger isoform 1 (NHE1) is a novel member of the calmodulin-binding proteins. Identification and characterization of calmodulin-binding sites. *J Biol Chem.* 1994;269(18):13703–13709.

336. Grinstein S, Woodside M, Sardet C, Pouyssegur J, Rotin D. Activation of the Na^+/H^+ antiporter during cell volume regulation. Evidence for a phosphorylation-independent mechanism. *J Biol Chem.* 1992;267(33): 23823–23828.

337. Fafournoux P, Noel J, Pouyssegur J. Evidence that Na^+/H^+ exchanger isoforms NHE1 and NHE3 exist as stable dimers in membranes with a high degree of specificity for homodimers. *J Biol Chem.* 1994; 269(4):2589–2596.

338. Grillo FG, Aronson PS. Inactivation of the renal microvillus membrane Na^+-H^+ exchanger by histidine-specific reagents. *J Biol Chem.* 1986;261(3):1120–1125.

339. Igarashi P, Aronson PS. Covalent modification of the renal Na^+/H^+ exchanger by N,N'-dicyclohexylcarbodiimide. *J Biol Chem.* 1987;262(2):860–868.

340. Landau M, Herz K, Padan E, Ben-Tal N. Model Structure of the Na^+/H^+ Exchanger 1 (NHE1). *J Biol Chem.* 2007;282(52):37854–37863.

341. Nygaard EB, Lagerstedt JO, Bjerre GP, Shi B, Budamagunta M, Poulsen KA *et al.* Structural modelling and electron paramagnetic resonance spectroscopy of the human Na^+/H^+ exchanger isoform 1, NHE1. *J Biol Chem* 2011;286:634–648.

342. Borgese F, Sardet C, Cappadoro M, Pouyssegur J, Motais R. Cloning and expression of a cAMP-activated Na^+/H^+ exchanger: evidence that the cytoplasmic domain mediates hormonal regulation. *Proc Natl Acad Sci USA.* 1992;89(15):6765–6769.

343. Motais R, Borgese F, Fievet B, Garcia-Romeu F. Regulation of Na^+/H^+ exchange and pH in erythrocytes of fish. *Comp Biochem and Physiol Part A: Physiology.* 1992;102(4):597–602.

344. Malapert M, Pellissier B, Borgese F. Asn49 is the unique glycosylation site of the trout red blood cell Na^+/H^+ exchanger. *Eur J Biochem.* 1998;257(1):228–235.

345. Malapert M, Guizouarn H, Fievet B, Jahns R, Garcia-Romeu F, Motais R et al. Regulation of Na$^+$/H$^+$ antiporter in trout red blood cells. *J Exp Biol.* 1997;200(2):353–360.
346. Nickerson JG, Dugan SG, Drouin G, Perry SF, Moon TW. Activity of the unique beta-adrenergic Na$^+$/H$^+$ exchanger in trout erythrocytes is controlled by a novel beta3-AR subtype. *Am J Physiol Reg Integ Comp Physiol.* 2003;285(3):R526–R535.
347. Nikinmaa M. The effects of adrenaline on the oxygen transport properties of Salmo gairdneri blood. *Comp Biochem Physiol A Comp Physiol.* 1982;71(2):353–356.
348. Nikinmaa M, Cech JJJ, Ryhänen EL, Salama A. Red cell function of carp (Cyprinus carpio) in acute hypoxia. *J Exp Biol.* 1987;47(1):53–58.
349. Mylvaganam SE, Bonaventura C, Bonaventura J, Getzoff ED. Structural basis for the Root effect in haemoglobin. *Nat Struct Mol Biol.* 1996;3(3):275–283.
350. Salama A, Nikinmaa M. The adrenergic responses of carp (Cyprinus carpio) red cells: effects of PO$_2$ and pH. *J Exp Biol.* 1988;136(1):405–416.
351. Paajaste M, Nikinmaa M. Effect of noradrenaline on the methaemoglobin concentration of rainbow trout red cells. *J Exp Zool.* 1991;260(1):28–32.
352. Nikinmaa M, Bogdanova A, Lecklin T. Oxygen dependency of the adrenergic Na/H exchange in rainbow trout erythrocytes is diminished by a hydroxyl radical scavenger. *Acta Physiol Scand.* 2003;178(2):149–154.
353. McLean LA, Zia S, Gorin FA, Cala PM. Cloning and expression of the Na$^+$/H$^+$ exchanger from Amphiuma RBCs: resemblance to mammalian NHE1. *Am J Physiol Cell Physiol.* 1999;276(5):C1025–C1037.
354. Pedersen SF, King SA, Rigor RR, Zhuang Z, Warren JM, Cala PM. Molecular cloning of NHE1 from winter flounder RBCs: activation by osmotic shrinkage, cAMP, and calyculin A. *Am J Physiol Cell Physiol.* 2003;284(6):C1561–C1576.
355. Pedersen SF, King SA, Nygaard EB, Rigor RR, Cala PM. NHE1 Inhibition by Amiloride- and Benzoylguanidine-type Compounds. *J Biol Chem.* 2007;282(27):19716–19727.
356. Rutherford P, Pizzonia J, Abu-Alfa A, Biemesderfer D, Reilly R, Aronson P. Sodium-hydrogen exchange isoform expression in blood cells: implications for studies in diabetes mellitus. *Exp Clin Endocrinol Diab.* 1997;105(S 02):13–16.

357. Khan I. Topology of the C-terminus of sodium hydrogen exchanger isoform-1: presence of an extracellular epitope. *Arch Biochem Biophys.* 2001;391(1):25–29.
358. Sarangarajan R, Dhabia N, Soleimani M, Baird N, Joiner C. NHE-1 is the sodium-hydrogen exchanger isoform present in erythroid cells. *Biochim Biophys Acta — Biomembranes.* 1998;1374(1–2):56–62.
359. Zerbini G, Maestroni A, Mangili R, Pozza G. Amiloride-insensitive Na^+-H^+ exchange: a candidate mediator of erythrocyte Na^+-Li^+ countertransport. *J Am Soc Nephrol.* 1998;9(12):2203–2211.
360. Escobales N, Canessa M. Ca^{2+}-activated Na^+ fluxes in human red cells. Amiloride sensitivity. *J Biol Chem.* 1985;260(22):11914–11923.
361. Escobales N, Canessa M. Amiloride-sensitive Na^+ transport in human red cells: Evidence for a Na/H exchange system. *J Membr Biol.* 1986;90(1):21–28.
362. Canessa M, Adragna N, Solomon HS, Connolly TM, Tosteson DC. Increased sodium-lithium countertransport in red cells of patients with essential hypertension. *New Eng J Med.* 1980;302(14):772–776.
363. Adragna N, Canessa M, Solomon H, Slater E, Tosteson D. Red cell lithium-sodium countertransport and sodium-potassium cotransport in patients with essential hypertension. *Hypertension.* 1982;4(6):795–804.
364. Matteucci E, Giampietro O. Building a bridge between clinical and basic research: the phenotypic elements of familial predisposition to type 1 diabetes. *Curr Medic Chem.* 2007;14:555–567.
365. Van Norren K, Thien T, Berden J, Elving L, De Pont J. Relevance of erythrocyte Na^+/Li^+ countertransport measurement in essential hypertension, hyperlipidaemia and diabetic nephropathy: a critical review. *Eur J Clin Inv.* 1998;28(5):339–352.
366. Canessa M. Erythrocyte sodium-lithium countertransport: another link between essential hypertension and diabetes. *Curr Op Nephrol Hypertens.* 1994;3(5):511–517.
367. Canessa M, Zerbini G, Laffel L. Sodium activation kinetics of red blood cell Na^+/Li^+ countertransport in diabetes: methodology and controversy. *J Am Soc Nephrol.* 1992;3(4):S41–S49.
368. Sechi LA, Melis A, Faedda R, Tedde R, Bartoli E. Heterogeneous derangement of cellular sodium metabolism in Bartter's syndrome. Description of two cases and review of the literature. *Panminerva Medica.* 1992;34(2):85–92.

369. Jennings ML, Adams-Lackey M, Cook KW. Absence of significant sodium-hydrogen exchange by rabbit erythrocyte sodium-lithium countertransporter. *Am J Physiol Cell Physiol.* 1985;249(1):C63–C68.
370. Jennings ML, Douglas SM, McAndrew PE. Amiloride-sensitive sodium-hydrogen exchange in osmotically shrunken rabbit red blood cells. *Am J Physiol Cell Physiol.* 1986;251(1):C32–C40.
371. Huot SJ, Aronson PS. Na^+-H^+ exchanger and its role in essential hypertension and diabetes mellitus. *Diabetes Care.* 1991;14(6):521–535.
372. Dudley C, Giuffra L, Raine A, Reeders S. Assessing the role of APNH, a gene encoding for a human amiloride-sensitive Na^+/H^+ antiporter, on the interindividual variation in red cell Na^+/Li^+ countertransport. *J Am Soc Nephrol.* 1991;2(4):937–943.
373. van Norren K, Gorissen R, Borgese F, Borggreven JMPM, De Pont JJHHM. The Na^+/H^+ exchanger present in trout erythrocyte membranes is not responsible for the amiloride-insensitive Na^+/Li^+ exchange activity. *J Membr Biol.* 1997;160(3):193–199.
374. Orlov SN, Kuznetsov SR, Pokudin NI, Tremblay J, Hamet P. Can we use erythrocytes for the study of the activity of the ubiquitous Na^+/H^+ exchanger (NHE-1) in essential hypertension? *Am J Hypertens.* 1998;11(7):774–783.
375. Lifton R, Hunt S, Williams R, Pouyssegur J, Lalouel J. Exclusion of the Na^+-H^+ antiporter as a candidate gene in human essential hypertension. *Hypertension.* 1991;17(1):8–14.
376. Zerbini G, Maestroni A, Breviario D, Mangili R, Casari G. Alternative splicing of NHE-1 mediates Na-Li countertransport and associates with activity rate. *Diabetes.* 2003;52(6):1511–1518.
377. Serrani RE, Mujica G, Gioia IA, Corchs JL. Neonatal red blood cells: amiloride-insensitive Na^+-H^+ transport isoform would express Na^+-Li^+ exchange. *Acta Physiologica et Pharmacologica Bulgarica.* 2000;25(3–4):71–74.
378. Richter S, Hamann J, Kummerow D, Bernhardt I. The monovalent cation "leak" transport in human erythrocytes: an electroneutral exchange process. *Biophys J.* 1997;73(2):733–745.
379. Bernhardt I, Bogdanova AY, Kummerow D, Kiessling K, Hamann J, Ellory JC. Characterization of the $K^+(Na^+)/H^+$ monovalent cation exchanger in the human red blood cell membrane: effects of transport inhibitors. *Gen Physiol Biophys.* 1999;18(2):119–137.

380. Kummerow D, Hamann J, Browning JA, Wilkins R, Ellory JC, Bernhardt I. Variations of intracellular pH in human erythrocytes via $K^+(Na^+)/H^+$ exchange under low ionic strength conditions. *J Membr Biol.* 2000; 176(3):207–216.

381. Thomas TH, West IC, Wilkinson R. Modification of erythrocyte Na^+/Li^+ countertransport kinetics by two types of thiol group. *Biochim Biophys Acta — Biomembranes.* 1995;1235(2):317–322.

382. Vareesangthip K, Panthongdee W, Shayakul C, Nitiyanant W, Ong-Aj-Yooth L. Abnormal kinetics of erythrocyte sodium lithium countertransport in patients with diabetic nephropathy in Thailand. *Journal of the Medical Association of Thailand.* 2006;89(Suppl 2):S48–S53.

383. Kędzierska K, Bober J, Ciechanowski K, Gołembiewska E, Kwiatkowska E, Nocen I et al. Copper modifies the activity of sodium-transporting systems in erythrocyte membrane in patients with essential hypertension. *Biol Trace Element Res.* 2005;107(1):21–32.

384. Morgan K, Spurlock G, Collins PA, Mir MA. Interaction of inhibitin with the human erythrocyte $Na^+(Li^+)_i/Na_o^+$ exchanger. *Biochim Biophys Acta.* 1989;979(1):53–61.

385. Ryu K, Adragna N, Lauf P. Kinetics of Na-Li exchange in high and low K sheep red blood cells. *Am J Physiol Cell Physiol.* 1989;257:C58–C64.

386. Pontremoli R, Zerbini G, Rivera A, Canessa M. Insulin activation of red blood cell Na^+/H^+ exchange decreases the affinity of sodium sites. *Kidney Int.* 1994;46(2):365–375.

387. Ceolotto G, Conlin P, Clari G, Semplicini A, Canessa M. Protein kinase C and insulin regulation of red blood cell Na^+/H^+ exchange. *Am J Physiol Cell Physiol.* 1997;272(3):C818–C826.

388. Juel C, Lundby C, Sander M, Calbet JAL, van Hall G. Human skeletal muscle and erythrocyte proteins involved in acid-base homeostasis: adaptations to chronic hypoxia. *J Physiol.* 2003;548(2):639–648.

389. Parker JC. Volume-activated cation transport in dog red cells: detection and transduction of the volume stimulus. *Comp Biochem Physiol Part A: Physiology.* 1992;102(4):615–618.

390. Parker JC, McManus TJ, Starke LC, Gitelman HJ. Coordinated regulation of Na/H exchange and K-Cl cotransport in dog red cells. *J Gen Physiol.* 1990;96(6):1141–1152.

391. Parker JC, Dunham PB, Minton AP. Effects of ionic strength on the regulation of Na/H exchange and K-Cl cotransport in dog red blood cells. *J Gen Physiol*. 1995;105(6):677–699.
392. Dunham PB. Cell shrinkage activates Na^+/H^+ exchange in dog red cells by relieving inhibition of exchange by Na^+ in isotonic medium. *Blood Cells Mol Dis*. 2004;32(3):389–393.
393. Dunham PB, Kelley SJ, Logue PJ, Mutolo MJ, Milanick MA. Na^+-inhibitory sites of the Na^+/H^+ exchanger are Li^+ substrate sites. *Am J Physiol Cell Physiol*. 2005 2005;289(2):C277–C282.
394. Morgan K, Canessa M. Interactions of external and internal H^+ and Na^+ with Na^+/Na^+ and Na^+/H^+ exchange of rabbit red cells: evidence for a common pathway. *J Membr Biol*. 1990;118(3):193–214.
395. Cala PM. Volume regulation by Amphiuma red blood cells. The membrane potential and its implications regarding the nature of the ion-flux pathways. *J Gen Physiol* 1980;76(6):683–708.
396. Cala PM, Maldonado H, Anderson SE. Cell volume and pH regulation by the Amphiuma red blood cell: a model for hypoxia-induced cell injury. *Comp Biochem Physiol Part A: Physiology*. 1992;102(4):603–608.
397. Pedersen SF, Cala PM. Comparative biology of the ubiquitous Na^+/H^+ exchanger, NHE1: lessons from erythrocytes. *J Exp Zool Part A*: 2004;301A(7):569–578.
398. Cala PM. Cell volume regulation by Amphiuma red blood cells. The role of Ca^{+2} as a modulator of alkali metal/H^+ exchange. *J Gen Physiol*. 1983; 82(6):761–784.
399. Cala PM, Maldonado HM. pH regulatory Na/H exchange by Amphiuma red blood cells. *J Gen Physiol*. 1994;103(6):1035–1053.
400. Ortiz-Acevedo A, Rigor RR, Maldonado HM, Cala PM. Activation of Na^+/H^+ and K^+/H^+ exchange by calyculin A in Amphiuma tridactylum red blood cells: implications for the control of volume-induced ion flux activity. *Am J Physiol Cell Physiol*. 2008;295(5):C1316–C1325.
401. Ortiz-Acevedo A, Rigor RR, Maldonado HM, Cala PM. Coordinated control of volume regulatory Na^+/H^+ and K^+/H^+ exchange pathways in Amphiuma red blood cells. *Am J Physiol Cell Physiol*. 2010;298(3):C510–C520.
402. Canessa M, Fabry ME, Suzuka SM, Morgan K, Nagel RL. Na^+/H^+ exchange is increased in sickle cell anemia and young normal red cells. *J Membr Biol*. 1990;116(2):107–115.

403. Canessa M. Red cell volume-related ion transport systems in hemoglobinopathies. *Hematol Oncol Clin North Am.* 1991;5(3):495–516.
404. Canessa M, Fabry ME, Blumenfeld N, Nagel RL. Volume-stimulated, Cl$^-$-dependent K$^+$ efflux is highly expressed in young human red cells containing normal hemoglobin or HbS. *J Membr Biol.* 1987;97(2):97–105.
405. Nagel RL, Lawrence C. The distinct pathobiology of sickle cell-hemoglobin C disease. Therapeutic implications. *Hematol Oncol Clin North Am.* 1991;5(3):433–451.
406. Brugnara C. Erythrocyte membrane transport physiology. *Curr Op Hemat.* 1997;4(2):122–127.
407. Baumgarth M, Beier N, Gericke R. (2-Methyl-5-(methylsulfonyl) benzoyl) guanidine Na$^+$/H$^+$ antiporter inhibitors. *J Med Chem.* 1997;40(13):2017–2034.
408. Rizvi SI, Zaid MA. Impairment of sodium pump and Na/H exchanger in erythrocytes from non-insulin dependent diabetes mellitus patients: effect of tea catechins. *Clinica Chimica Acta.* 2005;354(1–2):59–67.
409. Canessa M, Bize I, Solomon H, Adragna N, Tosteson DC, Dagher G *et al*. Na countertransport and cotransport in human red cells: function, dysfunction, and genes in essential hypertension. *Clin Exp Hypert.* 1981;3(4):783–795.
410. Garay R, Nazaret C, Hannaert P, Price M, Diez J, Dagher G *et al*. A classification of essential hypertensive patients according to the erythrocyte Na transport abnormalities: an application for monitoring the antihypertensive response to cicletanide. *Klinische Wochenschrift.* 1985;63(Suppl 3):30–32.
411. Garay RP. Kinetic aspects of red blood cell sodium transport systems in essential hypertension. *Hypertension.* 1987;10:111–114.
412. Canessa M, Brugnara C, Escobales N. The Li$^+$-Na$^+$ exchange and Na$^+$-K$^+$-Cl$^-$ cotransport systems in essential hypertension. *Hypertension.* 1987;10(5 Pt 2):4–10.
413. Narayanan G, Weeks S, Spurlock G, Mir MA, Newcombe R. Relationship between red cell sodium transport, blood pressure, and family history of hypertension. *Am J Hypertens.* 1988;1(2):187–189.
414. Barzilay J, Warram JH, Bak M, Laffel LMB, Canessa M, Krolewski AS. Predisposition to hypertension: Risk factor for nephropathy and hypertension in IDDM. *Kidney Int.* 1992;41(4):723–730.
415. de la Sierra A, Insa R, Compte M, Martinez-Amenós A, Sierra C, Hernández-Herrero G *et al*. Effect of long-term antihypertensive therapy

with angiotensin converting enzyme inhibitors on red cell sodium transport. *Am J. Hypertens.* 1995;8(6):622–625.
416. Chiarelli F, Verrotti A, Kalter-Leibovici O, Laron Z, Morgese G. Genetic predisposition to hypertension (as detected by Na^+/Li^+ countertransport) and risk of diabetic nephropathy in childhood diabetes. *J Paediat Child Health.* 1994;30(6):547–549.
417. Lijnen P, Petrov V, Amery A. Pravastatin has no direct effect on transmembrane cationic transport systems in human erythrocytes and platelets. *Eur J Clin Pharmacol.* 1994;47(3):281–283.
418. Orlov SN, Pausova Z, Gossard F, Gaudet D, Tremblay J, Kotchen T *et al.* Sibling resemblance of erythrocyte ion transporters in French-Canadian sibling-pairs affected with essential hypertension. *J Hypertens.* 1999; 17(12):1859–1865.
419. Strazzullo P, Cappuccio FP, Trevisan M, Iacoviello L, Iacone R, Barba G *et al.* Red blood cell sodium-lithium countertransport, blood pressure, and uric acid metabolism in untreated healthy men. *Am J Hypertens.* 1989; 2(8):634–636.
420. Corry DB, Tuck ML, Nicholas S, Weinman EJ. Increased Na//H antiport activity and abundance in uremic red blood cells. *Kidney Int.* 1993; 44(3):574–578.
421. Rota R, Timsit J, Hannedouche T, Ikeni A, Boitard C, Guicheney P. Erythrocyte Na^+/Li^+ countertransport and glomerular hyperfiltration in insulin-dependent diabetics. *Am J Hypertens.* 1993;6(6 Pt 1):534–537.
422. Canessa Ml, Morgan K, Semplicini A. Genetic differences in lithium-sodium exchange and regulation of the sodium-hydrogen exchanger in essential hypertension. *Cardiovasc Pharmacol.* 1988;12(Suppl 3): S92–S98.
423. Semplicini A, Canessa M, Mozzato MG, Ceolotto G, Marzola M, Buzzaccarini F *et al.* Red blood cell Na^+/H^+ and Li^+/Na^+ exchange in patients with essential hypertension. *Am J Hypertens.* 1989;2(12 Pt 1):903–908.
424. Garay R, Dagher G, Meyer P. An inherited sodium ion-potassium ion cotransport defect in essential hypertension. *Clin Sci (Lond).* 1980;6:191s–193s.
425. Tosteson DC, Adragna N, Bize I, Solomon H, Canessa M. Membranes, ions and hypertension. *Clin Sci (Lond).* 1981;61:5s–10s.
426. Dagher G, Canessa M. The Li-Na exchange in red cells of essential hypertensive patients with low Na-K cotransport. *J Hypertens.* 1984;2(2):195–201.

427. Canessa M, Spalvins A, Adragna N, Falkner B. Red cell sodium countertransport and cotransport in normotensive and hypertensive blacks. *Hypertension.* 1984;6(3):344–351.
428. Canessa M, Redgrave J, Laski C, Williams GH. Does sodium intake modify red cell Na$^+$ transporters in normal and hypertensive subjects? *Am J Hypertens.* 1989;2(7):515–523.
429. Canessa ML. The Na-K-Cl cotransport in essential hypertension: cellular functions and genetic environment interactions. *Int J Cardiol.* 1989; 25(Suppl 1):S37–S45.
430. Ellory JC, Flatman PW, Stewart GW. Inhibition of human red cell sodium and potassium transport by divalent cations. *J Physiol.* 1983;340(1):1–17.
431. Bianchi G, Salvati P, Ferrari P, Vezzoli G. Renal mechanisms and calcium in the pathogenesis of a type of genetic hypertension. *J Cardiovasc Pharmacol.* 1986;8:S124–S129.
432. Redgrave J, Canessa M, Gleason R, Hollenberg N, Williams G. Red blood cell lithium-sodium countertransport in non-modulating essential hypertension. *Hypertension.* 1989;13(6):721–726.
433. Pontremoli R, Spalvins A, Menachery A, Torielli L, Canessa M. Red cell sodium-proton exchange is increased in Dahl salt-sensitive hypertensive rats. *Kidney Int.* 1992;42(6):1355–1362.
434. Adragna NC, Chang JL, Morey MC, Williams RS. Effect of exercise on cation transport in human red cells. *Hypertension.* 1985;7:132–139.
435. Hespel P, Lijnen P, Fiocchi R, Denys B, Lissens W, M'Buyamba-Kabangu JR *et al.* Cationic concentrations and transmembrane fluxes in erythrocytes of humans during exercise. *J Appl Physiol.* 1986;61(1):37–43.
436. Saito T, Onuma N, Yamamoto M, Kai N, Yamamoto K, Iwata J *et al.* Exercise-loaded blood pressure and Li-Na countertransport system in the erythrocyte membrane as predictors of mild essential hypertension prognosis. *Clin Exp Pharm Physiol.* 1991;18(9):611–617.
437. Hespel P, Lijnen P, Fagard R, M'Buyamba-Kabangu J-R, Hoof RV, Lissens W *et al.* Changes in erythrocyte sodium and plasma lipids associated with physical training. *J. Hypertens.* 1988;6(2):159–166.
438. Krolewski AS, Canessa M, Warram JH, Laffel LMB, Christlieb R, Knowler WC *et al.* Predisposition to hypertension and susceptibility to renal disease in insulin-dependent diabetes mellitus. *New Eng J Med.* 1988;318(3): 140–145.

439. Johnson BA, Sowers JR, Zemel PC, Luft FC, Zemel MB. Increased sodium-lithium countertransport in black non-insulin-dependent diabetic hypertensives. *Am J Hypertens.* 1990;3(7):563–565.
440. Doria A, Fioretto P, Avogaro A, Carraro A, Morocutti A, Trevisan R et al. Insulin resistance is associated with high sodium-lithium countertransport in essential hypertension. *Am J Physiol Endocrinol Metab.* 1991; 261(6):E684–E691.
441. Viberti GC, Messent J. Risk factors for renal and cardiovascular disease in diabetic patients. *Cardiology.* 1991;79(Suppl 1):55–61.
442. Gall M-A, Rossing P, Jensen JS, Funder J, Parving H-H. Red cell $Na^+//Li^+$ countertransport in non-insulin-dependent diabetics with diabetic nephropathy. *Kidney Int.* 1991;39(1):135–140.
443. Nosadini R, Cipollina M, Solini A, Sambataro M, Morocutti A, Doria A, et al. Close relationship between microalbuminuria and insulin resistance in essential hypertension and non-insulin dependent diabetes mellitus. *J Am Soc Nephrol.* 1992;3(4):S56–S63.
444. Seely EW, Canessa ML, Graves SW. Impact of diabetes on sodium-lithium countertransport in pregnancy-induced hypertension. *Am J Hypertens.* 1993;6:422–426.
445. Canessa M, Falkner B, Hulman S. Red blood cell sodium-proton exchange in hypertensive blacks with insulin-resistant glucose disposal. *Hypertension.* 1993;22(2):204–213.
446. Falkner B, Canessa M, Anzalone D. Effect of angiotensin converting enzyme inhibitor (lisinopril) on insulin sensitivity and sodium transport in mild hypertension. *Am J Hypertens.* 1995;8(5):454–460.
447. Zerbini G, Ceolotto G, Gaboury C, Mos L, Pessina AC, Canessa M et al. Sodium-lithium countertransport has low affinity for sodium in hyperinsulinemic hypertensive subjects. *Hypertension.* 1995;25(5):986–993.
448. Falkner B, Kushner H, Levison S, Canessa M. Albuminuria in association with insulin and sodium-lithium countertransport in young African Americans with borderline hypertension. *Hypertension.* 1995;25(6):1315–1321.
449. Falkner B, Canessa M, Levison S, Kushner H. Sodium-lithium countertransport is associated with insulin resistance and urinary albumin excretion in young African-Americans. *Am J Kidney Dis.* 1997;29:45–53.
450. Giordano M, Castellino P, Solini A, Canessa M, DeFronzo RA. Na^+/Li^+ and Na^+/H^+ countertransport activity in hypertensive non-insulin-dependent

diabetic patients: role of insulin resistance and antihypertensive treatment. *Metabolism.* 1997;46:1316–1323.
451. Brent GA, Canessa M, Dluhy RG. Reversible alteration of red cell lithium-sodium countertransport in patients with thyroid disease. *J Clin Endocrinol Metab.* 1989;68(2):322–328.
452. Kelly RA, Canessa ML, Steinman TI, Mitch WE. Hemodialysis and red cell cation transport in uremia: role of membrane free fatty acids. *Kidney Int.* 1989;35(2):595–603.
453. Fabbri A, Boero R, Degli Esposti E, Guarena C, Lucatello A, Sturani A *et al.* Aggregation of erythrocyte sodium/lithium countertransport activity in families of patients with immunoglobulin A nephropathy. *Clin Sci (Lond).* 1992;83(2):241–245.
454. Rutherford PA, Thomas TH, O'Kelly J, West IC, Wilkinson R. Thiol group control of sodium-lithium countertransport kinetics in uraemia: evidence of a membrane abnormality affected by haemodialysis. *Nephron.* 1996; 72(2):184–188.
455. Vareesangthip K, Thomas TH, Wilkinson R. Abnormal effect of thiol groups on erythrocyte Na/Li countertransport kinetics in adult polycystic kidney disease. *Nephrol Dial Transplant.* 1995;10(12):2219–2223.
456. Lima PRM, Gontijo JAR, Lopes de Faria JB, Costa FF, Saad STO. Band 3 campinas: a novel splicing mutation in the band 3 gene (ae1) associated with hereditary spherocytosis, hyperactivity of Na^+/Li^+ countertransport and an abnormal renal bicarbonate handling. *Blood.* 1997;90(7):2810–2818.
457. Strazzullo P, Cappuccio FP, Trevisan M, Iacoviello L, Iacone R, Jossa F *et al.* Erythrocyte sodium/lithium countertransport and renal lithium clearance in a random sample of untreated middle-aged men. *Clin Sci (Lond).* 1989;77(3):337–342.
458. Ng LL, Quinn PA, Baker F, Carr SJ. Red cell Na^+/Li^+ countertransport and Na^+/H^+ exchanger isoforms in human proximal tubules. *Kidney Int.* 2000;58(1):229–235.
459. Hoffman JF, Kregenow FM. The characterization of new energy dependent cation transport processes in red blood cells. *Ann New York Acad Sci.* 1966;137(2):566–576.
460. Garay R, Adragna N, Canessa M, Tosteson D. Outward sodium and potassium cotransport in human red cells. *J Membr Biol.* 1981;62(3):169–174.

461. Canessa M, Bize I, Adragna N, Tosteson D. Cotransport of lithium and potassium in human red cells. *J Gen Physiol.* 1982;80(1):149–168.
462. Lauf PK, McManus TJ, Haas M, Forbush B, 3rd, Duhm J, Flatman PW et al. Physiology and biophysics of chloride and cation cotransport across cell membranes. *Fed Proc.* 1987;46(7):2377–2394.
463. Lytle C. Na^+-K^+-$2Cl^-$ Cotransport. In: Bernhardt I, Ellory CJ, editors. *Red Cell Membrane Transport in Health and Disease.* Berlin, Heidelberg, New York, Hong Kong, London, Milan, Paris, Tokyo: Springer; 2003, pp. 173–190.
464. Hebert SC, Gamba G, Kaplan M. The electroneutral Na^+-K^+-Cl^- cotransport family. *Kidney Int.* 1996;49(6):1638–1641.
465. Kaplan MR, Mount DB, Delpire E. Molecular mechanisms of Na Cl cotransport. *Annu Rev Physiol.* 1996;58:649–668.
466. Haas M. The Na-K-Cl cotransporters. *Am J Physiol Cell Physiol.* 1994;267(4):C869–C885.
467. Park JH, Saier JMH. Phylogenetic, structural and functional characteristics of the Na-K-Cl cotransporter family. *J Membr Biol.* 1996;149(3):161–168.
468. Haas M, Forbush B, 3rd. The Na-K-Cl cotransporters. *J Bioenerg Biomembr.* 1998;30(2):161–172.
469. Haas M, Forbush B, 3rd. The Na-K-Cl cotransporter of secretory epithelia. *Annu Rev Physiol.* 2000;62:515–534.
470. Russell JM. Sodium-potassium-chloride cotransport. *Physiol Rev.* 2000;80(1):211–276.
471. Hebert S, Mount D, Gamba G. Molecular physiology of cation-coupled Cl^- cotransport: the SLC12 family. *Pflügers Arch Eur J Physiol.* 2004;447:580–593.
472. Gamba G. Molecular physiology and pathophysiology of electroneutral cation-chloride cotransporters. *Physiol Rev.* 2005;85:423–493.
473. Delpire E, Kaplan MR, Plotkin MD, Hebert SC. The Na-(K)-Cl cotransporter family in the mammalian kidney: molecular identification and function(s). *Nephrol Dial Transpl.* 1996;11(10):1967–1973.
474. Delpire E, Mount DB. Human and murine phenotypes associated with defects in cation-chloride cotransport. *Annu Rev Physiol.* 2002;64(1):803–843.

475. Isenring P, Jacoby SC, Payne JA, Forbush B, 3rd. Comparison of Na-K-Cl cotransporters. NKCC1, NKCC2, and the HEK cell Na-K-Cl cotransporter. *J Biol Chem.* 1998;273(18):11295–11301.
476. Delpire E, Rauchman MI, Beier DR, Hebert SC, Gullans SR. Molecular cloning and chromosome localization of a putative basolateral Na^+-K^+-$2Cl^-$ cotransporter from mouse inner medullary collecting duct (mIMCD-3) cells. *J Biol Chem.* 1994;269(41):25677–25683.
477. Soybel DI, Gullans SR, Maxwell F, Delpire E. Role of basolateral Na^+-K^+-Cl^- cotransport in HCl secretion by amphibian gastric mucosa. *Am J Physiol Cell Physiol.* 1995;269(1):C242–C249.
478. Payne JA, Xu J-C, Haas M, Lytle CY, Ward D, Forbush B. Primary structure, functional expression, and chromosomal localization of the bumetanide-sensitive Na-K-Cl cotransporter in human colon. *J Biol Chem.* 1995;270(30):17977–17985.
479. Di Fulvio M, Alvarez-Leefmans FJ. The NKCC and NCC Genes. An in silico view. In: Alvarez-Leefmans FJ, Delpire E, editors. *Physiology and Pathology of Chloride Transporters and Channels in the Nervous System.* Amsterdam, Boston, Heidelberg, London, New York, Oxford, Paris, San Diego, San Francisco, Singapore, Sydney, Tokyo. Academic Press; 2009, pp. 169–209.
480. Moore-Hoon ML, Turner RJ. The structural unit of the secretory Na^+-K^+-$2Cl^-$ cotransporter (NKCC1) is a homodimer. *Biochemistry.* 2000;39(13):3718–3724.
481. Isenring P, Jacoby SC, Chang J, Forbush B, 3rd. Mutagenic mapping of the Na-K-Cl cotransporter for domains involved in ion transport and bumetanide binding. *J Gen Physiol.* 1998;112:549–558.
482. Gagnon KBE, England R, Delpire E. A single binding motif is required for SPAK activation of the Na-K-2Cl cotransporter. *Cell Physiol Biochem.* 2007;20(1–4):131–142.
483. Mount DB, Delpire E, Gamba G, Hall AE, Poch E, Hoover RS *et al.* The electroneutral cation-chloride cotransporters. *J Exp Biol.* 1998;201 (Pt 14):2091–2102.
484. Flatman PW. Regulation of Na^+-K^+-$2Cl^-$ cotransport by phosphorylation and protein-protein interactions. *Biochim Biophys Acta — Biomembr.* 2002;1566(1–2):140–151.

485. Cossins AR, Gibson JS. Volume-sensitive transport systems and volume homeostasis in vertebrate red blood cells. *J Exp Biol*. 1997;200(Pt 2): 343–352.
486. Muzyamba MC, Cossins AR, Gibson JS. Regulation of Na^+-K^+-$2Cl^-$ cotransport in turkey red cells: the role of oxygen tension and protein phosphorylation. *J Physiol*. 11999;517(2):421–429.
487. Mairbaeurl H, Hoffman JF. Internal magnesium, 2,3-diphosphoglycerate, and the regulation of the steady-state volume of human red blood cells by the Na/K/2Cl cotransport system. *J Gen Physiol*. 1992; 99(5):721–746.
488. Brugnara C, Canessa M, Cusi D, Tosteson DC. Furosemide-sensitive Na and K fluxes in human red cells. Net uphill Na extrusion and equilibrium properties. *J Gen Physiol*. 1986;87(1):91–112.
489. Gagnon KB, Delpire E. Molecular determinants of hyperosmotically activated NKCC1-mediated K^+/K^+ exchange. *J Physiol*. 2010;588(18): 3385–3396.
490. Canessa M, Brugnara C, Cusi D, Tosteson DC. Modes of operation and variable stoichiometry of the furosemide-sensitive Na and K fluxes in human red cells. *J Gen Physiol*. 1986;87(1):113–142.
491. Duhm J, Heller J, Zicha J. Kinetics of red cell Na^+ and K^+ transport in Prague hypertensive rats. *Clin Exp Hypert*. 1990;a12(7): 1203–1222.
492. Isenring P, Forbush B. Ion transport and ligand binding by the Na-K-Cl cotransporter, structure-function studies. *Comp Biochem Physiol — Part A: Mol Integr Physiol*. 2001;130(3):487–497.
493. Ellory JC, Hall AC, Ody SA. Factors affecting the activation and inactivation of KCl cotransport in "young" human red cells. *Biomed Biochim Acta*. 1990;49:S64–S69.
494. Lim J, Gasson C, Kaji DM. Urea inhibits NaK2Cl cotransport in human erythrocytes. *J Clin Inv*. 1995;96(5):2126–2132.
495. Zimmerman SB, Minton AP. Macromolecular crowding: biochemical, biophysical, and physiological consequences. *Annu Rev Biophys Biomol Struct*. 1993;22(1):27–65.
496. Delpire E, Gullans SR. Cell volume and K^+ transport during differentiation of mouse erythroleukemia cells. *Am J Physiol Cell Physiol*. 1994;266(2): C515–C523.

497. Ellory JC, Wolowyk MW. Evidence for bumetanide-sensitive, Na^+-dependent, partial Na-K-Cl co-transport in red blood cells of a primitive fish. *Can J Physiol Pharmacol.* 1991;69:588–591.
498. Milanick MA. Ferret red cells: Na/Ca exchange and Na-K-Cl cotransport. *Comp Biochem Physiol Part A: Physiol.* 1992;102(4):619–624.
499. Delpire E. The mammalian family of sterile 20p-like protein kinases. *Pflügers Arch Eur J Physiology.* 2009;458(5):953–967.
500. Delpire E, Gagnon KEB. SPAK and OSR1: STE20 kinases involved in the regulation of ion homoeostasis and volume control in mammalian cells. *Biochem J.* 2008;409:321–331.
501. Delpire E, Gagnon KEB. Kinetics of hyperosmotically-stimulated Na-K-2Cl cotransporter in Xenopus laevis oocytes. *J Physiol.* 2010, 301(5):C1074–85.
502. Piechotta K, Lu J, Delpire E. Cation chloride cotransporters interact with the stress-related kinases Ste20-related proline-alanine-rich kinase (SPAK) and oxidative stress response 1 (OSR1). *J Biol Chem.* 2002;277(52): 50812–50819.
503. Gagnon KBE, England R, Delpire E. Volume sensitivity of cation-Cl^- cotransporters is modulated by the interaction of two kinases: Ste20-related proline-alanine-rich kinase and WNK4. *Am J Physiol Cell Physiol.* 2006;290(1):C134–C142.
504. Garzon-Muvdi T, Pacheco-Alvarez D, Gagnon KBE, Vazquez N, Ponce-Coria J, Moreno E et al. WNK4 kinase is a negative regulator of K^+-Cl^- cotransporters. *Am J Physiol Renal Physiol.* 2007;292(4): F1197–F1207.
505. Piechotta K, Garbarini N, England R, Delpire E. Characterization of the interaction of the stress kinase SPAK with the Na^+-K^+-$2Cl^-$ cotransporter in the nervous system. *J Biol Chem.* 2003;278(52):52848–52856.
506. Gagnon KB, Delpire E. On the substrate recognition and negative regulation of SPAK, a kinase modulating Na^+-K^+-$2Cl^-$, cotransport activity. *Am J Physiol Cell Physiol.* 2010; 299(3):C614–C620.
507. Garay R, Senn N, Ollivier JP. Erythrocyte ion transport as indicator of sensitivity to antihypertensive drugs. *Am J Med Sci.* 1994;307(Suppl 1): S120–S125.
508. Mendonça M, Grichois ML, Dagher G, Garay R, Meyer P. Modulation of Na^+ transport systems in Wistar rat erythrocytes by excess dietary Na^+ intake. *Pflügers Arch Eur J Physiol.* 1983;398(1):64–68.

509. Orlov SN, Adragna NC, Adarichev VA, Hamet P. Genetic and biochemical determinants of abnormal monovalent ion transport in primary hypertension. *Am J Physiol Cell Physiol.* 1999;276(3):C511–C536.
510. De Mendonca M, Knorr A, Grichois ML, Ben-Ishay D, Garay RP, Meyer P. Erythrocytic sodium ion transport systems in primary and secondary hypertension of the rat. *Kidney Int Suppl.* 1982;11:S69–S75.
511. Ferrari P, Barber BR, Torielli L, Ferrandi M, Salardi S, Bianchi G. The Milan hypertensive rat as a model for studying cation transport abnormality in genetic hypertension. *Hypertension.* 1987;10(5 Pt 2):132–136.
512. Rosati C, Meyer P, Garay R. Sodium transport kinetics in erythrocytes from spontaneously hypertensive rats. *Hypertension.* 1988;11(1):41–48.
513. Bianchi G, Ferrari P, Cusi D, Tripodi G, Barber B. Genetic aspects of ion transport systems in hypertension. *J Hypertens Suppl.* 1990; 8(7):S213–S218.
514. Falkner B, Kushner H, Khalsa DK, Canessa M, Katz S. Sodium sensitivity, growth and family history of hypertension in young blacks. *J Hypertens Suppl.* 1986;4(5):S381–S383.
515. Garay RP, Alvarez-Guerra M, Alda JO, Nazaret C, Soler A, Vargas F. Regulation of renal Na-K-Cl cotransporter NKCC2 by humoral natriuretic factors: relevance in hypertension. *Clin Exp Hypertens.* 1998; 20(5–6):675–682.
516. Flatman PW. Cotransporters, WNKs and hypertension: important leads from the study of monogenetic disorders of blood pressure regulation. *Clin Sci.* 2007;112(4):203–216.
517. Lauf PK, Bauer J, Adragna NC, Fujise H, Zade-Oppen AM, Ryu KH et al. Erythrocyte K-Cl cotransport: properties and regulation. *Am J Physiol Cell Physiol.* 1992;263(5):C917–932.
518. Lauf PK, Adragna NC. A thermodynamic study of electroneutral K-Cl cotransport in pH- and volume-clamped low K sheep erythrocytes with normal and low internal magnesium. *J Gen Physiol.* 1996;108(4): 341–350.
519. Kaji DM. Effect of membrane potential on K-Cl transport in human erythrocytes. *Am J Physiol Cell Physiol.* 1993;264(2):C376–C382.
520. Jennings ML, Adame MF. Direct estimate of 1:1 stoichiometry of K^+-Cl^- cotransport in rabbit erythrocytes. *Am J Physiol Cell Physiol.* 2001;281(3):C825–C832.

521. Lauf PK. K:Cl cotransport: emerging molecular aspects of a ouabain-resistant, volume-responsive transport system in red blood cells. *Kidney Blood Press Res.* 1988;11(3–5):248–259.
522. al-Rohil N, Jennings ML. Volume-dependent K^+ transport in rabbit red blood cells comparison with oxygenated human SS cells. *Am J Physiol Cell Physiol.* 1989;257(1):C114–C121.
523. Flatman PW, Adragna NC, Lauf PK. Role of protein kinases in regulating sheep erythrocyte K-Cl cotransport. *Am J Physiol Cell Physiol.* 1996; 271(1):C255–C263.
524. Scott CC, Suzan MH, Richard P, Rettig RK, Guo-Ping Z, Patrick GG *et al.* Multiple isoforms of the KC1 cotransporter are expressed in sickle and normal erythroid cells. *Exp Hematol.* 2005;33(6):624–631.
525. Rinehart J, Maksimova YD, Tanis JE, Stone K, Hodson CA, Zhang J *et al.* Sites of regulated phosphorylation that control K-Cl cotransporter activity. *Cell* 2009;138:525–536.
526. Gillen CM, Brill S, Payne JA, Forbush B. Molecular cloning and functional expression of the K-Cl cotransporter from rabbit, rat, and human. *J Biol Chem.* 1996;271(27):16237–16244.
527. Payne JA, Stevenson TJ, Donaldson LF. Molecular characterization of a putative K-Cl cotransporter in Rat Brain. *J Biol Chem.* 1996;271(27): 16245–16252.
528. Mount DB, Mercado A, Song L, Xu J, George AL, Delpire E *et al.* Cloning and characterization of KCC3 and KCC4, new members of the cation-chloride cotransporter gene family. *J Biol Chem* 1999;274(23): 16355–16362.
529. Holtzman EJ, Kumar S, Faaland CA, Warner F, Logue PJ, Erickson SJ *et al.* Cloning, characterization, and gene organization of K-Cl cotransporter from pig and human kidney and C. elegans. *Am J Physiol Renal Physiol.* 1998;275(4):F550–F564.
530. Hiki K, D'Andrea RJ, Furze J, Crawford J, Woollatt E, Sutherland GR *et al.* Cloning, characterization, and chromosomal location of a novel human K^+-Cl^- cotransporter. *J Biol Chem* 1999;274(15):10661–10667.
531. Su W, Shmukler BE, Chernova MN, Stuart-Tilley AK, de Franceschi L, Brugnara C *et al.* Mouse K^+-Cl^- cotransporter KCC1: cloning, mapping, pathological expression, and functional regulation. *Am J Physiol Cell Physiol.* 1999;277(5):C899–C912.

532. Ochiai H, Higa K, Fujise H. Molecular identification of K-Cl cotransporter in dog erythroid progenitor cells. *J Biochem*. 2004;135(3):365–374.
533. Zhang JJ, Misri S, Adragna NC, Gagnon KB, Fyffe RE, Lauf PK. Cloning and expression of sheep renal K-CI cotransporter-1. *Cell Physiol Biochem*. 2005;16(1–3):87–98.
534. Pellegrino CM, Rybicki AC, Musto S, Nagel RL, Schwartz RS. Molecular identification and expression of erythroid K:Cl cotransporter in human and mouse erythroleukemic cells. *Blood Cells Mol Dis*. 1998;24(1):31–40.
535. Karadsheh MF, Delpire E. Neuronal restrictive silencing element is found in the KCC2 gene: molecular basis for KCC2-specific expression in neurons. *J Neurophysiol*. 2001;85(2):995–997.
536. Uvarov P, Ludwig A, Markkanen M, Pruunsild P, Kaila K, Delpire E et al. A novel N-terminal isoform of the neuron-specific K-Cl cotransporter KCC2. *J Biol Chem*. 2007;282(42):30570–30576.
537. Mercado A, Vazquez N, Song L, Cortes R, Enck AH, Welch R et al. NH2-terminal heterogeneity in the KCC3 K^+-Cl^- cotransporter. *Am J Physiol Renal Physiol*. 2005;289(6):F1246–1261.
538. Bergeron MJ, Gagnon E, Caron L, Isenring P. Identification of key functional domains in the C terminus of the K^+-Cl^- cotransporters. *J Biol Chem*. 2006;281(23):15959–15969.
539. Gagnon E, Bergeron MJ, Brunet GM, Daigle ND, Simard CF, Isenring P. Molecular mechanisms of Cl^- transport by the renal Na^+-K^+-Cl^- cotransporter. *J Biol Chem*. 2004;279(7):5648–5654.
540. Lauf PK. Thiol-dependent passive K/Cl transport in sheep red cells: IV. furosemide inhibition as a function of external Rb^+, Na^+, and Cl^-. *J Membr Biol* 1984;77(1):57–62.
541. Delpire E, Lauf PK. Kinetics of DIDS inhibition of swelling-activated K-Cl contrasport in low K sheep erythrocytes. *J Membr Biol*. 1992;126(1):89–96.
542. Kregenow FM. Osmoregulatory salt transporting mechanisms: control of cell volume in anisotonic media. *Annu Rev Physiol*. 1981;43(1):493–505.
543. Bauer J, Lauf PK. Inactivation of regulatory volume decrease in human peripheral blood lymphocytes by N-ethylmaleimide. *Biochem Biophys Res Commun*. 1983;117(1):154–160.
544. Lauf PK, Mangor-Jensen A. Effects of A23187 and Ca^{2+} on volume- and thiol-stimulated, ouabain-resistant K^+Cl^- fluxes in low K^+ sheep erythrocytes. *Biochem Biophys Res Commun*. 1984;125(2):790–796.

545. Lauf PK, Adragna NC, Garay RP. Activation by N-ethylmaleimide of a latent K^+-Cl^- flux in human red blood cells. *Am J Physiol Cell Physiol.* 1984;246(5):C385–C390.
546. Lauf PK. K-Cl cotransport: sulfhydryls, divalent cations, and the mechanism of volume activation in a red cell. *J Membr Biol.* 1985;88(1):1–13.
547. Lauf PK. Passive K^+-Cl^- fluxes in low-K^+ sheep erythrocytes: modulation by A23187 and bivalent cations. *Am J Physiol Cell Physiol.* 1985;249(3):C271–C278.
548. Lauf PK. Thiol-dependent K-Cl transport in sheep red cells: VIII. Activation through metabolically and chemically reversible oxidation by diamide. *J Membr Biol.* 1988;101(1):179–188.
549. Lauf PK. Kinetic comparison of ouabain-resistant K:Cl fluxes (K:Cl [Co]-transport) stimulated in sheep erythrocytes by membrane thiol oxidation and alkylation. *Mol Cell Biochem.* 1988;82(1):97–106.
550. Lauf PK. Volume and anion dependency of ouabain-resistant K-Rb fluxes in sheep red blood cells. *Am J Physiol Cell Physiol.* 1988;255(3):C331–C339.
551. Lauf PK. Thiol-dependent passive K:Cl transport in sheep red blood cells: X. A hydroxylamine-oxidation induced K:Cl flux blocked by diethylpyrocarbonate. *J Membr Biol.* 1990;118(2):153–159.
552. Lauf PK, Erdmann A, Adragna NC. K-Cl cotransport, pH, and role of Mg in volume-clamped low-K sheep erythrocytes: three equilibrium states. *Am J Physiol Cell Physiol.* 1994;266(1):C95–C103.
553. Bize I, Dunham PB. H_2O_2 activates red blood cell K-Cl cotransport *via* stimulation of a phosphatase. *Am J Physiol Cell Physiol.* 1995;269(4):C849–C855.
554. Kelley SJ, Dunham PB. Mechanism of swelling activation of K-Cl cotransport in inside-out vesicles of LK sheep erythrocyte membranes. *Am J Physiol Cell Physiol.* 1996;270(4):C1122–C1130.
555. Adragna NC, Lauf PK. Oxidative activation of K-Cl cotransport by diamide in erythrocytes from humans with red cell disorders, and from several other mammalian species. *J Membr Biol.* 1997;155(3):207–217.
556. Lauf PK, Adragna NC. K-Cl cotransport: properties and molecular mechanism. *Cell Physiol Biochem.* 2000;10(5–6):341–354.
557. Muzyamba MC, Speake PF, Gibson JS. Oxidants and regulation of K^+-Cl^- cotransport in equine red blood cells. *Am J Physiol Cell Physiol.* 2000;279(4):C981–C989.

558. Adragna NC, Lauf PK. K-Cl cotransport function and its potential contribution to cardiovascular disease. *Pathophysiol.* 2007;14(3): 135–146.
559. Kaji DM, Gasson C. Urea activation of K-Cl transport in human erythrocytes. *Am J Physiol Cell Physiol.* 1995;268(4):C1018–1025.
560. Adragna NC, Lauf PK. Role of nitrite, a nitric oxide derivative, in K-Cl cotransport activation of low-potassium sheep red blood cells. *J Membr Biology.* 1998;166(3):157–167.
561. Mallozzi C, De Franceschi L, Brugnara C, Di Stasi AMM. Protein phosphatase 1[alpha] is tyrosine-phosphorylated and inactivated by peroxynitrite in erythrocytes through the src family kinase fgr. *Free Radical Biol Med.* 2005;38(12):1625–1636.
562. Lauf PK. Thiol-dependent passive K/Cl transport in sheep red cells: II. Loss of Cl^- and N-ethylmaleimide sensitivity in maturing high K^+ cells. *J Membr Biol.* 1983;73(3):247–256.
563. Garay RP, Nazaret C, Hannaert PA, Cragoe EJ. Demonstration of a $[K^+,Cl^-]$-cotransport system in human red cells by its sensitivity to [(dihydroindenyl)oxy]alkanoic acids: regulation of cell swelling and distinction from the bumetanide-sensitive $[Na^+,K^+,Cl^+]$-cotransport system. *Mol Pharm.* 1988;33(6):696–701.
564. Lauf PK. Modulation of K:Cl cotransport by diethylpyrocarbonate. *Prog Clin Biol Res.* 1989;292:339–347.
565. Ryu KH, Lauf PK. Evidence for inhibitory SH groups in the thiol activated K:Cl cotransporter of low K sheep red blood cells. *Mol Cell Biochem.* 1990;99(2):135–140.
566. Adragna NC, Lauf PK. Quinine and quinidine inhibit and reveal heterogeneity of K-Cl cotransport in low K sheep erythrocytes. *J Membr Biol.* 1994;142(2):195–207.
567. Ferrell CM, Lauf PK, Wilson BA, Adragna NC. Lithium and protein kinase C modulators regulate swelling-activated K-Cl cotransport and reveal a complete phosphatidylinositol cycle in low K sheep erythrocytes. *J Membr Biol.* 2000;177(1):81–93.
568. Lauf PK, Adragna NC, Agar NS. Glutathione removal reveals kinases as common targets for K-Cl cotransport stimulation in sheep erythrocytes. *Am J Physiol Cell Physiol.* 1995;269(1):C234–C241.

569. Lauf PK, Adragna NC. Twenty-five years of K-Cl cotransport: from stimulation by a thiol reaction to cloning of the full-length KCCs. *Adv Exp Med Biol.* 2004;559:11–28.
570. Lauf PK. Incorporation of ^{3}H-N-ethylmaleimide into sheep red cell membrane thiol groups following protection by diamide-induced oxidation. *Mol Cell Biochem.* 1992;114(1):13–20.
571. Lauf PK, Adragna NC. Temperature-induced functional deocclusion of thiols inhibitory for sheep erythrocyte K-Cl cotransport. *Am J Physiol Cell Physiol.* 1995;269(5):C1167–1175.
572. Lauf PK, Bauer J. Direct evidence for chloride-dependent volume reduction in macrocytic sheep reticulocytes. *Biochem Biophys Res Commun.* 1987;144(2):849–855.
573. Gibson J, Ellory J, Culliford S, Fincham D. Volume-sensitive KCl co-transport and taurine fluxes in horse red blood cells. *Exp Physiol.* 1993;78(5):685–695.
574. Gibson J, Godart H, Ellory J, Staines H, Honess N, Cossins A. Modulation of K^+-Cl^- cotransport in equine red blood cells. *Exp Physiol.* 1994;79(6):997–1009.
575. Campbell EH, Cossins AR, Gibson JS. Oxygen-dependent K^+ influxes in Mg^{2+}-clamped equine red blood cells. *J Physiol.* 1999;515(2):431–437.
576. Gibson J, Cossins A, Ellory J. Oxygen-sensitive membrane transporters in vertebrate red cells. *J Exp Biol.* 2000;203(9):1395–1407.
577. Delpire E, Lauf PK. Kinetics of Cl-dependent K fluxes in hyposmotically swollen low K sheep erythrocytes. *J Gen Physiol.* 1991;97(2):173–193.
578. Lytle C, McManus TJ, Haas M. A model of Na-K-2Cl cotransport based on ordered ion binding and glide symmetry. *Am J Physiol Cell Physiol.* 1998;274(2):C299–C309.
579. Payne JA. Functional characterization of the neuronal-specific K-Cl cotransporter: implications for $[K^+]_o$ regulation. *Am J Physiol Cell Physiol.* 1997;273(5):C1516–C1525.
580. Thomas S, Egée S. Fish red blood cells: characteristics and physiological role of the membrane ion transporters. *Comp Biochem Physiol — Part A: Mol & Integr Physiol.* 1998;119(1):79–86.
581. Dunham PB. Ion transport in sheep red blood cells. *Comp Biochem Physiol Part A: Physiology.* 1992;102(4):625–630.

582. Armsby CC, Brugnara C, Alper SL. Cation transport in mouse erythrocytes: role of K^+-Cl^- cotransport in regulatory volume decrease. *Am J Physiol Cell Physiol*. 268(4):C894–902.
583. Jennings ML, al-Rohil N. Kinetics of activation and inactivation of swelling-stimulated K^+/Cl^- transport. The volume-sensitive parameter is the rate constant for inactivation. *J Gen Physiol*. 1990;95(6):1021–1040.
584. Jennings ML, Schulz RK. Okadaic acid inhibition of KCl cotransport. Evidence that protein dephosphorylation is necessary for activation of transport by either cell swelling or N-ethylmaleimide. *J Gen Physiol*. 1991;97(4):799–817.
585. Kaji DM, Tsukitani Y. Role of protein phosphatase in activation of KCl cotransport in human erythrocytes. *Am J Physiol Cell Physiol*. 1991; 260(1):C176–C180.
586. Starke LC, Jennings ML. K-Cl cotransport in rabbit red cells: further evidence for regulation by protein phosphatase type 1. *Am J Physiol*. 1993;264(1 Pt 1):C118–C124.
587. Namboodiripad AN, Jennings ML. Permeability characteristics of erythrocyte membrane to okadaic acid and calyculin A. *Am J Physiol*. 1996;270 (2 Pt 1):C449–C456.
588. Jennings ML. Volume-sensitive K^+/Cl^- cotransport in rabbit erythrocytes. *J Gen Physiol*. 1999;114(6):743–758.
589. Delpire E, Days E, Lewis LM, Mi D, Kim K, Lindsley CW et al. Small-molecule screen identifies inhibitors of the neuronal K-Cl cotransporter KCC2. *Proc Natl Acad Sci USA*. 2009;106(13):5383–5388.
590. Bize I, Dunham PB. Staurosporine, a protein kinase inhibitor, activates K-Cl cotransport in LK sheep erythrocytes. *Am J Physiol Cell Physiol*. 1994;266(3):C759–C770.
591. Bize I, Munoz P, Canessa M, Dunham PB. Stimulation of membrane serine-threonine phosphatase in erythrocytes by hydrogen peroxide and staurosporine. *Am J Physiol Cell Physiol*. 1998;274(2):C440–C446.
592. Bize I, Guvenc B, Robb A, Buchbinder G, Brugnara C. Serine/threonine protein phosphatases and regulation of K-Cl cotransport in human erythrocytes. *Am J Physiol Cell Physiol*. 1999;277(5):C926–C936.
593. Bize I, Güvenç B, Buchbinder G, Brugnara C. Stimulation of human erythrocyte K-Cl cotransport and protein phosphatase type 2A by n-ethylmaleimide: role of intracellular Mg^{++}. *J Membr Biol*. 2000;177(2):159–168.

594. Kelley SJ, Thomas R, Dunham PB. Candidate inhibitor of the volume-sensitive kinase regulating K-Cl cotransport: the myosin light chain kinase inhibitor ML-7. *J Membr Biol.* 2000;178(1):31–41.
595. de Franceschi L, Fumagalli L, Olivieri O, Corrocher R, Lowell CA, Berton G. Deficiency of Src family kinases Fgr and Hck results in activation of erythrocyte K/Cl cotransport. *J Clin Invest.* 1997;99(2):220–227.
596. De Franceschi L, Villa-Moruzzi E, Biondani A, Siciliano A, Brugnara C, Alper S et al. Regulation of K-Cl cotransport by protein phosphatase 1α in mouse erythrocytes. *Pflügers Arch Eur J Physiol.* 2006;451(6):760–768.
597. Agalakova N, Gusev G. Effects of phorbol 12-myristate 13-acetate on potassium transport in the red blood cells of frog *Rana temporaria*. *J Comp Physiol B.* 2009;179(4):443–450.
598. Gusev G, Agalakova N. Regulation of K-Cl cotransport in erythrocytes of frog Rana temporaria by commonly used protein kinase and protein phosphatase inhibitors. *J Comp Physiol B.* 2010;180(3):385–391.
599. Ochiai H, Higa K, Hisamatsu S, Fujise H. Comparison of K-Cl cotransport expression in high and low K dog erythrocytes. *Exp Anim.* 2006;55(1):57–63.
600. Zade-Oppen AM, Lauf PK. Thiol-dependent passive K:Cl transport in sheep red blood cells: IX. Modulation by pH in the presence and absence of DIDS and the effect of NEM. *J Membr Biol.* 1990;118(2):143–145.
601. Bergeron MJ, Gagnon E, Wallendorff B, Lapointe J-Y, Isenring P. Ammonium transport and pH regulation by K^+-Cl^- cotransporters. *Am J Physiol Renal Physiol.* 2003;285(1):F68–F78.
602. Lauf PK. Thiol-dependent passive K^+Cl^- transport in sheep red blood cells: VI. Functional heterogeneity and immunologic identity with volume-stimulated $K^+(Rb^+)$ fluxes. *J Membr Biol.* 1984;82(2):167–178.
603. Godart H, Dormandy A, Ellory JC. Do HbSS erythrocytes lose KCl in physiological conditions? *Brit J Haematol.* 1997;98(1):25–32.
604. Lauf PK. Foreign anions modulate volume set point of sheep erythrocyte K-Cl cotransport. *Am J Physiol Cell Physiol.* 1991;260(3):C503–C512.
605. Sachs JR. Soluble polycations and cationic amphiphiles inhibit volume-sensitive K-Cl cotransport in human red cell ghosts. *Am J Physiol Cell Physiol.* 1994;266(4):C997–C1005.
606. Smith DK, Lauf PK. Effects of N-ethylmaleimide on ouabain-insensitive cation fluxes in human red cell ghosts. *Biochim Biophys Acta — Biomembranes.* 1985;818(2):251–259.

607. Lauf PK. A chemically unmasked, chloride dependent K^+ transport in low K^+ sheep red cells: genetic and evolutionary aspects. *Prog Clin Biol Res.* 1981;56:13–34.
608. Lauf PK, Zeidler RB, Kim HD. Pig reticulocytes. V. Development of Rb^+ influx during *in vitro* maturation. *J Cell Physiol.* 1984;121(2):284–290.
609. Hall AC, Ellory JC. Evidence for the presence of volume-sensitive KCl transport in "young" human red cells. *Biochim Biophys Acta — Biomembranes.* 1986;858(2):317–320.
610. Ellory JC, Hall AC, Amess JA. Passive potassium transport in human red cells. *Biomed Biochim Acta.* 1987;46:S31–S35.
611. Casula S, Zolotarev AS, Stuart-Tilley AK, Wilhelm S, Shmukler BE, Brugnara C et al. Chemical crosslinking studies with the mouse KCC1 K-Cl cotransporter. *Blood Cells Mol Dis.* 2009;42(3):233–240.
612. Simard CF, Bergeron MJ, Frenette-Cotton R, Carpentier GA, Pelchat M-E, Caron L et al. Homooligomeric and heterooligomeric associations between K^+-Cl^- cotransporter isoforms and between K^+-Cl^- and Na^+-K^+-Cl^- cotransporters. *J Biol Chem.* 2007;282(25):18083–18093.
613. Godart H, Ellory JC. KCl cotransport activation in human erythrocytes by high hydrostatic pressure. *J Physiol.* 1996;491(Pt 2):423–434.
614. Willis JS, Anderson GL. Activation of K-Cl cotransport by mild warming in guinea pig red cells. *J Membr Biol.* 1998;163(3):193–203.
615. Orlov SN, Kolosova IA, Cragoe EJ, Gurlo TG, Mongin AA, Aksentsev SL et al. Kinetics and peculiarities of thermal inactivation of volume-induced Na^+/H^+ exchange, Na^+, K^+, $2Cl^-$ cotransport and K^+, Cl^- cotransport in rat erythrocytes. *Biochim Biophys Acta — Biomembranes.* 1993;1151(2):186–192.
616. Canessa M, Spalvins A, Nagel RL. Volume-dependent and NEM-stimulated K^+Cl^- transport is elevated in oxygenated SS, SC and CC human red cells. *FEBS Letters.* 1986;200(1):197–202.
617. Fabry M, Romero J, Buchanan I, Suzuka S, Stamatoyannopoulos G, Nagel R et al. Rapid increase in red blood cell density driven by K:Cl cotransport in a subset of sickle cell anemia reticulocytes and discocytes. *Blood.* 1991;78(1):217–225.
618. Joiner CH. Cation transport and volume regulation in sickle red blood cells. *Am J Physiol Cell Physiol.* 1993;264(2):C251–270.

619. Brugnara C. Erythrocyte dehydration in pathophysiology and treatment of sickle cell disease. *Curr Op Hemat.* 1995;2(2):132–138.
620. Kaul DK, Fabry ME, Nagel RL. The pathophysiology of vascular obstruction in the sickle syndromes. *Blood Rev.* 1996;10(1):29–44.
621. Brugnara C. Therapeutic strategies for prevention of sickle cell dehydration. *Blood Cells Mol Dis.* 2001;27(1):71–80.
622. Joiner CH, Franco RS. The activation of KCL cotransport by deoxygenation and its role in sickle cell dehydration. *Blood Cells Mol Dis.* 2001;27(1):158–164.
623. Bookchin RM, Lew VL. Sickle red cell dehydration: mechanisms and interventions. *Curr Opin Hemat.* 2002;9(2):107–110.
624. Brugnara C. Sickle cell disease: from membrane pathophysiology to novel therapies for prevention of erythrocyte dehydration. *J Ped Hemat/Oncol.* 2003;25(12):927–933.
625. Nagel RL, Fabry ME, Steinberg MH. The paradox of hemoglobin SC disease. *Blood Rev.* 2003;17(3):167–178.
626. Brugnara C. Membrane transport of Na and K and cell dehydration in sickle erythrocytes. *Cell Mol Life Sci.* 1993;49(2):100–109.
627. Canessa M, Romero J, Lawrence C, Nagel R, Fabry M. Rate of activation and deactivation of K-Cl cotransport by changes in cell volume in hemoglobin SS, CC and AA red cells. *J Membr Biol.* 1994;142(3):349–362.
628. Romero JR, Fabry ME, Suzuka SM, Costantini F, Nagel RL, Canessa M. K:Cl cotransport in red cells of transgenic mice expressing high levels of human hemoglobin S. *Am J Hemat.* 1997;55(2):112–114.
629. Romero JR, Suzuka SM, Nagel RL, Fabry ME. Expression of HbC and HbS, but not HbA, results in activation of K-Cl cotransport activity in transgenic mouse red cells. *Blood.* 2004;103(6):2384–2390.
630. Brugnara C, De Franceschi L, Bennekou P, Alper SL, Christophersen P. Novel therapies for prevention of erythrocyte dehydration in sickle cell anemia. *Drug News Perspect.* 2001;14(4):208–220.
631. Mueller BU, Brugnara C. Prevention of red cell dehydration: a possible new treatment for sickle cell disease. *Fetal Pediat Path.* 2001;20(1):15–25.
632. Steinberg MH, Brugnara C. Pathophysiological-based approaches to treatment of sickle cell disease. *Annu Rev Med.* 2003;54:89–112.

633. de Franceschi L, Turrini F, Honczarenko M, Ayi K, Rivera A, Fleming M et al. In vivo reduction of erythrocyte oxidant stress in a murine model of beta-thalassemia. *Haematologica* 2004;89(11):1287–1298.
634. Rust MB, Alper SL, Rudhard Y, Shmukler BE, Vicente R, Brugnara C et al. Disruption of erythroid K-Cl cotransporters alters erythrocyte volume and partially rescues erythrocyte dehydration in SAD mice. *J Clin Invest.* 2007;117(6):1708–1717.
635. Hankins JS, Wynn LW, Brugnara C, Hillery CA, Li CS, Wang WC. Phase I study of magnesium pidolate in combination with hydroxycarbamide for children with sickle cell anaemia. *Br J Haemat.* 2008;140(1):80–85.
636. Olivieri O, de Franceschi L, de Gironcoli M, Girelli D, Corrocher R. Potassium loss and cellular dehydration of stored erythrocytes following incubation in autologous plasma: role of the KCl cotransport system. *Vox Sang.* 1993;65(2):95–102.
637. Sun Y-T, Lin T-S, Tzeng S-F, Delpire E, Shen M-R. Deficiency of electroneutral K^+-Cl^- cotransporter 3 causes a disruption in impulse propagation along peripheral nerves. *Glia.* 2010;58(13):1544–1552.
638. Shen M-R, Chou C-Y, Ellory JC. Volume-sensitive KCl cotransport associated with human cervical carcinogenesis. *Pflügers Arch Eur J Physiol.* 2000;440(5):751–760.
639. Shen M-R, Chou C-Y, Hsu K-F, Liu H-S, Dunham PB, Holtzman EJ et al. The KCl cotransporter isoform KCC3 can play an important role in cell growth regulation. *Proc Natl Acad Sci. USA* 2001;98(25):14714–14719.
640. Shen M-R, Chou C-Y, Hsu K-F, Hsu Y-M, Chiu W-T, Tang M-J et al. KCl cotransport is an important modulator of human cervical cancer growth and invasion. *J Biol Chem.* 2003;278(41):39941–39950.
641. Hsu YM, Chou CY, Chen HH, Lee WY, Chen YF, Lin PW et al. IGF-1 upregulates electroneutral K-Cl cotransporter KCC3 and KCC4 which are differentially required for breast cancer cell proliferation and invasiveness. *J Cell Physiol.* 2007;210(3):626–636.
642. Hsu Y-M, Chen Y-F, Chou C-Y, Tang M-J, Chen JH, Wilkins RJ et al. KCl cotransporter-3 down-regulates E-cadherin/beta-catenin complex to promote epithelial-mesenchymal transition. *Cancer Res.* 2007;67(22): 11064–11073.
643. Chen YF, Chou CY, Ellory JC, Shen MR. The emerging role of KCl cotransport in tumor biology. *Am J Transl Res.* 2010;2(4):354–355.

644. Boettger T, Hubner CA, Maier H, Rust MB, Beck FX, Jentsch TJ. Deafness and renal tubular acidosis in mice lacking the K-Cl co-transporter KCC4. *Nature.* 2002;416(6883):874-878.

645. Gardos G. Effect of ethylenediaminetetraacetate on the permeability of human erythrocytes. *Acta Physiol Hung.* 1958;14(1):1–5.

646. Gardos G. The function of calcium in the potassium permeability of human erythrocytes. *Biochim Biophys Acta.* 1958;30(3):653–654.

647. Tharp DL, Bowles DK. The intermediate-conductance Ca^{2+}-activated K^+ channel (KCa3.1) in vascular disease. *Cardiovasc Hematol Agents Med Chem.* 2009;7(1):1–11.

648. Ishii TM, Silvia C, Hirschberg B, Bond CT, Adelman JP, Maylie J. A human intermediate conductance calcium-activated potassium channel. *Proc Natl Acad Sci USA.* 1997;94(21):11651–11656.

649. Vandorpe DH, Shmukler BE, Jiang L, Lim B, Maylie J, Adelman JP et al. cDNA cloning and functional characterization of the mouse Ca^{2+}-gated K^+ channel, mIK1. Roles in regulatory volume decrease and erythroid differentiation. *J Biol Chem.* 1998;273(34):21542–21553.

650. Ghanshani S, Coleman M, Gustavsson P, Wu AC, Gargus JJ, Gutman GA et al. Human calcium-activated potassium channel gene KCNN4 maps to chromosome 19q13.2 in the region deleted in diamond-blackfan anemia. *Genomics.* 1998;51(1):160–161.

651. Jensen BS, Strobaek D, Christophersen P, Jorgensen TD, Hansen C, Silahtaroglu A et al. Characterization of the cloned human intermediate-conductance Ca^{2+}-activated K^+ channel. *Am J Physiol.* 1998;275(3 Pt 1): C848–856.

652. Neylon CB, Lang RJ, Fu Y, Bobik A, Reinhart PH. Molecular cloning and characterization of the intermediate-conductance Ca^{2+}-activated K^+ channel in vascular smooth muscle: relationship between K_{Ca} channel diversity and smooth muscle cell function. *Circ Res.* 1999;85(9):e33–e43.

653. Lauf PK, Misri S, Chimote AA, Adragna NC. Apparent intermediate K conductance channel hyposmotic activation in human lens epithelial cells. *Am J Physiol Cell Physiol.* 2008;294(3):C820–832.

654. Begenisich T, Nakamoto T, Ovitt CE, Nehrke K, Brugnara C, Alper SL et al. Physiological roles of the intermediate conductance, Ca^{2+}-activated potassium channel KCNN4. *J Biol Chem.* 2004;279(46): 47681–47687.

655. Klein H, Garneau L, Banderali U, Simoes M, Parent L, Sauvé R. Structural determinants of the closed KCa3.1 channel pore in relation to channel gating: results from a substituted cysteine accessibility analysis. *J Gen Physiol*. 2007;129(4):299–315.
656. Brugnara C, De Franceschi L, Alper SL. Ca^{2+}-activated K^+ transport in erythrocytes. Comparison of binding and transport inhibition by scorpion toxins. *J Biol Chem*. 1993;268(12):8760–8768.
657. Heinz A, Hoffman JF. Membrane sidedness and the interaction of H^+ and K^+ on Ca^{2+}-activated K^+ transport in human red blood cells. *Proc Natl Acad Sci USA*. 1990;87(5):1998–2002.
658. de Franceschi L, Saadane N, Trudel M, Alper SL, Brugnara C, Beuzard Y. Treatment with oral clotrimazole blocks Ca^{2+}-activated K^+ transport and reverses erythrocyte dehydration in transgenic SAD mice. A model for therapy of sickle cell disease. *J Clin Invest*. 1994;93(4):1670–4676.
659. Brugnara C, Gee B, Armsby CC, Kurth S, Sakamoto M, Rifai N *et al.* Therapy with oral clotrimazole induces inhibition of the Gardos channel and reduction of erythrocyte dehydration in patients with sickle cell disease. *J Clin Invest*. 1996;97(5):1227–1234.
660. de Franceschi L, Brugnara C, Rouyer-Fessard P, Jouault H, Beuzard Y. Formation of dense erythrocytes in SAD mice exposed to chronic hypoxia: evaluation of different therapeutic regimens and of a combination of oral clotrimazole and magnesium therapies. *Blood*. 1999;94(12):4307–4313.
661. Brugnara C, Armsby CC, Sakamoto M, Rifai N, Alper SL, Platt O. Oral administration of clotrimazole and blockade of human erythrocyte Ca^{++}-activated K^+ channel: the imidazole ring is not required for inhibitory activity. *J Pharmacol Exp Ther*. 1995;273(1):266–272.
662. Stocker JW, De Franceschi L, McNaughton-Smith GA, Corrocher R, Beuzard Y, Brugnara C. ICA-17043, a novel Gardos channel blocker, prevents sickled red blood cell dehydration *in vitro* and *in vivo* in SAD mice. *Blood*. 2003;101(6):2412–2418.
663. Gibson JS, Stewart GW, Ellory JC. Effect of dimethyl adipimidate on K^+ transport and shape change in red blood cells from sickle cell patients. *FEBS Lett*. 2000;480(2–3):179–183.
664. Lew VL, Tiffert T, Etzion Z, Perdomo D, Daw N, Macdonald L *et al.* Distribution of dehydration rates generated by maximal Gardos-channel

activation in normal and sickle red blood cells. *Blood.* 2005;105(1): 361–367.
665. Baunbaek M, Bennekou P. Evidence for a random entry of Ca^{2+} into human red cells. *Bioelectrochem.* 2008;73(2):145–150.
666. Barksmann TL, Kristensen BI, Christophersen P, Bennekou P. Pharmacology of the human red cell voltage-dependent cation channel; Part I. Activation by clotrimazole and analogues. *Blood Cells Mol Dis.* 2004;32(3):384–388.
667. Browning JA, Ellory JC, Gibson JS. Pathophysiology of red cell volume. *Contrib Nephrol.* 2006;152:241–268.
668. Bennekou P, Kristensen BI, Christophersen P. The human red cell voltage-regulated cation channel. The interplay with the chloride conductance, the Ca^{2+}-activated K^+ channel and the Ca^{2+} pump. *J Membr Biol.* 2003; 195(1):1–8.
669. Varecka L, Peterajova E, Pisova E. Properties of the Ca^{2+} influx reveal the duality of events underlying the activation by vanadate and fluoride of the Gardos effect in human red blood cells. *FEBS Lett.* 1998; 433(1–2):157–160.
670. Ellory JC, Culliford SJ, Smith PA, Wolowyk MW, Knaus EE. Specific inhibition of Ca-activated K channels in red cells by selected dihydropyridine derivatives. *Br J Pharmacol.* 1994;111(3):903–905.
671. Ellory JC, Kirk K, Culliford SJ, Nash GB, Stuart J. Nitrendipine is a potent inhibitor of the Ca^{2+}-activated K^+ channel of human erythrocytes. *FEBS Letters.* 1992;296(2):219–221.
672. Vandorpe DH, Xu C, Shmukler BE, Otterbein LE, Trudel M, Sachs F *et al.* Hypoxia activates a Ca^{2+}-permeable cation conductance sensitive to carbon monoxide and to GsMTx-4 in human and mouse sickle erythrocytes. *PLoS One.* 2010;5(1):e8732.
673. Pinet C, Antoine S, Filoteo AG, Penniston JT, Coulombe A. Reincorporated plasma membrane $Ca2^+$-ATPase can mediate B-Type Ca^{2+} channels observed in native membrane of human red blood cells. *J Membr Biol.* 2002;187(3):185–201.
674. Shartava A, Shah AK, Goodman SR. N-acetylcysteine and clotrimazole inhibit sickle erythrocyte dehydration induced by 1-chloro-2,4-dinitrobenzene. *Am J Hematol.* 1999;62(1):19–24.

675. Gibson JS, Muzyamba MC. Modulation of Gardos channel activity by oxidants and oxygen tension: effects of 1-chloro-2,4-dinitrobenzene and phenazine methosulphate. *Bioelectrochem.* 2004;62(2):147–152.
676. Maher AD, Kuchel PW. The Gardos channel: a review of the Ca^{2+}-activated K^+ channel in human erythrocytes. *Int J Biochem Cell Biol.* 2003;35(8):1182–1197.
677. Li Q, Jungmann V, Kiyatkin A, Low PS. Prostaglandin E2 stimulates a Ca^{2+}-dependent K^+ channel in human erythrocytes and alters cell volume and filterability. *J Biol Chem.* 1996;271(31):18651–18656.
678. Kaestner L, Tabellion W, Lipp P, Bernhardt I. Prostaglandin E2 activates channel-mediated calcium entry in human erythrocytes: an indication for a blood clot formation supporting process. *Thromb Haemost.* 2004;92(6): 1269–1272.
679. Makhro A, Wang J, Vogel J, Boldyrev AA, Gassmann M, Kaestner L et al. Functional NMDA receptors in rat erythrocytes. *Am J Physiol Cell Physiol.* 2010;298(6):C1315–1325.
680. Adragna NC. Homocysteine: to measure or not to measure? Focus on "functional NMDA receptors in rat erythrocytes". *Am J Physiol Cell Physiol.* 2010;298(6):C1301–1302.
681. Lang PA, Kaiser S, Myssina S, Wieder T, Lang F, Huber SM. Role of Ca^{2+}-activated K^+ channels in human erythrocyte apoptosis. *Am J Physiol Cell Physiol.* 2003;285(6):C1553–1560.
682. Lang F, Birka C, Myssina S, Lang KS, Lang PA, Tanneur V et al. Erythrocyte ion channels in regulation of apoptosis. *Adv Exp Med Biol.* 2004;559:211–217.
683. Skals M, Jensen UB, Ousingsawat J, Kunzelmann K, Leipziger J, Praetorius HA. Escherichia coli alpha-hemolysin triggers shrinkage of erythrocytes via K(Ca)3.1 and TMEM16A channels with subsequent phosphatidylserine exposure. *J Biol Chem.* 2010 May 14;285(20):15557–15565.
684. Myssina S, Lang PA, Kempe DS, Kaiser S, Huber SM, Wieder T et al. Cl-channel blockers NPPB and niflumic acid blunt Ca^{2+}-induced erythrocyte "apoptosis". *Cell Physiol Biochem.* 2004;14(4–6):241–248.
685. Kucherenko YV, Morsdorf D, Lang F. Acid-sensitive outwardly rectifying anion channels in human erythrocytes. *J Membr Biol.* 2009;230(1):1–10.
686. Föller M, Huber SM, Lang F. Erythrocyte programmed cell death. *IUBMB Life.* 2008;60(10):661–668.

687. de Franceschi L, Rivera A, Fleming MD, Honczarenko M, Peters LL, Gascard P et al. Evidence for a protective role of the Gardos channel against hemolysis in murine spherocytosis. *Blood*. 2005;106(4):1454–1459.
688. Rivera A, Jarolim P, Brugnara C. Modulation of Gardos channel activity by cytokines in sickle erythrocytes. *Blood*. 2002;99(1):357–363.
689. Rivera A, Rotter MA, Brugnara C. Endothelins activate Ca^{2+}-gated K^+ channels via endothelin B receptors in CD-1 mouse erythrocytes. *Am J Physiol*. 1999;277(4 Pt 1):C746–754.
690. Rivera A. Reduced sickle erythrocyte dehydration *in vivo* by endothelin-1 receptor antagonists. *Am J Physiol Cell Physiol*. 2007;293(3):C960–966.
691. Foller M, Mahmud H, Qadri SM, Gu S, Braun M, Bobbala D et al. Endothelin B receptor stimulation inhibits suicidal erythrocyte death. *FASEB J*. 2010;24(9):3351–3359.
692. Zimmermann A, Schatzmann HJ. Calcium transport by red blood cell membranes from young and adult cattle. *Experientia*. 1985;41(6):743–745.
693. Wieth JO, Andersen OS, Brahm J, Bjerrum PJ, Borders CL, Jr. Chloride-bicarbonate exchange in red blood cells: physiology of transport and chemical modification of binding sites. *Philos Trans R Soc Lond B Biol Sci*. 1982;299(1097):383–399.
694. Jones GS, Knauf PA. Mechanism of the increase in cation permeability of human erythrocytes in low-chloride media. Involvement of the anion transport protein capnophorin. *J Gen Physiol*. 1985;86(5):721–738.
695. Simon JS, Deshmukh G, Couch FJ, Merajver SD, Weber BL, Van Vooren P et al. Chromosomal mapping of the rat Slc4a family of anion exchanger genes, Ae1, Ae2, and Ae3. *Mamm Genome*. 1996;7(5):380–382.
696. Showe LC, Ballantine M, Huebner K. Localization of the gene for the erythroid anion exchange protein, band 3 (EMPB3), to human chromosome 17. *Genomics*. 1987;1(1):71–76.
697. Love JM, Knight AM, McAleer MA, Todd JA. Towards construction of a high resolution map of the mouse genome using PCR-analysed microsatellites. *Nucleic Acids Res*. 1990;18(14):4123–4130.
698. Jacob HJ, Brown DM, Bunker RK, Daly MJ, Dzau VJ, Goodman A et al. A genetic linkage map of the laboratory rat, Rattus norvegicus. *Nat Genet*. 1995;9(1):63–69.
699. Kopito RR. Molecular biology of the anion exchanger gene family. *Int Rev Cytol*. 1990;123:177–199.

700. Wang DN, Sarabia VE, Reithmeier RA, Kuhlbrandt W. Three-dimensional map of the dimeric membrane domain of the human erythrocyte anion exchanger, Band 3. *EMBO J.* 1994;13(14):3230–3235.
701. Tam LY, Landolt-Marticorena C, Reithmeier RA. Glycosylation of multiple extracytosolic loops in Band 3, a model polytopic membrane protein. *Biochem J.* 1996;318(Pt 2):645–658.
702. Popov M, Li J, Reithmeier RA. Transmembrane folding of the human erythrocyte anion exchanger (AE1, Band 3) determined by scanning and insertional N-glycosylation mutagenesis. *Biochem J.* 1999;339(Pt 2):269–279.
703. Vince JW, Reithmeier RA. Structure of the band 3 transmembrane domain. *Cell Mol Biol (Noisy-le-grand).* 1996;42(7):1041–1051.
704. Landolt-Marticorena C, Charuk JH, Reithmeier RA. Two glycoprotein populations of band 3 dimers are present in human erythrocytes. *Mol Membr Biol.* 1998;15(3):153–158.
705. Cordat E, Li J, Reithmeier RA. Carboxyl-terminal truncations of human anion exchanger impair its trafficking to the plasma membrane. *Traffic.* 2003;4(9):642–651.
706. Thevenin BJ, Willardson BM, Low PS. The redox state of cysteines 201 and 317 of the erythrocyte anion exchanger is critical for ankyrin binding. *J Biol Chem.* 1989;264(27):15886–15892.
707. Stefanovic M, Markham NO, Parry EM, Garrett-Beal LJ, Cline AP, Gallagher PG et al. An 11-amino acid beta-hairpin loop in the cytoplasmic domain of band 3 is responsible for ankyrin binding in mouse erythrocytes. *Proc Natl Acad Sci USA.* 2007;104(35):13972–13977.
708. Franco T, Low PS. Erythrocyte adducin: a structural regulator of the red blood cell membrane. *Transf Clin Biol.* 2010;17(3):87–94.
709. Wang DN, Kuhlbrandt W, Sarabia VE, Reithmeier RA. Two-dimensional structure of the membrane domain of human band 3, the anion transport protein of the erythrocyte membrane. *EMBO J.* 1993;12(6):2233–2239.
710. Brahm J. Temperature-dependent changes of chloride transport kinetics in human red cells. *J Gen Physiol.* 1977;70(3):283–306.
711. Brahm J, Wimberley PD. Chloride and bicarbonate transport in fetal red cells. *J Physiol.* 1989;419:141–156.
712. Milanick MA, Gunn RB. Proton-sulfate co-transport: mechanism of H^+ and sulfate addition to the chloride transporter of human red blood cells. *J Gen Physiol.* 1982;79(1):87–113.

713. Gunn RB, Frohlich O. Asymmetry in the mechanism for anion exchange in human red blood cell membranes. Evidence for reciprocating sites that react with one transported anion at a time. *J Gen Physiol.* 1979;74(3):351–374.
714. Wieth JO, Bjerrum PJ. Titration of transport and modifier sites in the red cell anion transport system. *J Gen Physiol.* 1982;79(2):253–282.
715. Jennings ML, Whitlock J, Shinde A. Pre-steady state transport by erythrocyte band 3 protein: uphill countertransport induced by the impermeant inhibitor H_2DIDS. *Biochem Cell Biol.* 1998;76(5):807–813.
716. Knauf PA, Gasbjerg PK, Brahm J. The asymmetry of chloride transport at 38 degrees C in human red blood cell membranes. *J Gen Physiol.* 1996;108(6):577–589.
717. Knauf PA, Law FY, Marchant PJ. Relationship of net chloride flow across the human erythrocyte membrane to the anion exchange mechanism. *J Gen Physiol.* 1983;81(1):95–126.
718. Knauf PA, Law FY, Tarshis T, Furuya W. Effects of the transport site conformation on the binding of external NAP-taurine to the human erythrocyte anion exchange system. Evidence for intrinsic asymmetry. *J Gen Physiol.* 1984;83(5):683–701.
719. Janas T, Bjerrum PJ, Brahm J, Wieth JO. Kinetics of reversible DIDS inhibition of chloride self exchange in human erythrocytes. *Am J Physiol.* 1989;257(4 Pt 1):C601–606.
720. Salhany JM. Kinetic evidence for modulation by glycophorin A of a conformational equilibrium between two states of band 3 (SLC4A1) bound reversibly by the competitive inhibitor DIDS. *Blood Cells Mol Dis.* 2009;42(3):185–191.
721. Landolt-Marticorena C, Casey JR, Reithmeier RA. Transmembrane helix-helix interactions and accessibility of H_2DIDS on labelled band 3, the erythrocyte anion exchange protein. *Mol Membr Biol.* 1995;12(2): 173–182.
722. Salhany JM. Anion binding characteristics of the band 3/4,4-dibenzamidostilbene-2,2-disulfonate binary complex: evidence for both steric and allosteric interactions. *Biochem Cell Biol.* 1999;77(6):543–549.
723. Schopfer LM, Salhany JM. Characterization of the stilbenedisulfonate binding site on band 3. *Biochemistry.* 1995;34(26):8320–8329.
724. Wood PG, Muller H, Sovak M, Passow H. Role of Lys 558 and Lys 869 in substrate and inhibitor binding to the murine band 3 protein: a study of the

effects of site-directed mutagenesis of the band 3 protein expressed in the oocytes of Xenopus laevis. *J Membr Biol.* 1992;127(2):139–148.

725. Karbach D, Staub M, Wood PG, Passow H. Effect of site-directed mutagenesis of the arginine residues 509 and 748 on mouse band 3 protein-mediated anion transport. *Biochim Biophys Acta.* 1998;1371(1):114–122.

726. Sarabia VE, Casey JR, Reithmeier RA. Molecular characterization of the band 3 protein from Southeast Asian ovalocytes. *J Biol Chem.* 1993; 268(14):10676–10680.

727. Cheung JC, Reithmeier RA. Membrane integration and topology of the first transmembrane segment in normal and Southeast Asian ovalocytosis human erythrocyte anion exchanger 1. *Mol Membr Biol.* 2005;22(3):203–214.

728. Jennings ML, Smith SJ. Anion-proton cotransport through the human red blood cell band 3 protein. Role of glutamate 681. *J Biol Chem* 1992; 267(20):13964–13971.

729. Jennings ML, Gosselink PG. Anion exchange protein in Southeast Asian ovalocytes: heterodimer formation between normal and variant subunits. *Biochemistry.* 1995;34(11):3588–3595.

730. Knauf PA, Law FY, Leung TW, Atherton SJ. Relocation of the disulfonic stilbene sites of AE1 (band 3) on the basis of fluorescence energy transfer measurements. *Biochemistry.* 2004;43(38):11917–11931.

731. Gasbjerg PK, Knauf PA, Brahm J. Kinetics of bicarbonate transport in human red blood cell membranes at body temperature. *J Gen Physiol.* 1996;108(6):565–575.

732. Knauf PA, Law F-Y, Leung T-WV, Gehret AU, Perez ML. Substrate-dependent reversal of anion transport site orientation in the human red blood cell anion-exchange protein, AE1. *Proc Natl Acad Sci USA.* 2002;99(16):10861–10864.

733. Campanella ME, Chu H, Low PS. Assembly and regulation of a glycolytic enzyme complex on the human erythrocyte membrane. *Proc Natl Acad Sci USA.* 2005;102(7):2402–2407.

734. Romano L, Passow H. Characterization of anion transport system in trout red blood cell. *Am J Physiol.* 1984;246(3 Pt 1):C330–338.

735. Cass A, Dalmark M. Chloride transport by self-exchange and by KCl salt diffusion in gramicidin-treated red blood cells. *Acta Physiol Scand.* 1979;107(3):193–203.

736. Dalmark M. Effects of halides and bicarbonate on chloride transport in human red blood cells. *J Gen Physiol*. 1976;67(2):223–234.
737. Wieth JO. Effect of some monovalent anions on chloride and sulphate permeability of human red cells. *J Physiol*. 1970;207(3):581–609.
738. Kaplan JH, Pring M, Passow H. Band-3 protein-mediated anion conductance of the red cell membrane. Slippage vs ionic diffusion. *FEBS Lett*. 1983;156(1):175–179.
739. Campanella ME, Chu H, Wandersee NJ, Peters LL, Mohandas N, Gilligan DM et al. Characterization of glycolytic enzyme interactions with murine erythrocyte membranes in wild-type and membrane protein knockout mice. *Blood*. 2008;112(9):3900–3906.
740. Low PS, Rathinavelu P, Harrison ML. Regulation of glycolysis via reversible enzyme binding to the membrane protein, band 3. *J Biol Chem*. 1993;268(20):14627–14631.
741. Harrison ML, Rathinavelu P, Arese P, Geahlen RL, Low PS. Role of band 3 tyrosine phosphorylation in the regulation of erythrocyte glycolysis. *J Biol Chem*. 1991;266(7):4106–4111.
742. Harrison ML, Isaacson CC, Burg DL, Geahlen RL, Low PS. Phosphorylation of human erythrocyte band 3 by endogenous p72syk. *J Biol Chem*. 1994;269(2):955–959.
743. Walder JA, Chatterjee R, Steck TL, Low PS, Musso GF, Kaiser ET et al. The interaction of hemoglobin with the cytoplasmic domain of band 3 of the human erythrocyte membrane. *J Biol Chem*. 1984;259(16):10238–10246.
744. Moriyama R, Lombardo CR, Workman RF, Low PS. Regulation of linkages between the erythrocyte membrane and its skeleton by 2,3-diphosphoglycerate. *J Biol Chem*. 1993;268(15):10990–10996.
745. Vince JW, Reithmeier RA. Identification of the carbonic anhydrase II binding site in the Cl^-/HCO_3^- anion exchanger AE1. *Biochemistry*. 2000;39(18):5527–5533.
746. Sterling D, Reithmeier RA, Casey JR. A transport metabolon. Functional interaction of carbonic anhydrase II and chloride/bicarbonate exchangers. *J Biol Chem*. 2001;276(51):47886–47894.
747. Chu H, Low PS. Mapping of glycolytic enzyme-binding sites on human erythrocyte band 3. *Biochem J*. 2006;400(1):143–151.

748. Bruce LJ. Red cell membrane transport abnormalities. *Curr Opin Hematology.* 2008;15(3):184–190.
749. Jarolim P. Disorders of Band 3. In: Bernhardt I, Ellory CJ, editors. *Red Cell Membrane Transport in Health and Disease.* Berlin, Heidelberg, New York, Hong Kong, London, Milan, Paris, Tokyo, Springer; 2003, pp. 603–619.
750. Bruce LJ, C Robinson H, Guizouarn H, Borgese F, Harrison P, King M-J *et al.* Monovalent cation leaks in human red cells caused by single amino-acid substitutions in the transport domain of the band 3 chloride-bicarbonate exchanger, AE1. *Nat Genet.* 2005;37(11):1258–1263.
751. Egee S, Lapaix F, Decherf G, Staines HM, Ellory JC, Doerig C *et al.* A stretch-activated anion channel is up-regulated by the malaria parasite Plasmodium falciparum. *J Physiol.* 2002;542(Pt 3):795–801.
752. Naftalin RJ, Afzal I, Browning JA, Wilkins RJ, Ellory JC. Effects of high pressure on glucose transport in the human erythrocyte. *J Membr Biol.* 2002;186(3):113–129.
753. Kanki T, Young MT, Sakaguchi M, Hamasaki N, Tanner MJ. The N-terminal region of the transmembrane domain of human erythrocyte band 3. Residues critical for membrane insertion and transport activity. *J Biol Chem.* 2003;278(8):5564–5573.
754. Bruce LJ, Ring SM, Ridgwell K, Reardon DM, Seymour CA, Van Dort HM *et al.* South-east Asian ovalocytic (SAO) erythrocytes have a cold sensitive cation leak: implications for *in vitro* studies on stored SAO red cells. *Biochim Biophys Acta.* 1999;1416(1–2):258–270.
755. Ellory JC, Guizouarn H, Borgese F, Bruce LJ, Wilkins RJ, Stewart GW. Review. Leaky Cl–HCO_3^- exchangers: cation fluxes via modified AE1. *Philos Trans R Soc Lond B Biol Sci.* 2009;364(1514):189–194.
756. Stewart GW, Ellory JC, Klein RA. Increased human red cell cation passive permeability below 12 degrees C. *Nature.* 1980;286(5771):403–404.
757. Bruce LJ, Ring SM, Anstee DJ, Reid ME, Wilkinson S, Tanner MJ. Changes in the blood group Wright antigens are associated with a mutation at amino acid 658 in human erythrocyte band 3: a site of interaction between band 3 and glycophorin A under certain conditions. *Blood.* 1995;85(2):541–547.
758. Spring FA, Bruce LJ, Anstee DJ, Tanner MJ. A red cell band 3 variant with altered stilbene disulphonate binding is associated with the Diego (Dia) blood group antigen. *Biochem J.* 1992;288(Pt 3):713–716.

759. Bruce LJ, Zelinski T, Ridgwell K, Tanner MJ. The low-incidence blood group antigen, Wda, is associated with the substitution Val557 → Met in human erythrocyte band 3 (AE1). *Vox Sanginis.* 1996;71(2):118–120.
760. Inaba M, Yawata A, Koshino I, Sato K, Takeuchi M, Takakuwa Y et al. Defective anion transport and marked spherocytosis with membrane instability caused by hereditary total deficiency of red cell band 3 in cattle due to a nonsense mutation. *J Clin Invest.* 1996;97(8):1804–1817.
761. Akel A, Wagner CA, Kovacikova J, Kasinathan RS, Kiedaisch V, Koka S et al. Enhanced suicidal death of erythrocytes from gene-targeted mice lacking the Cl^-/HCO_3^- exchanger AE1. *Am J Physiol Cell Physiol.* 2007; 292(5):C1759–1767.
762. Lang KS, Lang PA, Bauer C, Duranton C, Wieder T, Huber SM et al. Mechanisms of suicidal erythrocyte death. *Cell Physiol Biochem.* 2005;15(5):195–202.
763. Nikinmaa M, Railo E. Anion movements across lamprey (*Lampetra fluviatilis*) red cell membrane. *Biochim Biophys Acta.* 1987;899(1): 134–136.
764. Cameron BA, Perry SF, 2nd, Wu C, Ko K, Tufts BL. Bicarbonate permeability and immunological evidence for an anion exchanger-like protein in the red blood cells of the sea lamprey, Petromyzon marinus. *J Comp Physiol B.* 1996;166(3):197–204.
765. Hagerstrand H, Danieluk M, Bobrowska-Hagerstrand M, Holmstrom T, Kralj-Iglic V, Lindqvist C et al. The lamprey (Lampetra fluviatilis) erythrocyte; morphology, ultrastructure, major plasma membrane proteins and phospholipids, and cytoskeletal organization. *Mol Membr Biol.* 1999; 16(2):195–204.
766. Nikinmaa M, Airaksinen S, Virkki LV. Haemoglobin function in intact lamprey erythrocytes: interactions with membrane function in the regulation of gas transport and acid-base balance. *J Exp Biol.* 1995;198(Pt 12): 2423–2430.
767. Nikinmaa M. Oxygen and carbon dioxide transport in vertebrate erythrocytes: an evolutionary change in the role of membrane transport. *J Exp Biol.* 1997;200(Pt 2):369–380.
768. Roudier N, Verbavatz J-M, Maurel C, Ripoche P, Tacnet F. Evidence for the presence of aquaporin-3 in human red blood cells. *J Biol Chem.* 1998 1998;273(14):8407–8412.

769. Liu Y, Promeneur D, Rojek A, Kumar N, Frokiaer J, Nielsen S et al. Aquaporin 9 is the major pathway for glycerol uptake by mouse erythrocytes, with implications for malarial virulence. *Proc Natl Acad Sci USA.* 2007;104(30):12560–12564.
770. Preston GM, Agre P. Isolation of the cDNA for erythrocyte integral membrane protein of 28 kilodaltons: member of an ancient channel family. *Proc Natl Acad Sci USA.* 1991;88(24):11110–11114.
771. Moon C, Preston GM, Griffin CA, Jabs EW, Agre P. The human aquaporin-CHIP gene. Structure, organization, and chromosomal localization. *J Biol Chem.* 1993;268(21):15772–15778.
772. Moon C, Williams JB, Preston GM, Copeland NG, Gilbert DJ, Nathans D et al. The mouse aquaporin-1gene. *Genomics.* 1995;30(2):354–357.
773. Preston GM, Carroll TP, Guggino WB, Agre P. Appearance of water channels in xenopus oocytes expressing red cell CHIP28 protein. *Science.* 1992;256(5055):385–387.
774. Preston GM, Jung JS, Guggino WB, Agre P. The mercury-sensitive residue at cysteine 189 in the CHIP28 water channel. *J Biol Chem.* 1993; 268(1):17–20.
775. Agre P, Preston GM, Smith BL, Jung JS, Raina S, Moon C et al. Aquaporin CHIP: the archetypal molecular water channel. *Am J Physiol Renal Physiol.* 1993;265(4):F463–476.
776. Walz T, Smith BL, Zeidel ML, Engel A, Agre P. Biologically active two-dimensional crystals of aquaporin CHIP. *J Biol Chem.* 1994; 269(3):1583–1586.
777. Walz T, Smith BL, Agre P, Engel A. The three-dimensional structure of human erythrocyte aquaporin CHIP. *EMBO J.* 1994;13(13):2985–2993.
778. Jung JS, Preston GM, Smith BL, Guggino WB, Agre P. Molecular structure of the water channel through aquaporin CHIP. The hourglass model. *J Biol Chem.* 1994;269(20):14648–14654.
779. Mathai JC, Mori S, Smith BL, Preston GM, Mohandas N, Collins M et al. Functional analysis of aquaporin-1 deficient red cells. The Colton-null phenotype. *J Biol Chem.* 1996;271(3):1309–1313.
780. Murata K, Mitsuoka K, Hirai T, Walz T, Agre P, Heymann JB et al. Structural determinants of water permeation through aquaporin-1. *Nature.* 2000;407(6804):599–605.

781. Saparov SM, Kozono D, Rothe U, Agre P, Pohl P. Water and ion permeation of aquaporin-1 in planar lipid bilayers. Major differences in structural determinants and stoichiometry. *J Biol Chem.* 2001;276(34):31515–31520.
782. Hazama A, Kozono D, Guggino WB, Agre P, Yasui M. Ion permeation of AQP6 water channel protein. Single channel recordings after Hg^{2+} activation. *J Biol Chem.* 2002;277(32):29224–29230.
783. Ikeda M, Beitz E, Kozono D, Guggino WB, Agre P, Yasui M. Characterization of aquaporin-6 as a nitrate channel in mammalian cells. Requirement of pore-lining residue threonine 63. *J Biol Chem.* 2002; 277(42):39873–3989.
784. Smith BL, Preston GM, Spring FA, Anstee DJ, Agre P. Human red cell aquaporin CHIP. I. Molecular characterization of ABH and Colton blood group antigens. *J Clin Invest.* 1994;94(3):1043–1049.
785. Preston G, Smith B, Zeidel M, Moulds J, Agre P. Mutations in aquaporin-1 in phenotypically normal humans without functional CHIP water channels. *Science.* 1994;265(5178):1585–1587.
786. Rabaud NE, Song L, Wang Y, Agre P, Yasui M, Carbrey JM. Aquaporin 6 binds calmodulin in a calcium-dependent manner. *Biochem Biophys Res Commun.* 2009;383(1):54–57.
787. Brahm J. Urea permeability of human red cells. *J Gen Physiol.* 1983;82(1):1–23.
788. Fröhlich O, Macey RI, Edwards-Moulds J, Gargus JJ, Gunn RB. Urea transport deficiency in Jk(a-b-) erythrocytes. *Am J Physiol Cell Physiol.* 1991;260(4):C778–C783.
789. Sands JM, Timmer RT, Gunn RB. Urea transporters in kidney and erythrocytes. *Am J Physiol Renal Physiol.* 1997;273(3):F321–F339.
790. Timmer RT, Klein JD, Bagnasco SM, Doran JJ, Verlander JW, Gunn RB *et al.* Localization of the urea transporter UT-B protein in human and rat erythrocytes and tissues. *Am J Physiol Cell Physiol.* 2001;281(4):C1318–C1325.
791. Gasbjerg PK, Brahm J. Glucose transport kinetics in human red blood cells. *Biochim Biophys Acta.* 1991;1062(1):83–93.
792. Naftalin R, J. Glucose Transport. In: Bernhardt I, Ellory JC, editors. *Red Cell Membrane Transport in Health and Disease.* Berlin, Heidelberg, New York, Hong Kong, London, Mailan, Paris, Tokyo. Springer; 2003, pp. 339–372.

793. Lauf PK, Poulik MD. Solubilization and structural integrity of the human red cell membrane. *Br J Haemat*. 1968;15(2):191–202.
794. Ballas SK, Clark MR, Mohandas N, Colfer HF, Caswell MS, Bergren MO *et al*. Red cell membrane and cation deficiency in Rh null syndrome. *Blood*. 1984;63(5):1046–1055.
795. Colin Y, Cherif-Zahar B, Le Van Kim C, Raynal V, Van Huffel V, Cartron JP. Genetic basis of the RhD-positive and RhD-negative blood group polymorphism as determined by Southern analysis. *Blood*. 1991;78(10):2747–2752.
796. Masouredis SP. Rho(D) genotype and red cell Rho(D) antigen content. *Science*. 1960;131:1442.
797. Bloy C, Blanchard D, Lambin P, Goossens D, Rouger P, Salmon C *et al*. Characterization of the D, C, E and G antigens of the Rh blood group system with human monoclonal antibodies. *Mol Immunol*. 1988;25(9):925–930.
798. Cartron JP. Rh blood group system and molecular basis of Rh-deficiency. *Baillieres Best Pract Res Clin Haematol*. 1999;12(4):655–689.
799. Gruswitz F, Chaudhary S, Ho JD, Schlessinger A, Pezeshki B, Ho C-M *et al*. Function of human Rh based on structure of RhCG at 2.1Å. *Proc Natl Acad Sci USA*. 2010;107(21):9638–9643.
800. Avent ND, Reid ME. The Rh blood group system: a review. *Blood*. 2000;95(2):375–387.
801. Vos GH, Vos D, Kirk RL, Sanger R. A sample of blood with no detectable Rh antigens. *Lancet*. 1961;1(7167):14–15.
802. Ishimori T, Hasekura H. A Japanese with no detectable Rh blood group antigens due to silent Rh alleles or deleted chromosomes. *Transfusion*. 1967;7(2):84–87.
803. Westhoff CM. The structure and function of the Rh antigen complex. *Semin Hematol*. 2007;44(1):42–50.
804. Liu Z, Peng J, Mo R, Hui C, Huang CH. Rh type B glycoprotein is a new member of the Rh superfamily and a putative ammonia transporter in mammals. *J Biol Chem*. 2001;276(2):1424–1433.
805. Liu Z, Chen Y, Mo R, Hui C, Cheng JF, Mohandas N *et al*. Characterization of human RhCG and mouse RhCG as novel nonerythroid Rh glycoprotein homologues predominantly expressed in kidney and testis. *J Biol Chem*. 2000;275(33):25641–28651.

806. Goossens D, Bony V, Gane P, Colin Y, Cartron JP. Generation of mice with inactivated Rh or Rhag genes. *Transfus Clin Biol.* 2006; 13(1–2):164–166.
807. Soupene E, Ramirez RM, Kustu S. Evidence that fungal MEP proteins mediate diffusion of the uncharged species NH_3 across the cytoplasmic membrane. *Mol Cell Biol.* 2001;21(17):5733–5741.
808. Li X, Jayachandran S, Nguyen HH, Chan MK. Structure of the Nitrosomonas europaea Rh protein. *Proc Natl Acad Sci USA.* 2007; 104(49):19279–19284.
809. Westhoff CM, Ferreri-Jacobia M, Mak D-OD, Foskett JK. Identification of the erythrocyte Rh Blood group glycoprotein as a mammalian ammonium transporter. *J Biol Chem.* 2002;277(15):12499–12502.
810. Mak DO, Dang B, Weiner ID, Foskett JK, Westhoff CM. Characterization of ammonia transport by the kidney Rh glycoproteins RhBG and RhCG. *Am J Physiol Renal Physiol.* 2006;290(2):F297–F305.
811. Westhoff CM, Siegel DL, Burd CG, Foskett JK. Mechanism of genetic complementation of ammonium transport in yeast by human erythrocyte Rh-associated glycoprotein. *J Biol Chem.* 2004;279(17):17443–17448.
812. Labotka RJ, Lundberg P, Kuchel PW. Ammonia permeability of erythrocyte membrane studied by 14N and 15N saturation transfer NMR spectroscopy. *Am J Physiol.* 1995;268(3 Pt 1):C686–C699.
813. Bruce LJ. Hereditary stomatocytosis and cation leaky red cells — recent developments. *Blood Cells Mol Dis.* 2009;42(3):216–222.
814. Bruce LJ, Guizouarn H, Burton NM, Gabillat N, Poole J, Flatt JF et al. The monovalent cation leak in overhydrated stomatocytic red blood cells results from amino acid substitutions in the Rh-associated glycoprotein. *Blood.* 2009;113(6):1350–1357.
815. Silverman DN, Tu C, Wynns GC. Depletion of ^{18}O from $C^{18}O_2$ in erythrocyte suspensions. The permeability of the erythrocyte membrane to CO_2. *J Biol Chem.* 1976;251(14):4428–4435.
816. Gutknecht J, Bisson MA, Tosteson FC. Diffusion of carbon dioxide through lipid bilayer membranes: effects of carbonic anhydrase, bicarbonate, and unstirred layers. *J Gen Physiol.* 1977;69(6):779–794.
817. Endeward V, Cartron JP, Ripoche P, Gros G. RhAG protein of the Rhesus complex is a CO_2 channel in the human red cell membrane. *FASEB J.* 2008;22(1):64–73.

818. Duranton C, Huber SM, Lang F. Oxidation induces a Cl(−)-dependent cation conductance in human red blood cells. *J Physiol.* 2002;539(Pt 3): 847–855.
819. Bernhardt I, Hall AC, Ellory JC. Effects of low ionic strength media on passive human red cell monovalent cation transport. *J Physiol.* 1991; 434:489–506.
820. Browning JA, Robinson HC, Ellory JC, Gibson JS. Deoxygenation-induced non-electrolyte pathway in red cells from sickle cell patients. *Cell Physiol Biochem.* 2007;19(1–4):165–174.
821. Ellory JC, Sequeira R, Constantine A, Wilkins RJ, Gibson JS. Non-electrolyte permeability of deoxygenated sickle cells compared. *Blood Cells Mol Dis.* 2008;41(1):44–49.
822. Fugelli K. Regulation of cell volume in flounder (*Pleuronectes flesus*) erythrocytes accompanying a decrease in plasma osmolarity. *Comp Biochem Physiol.* 1967;22(1):253–260.
823. Kiessling K, Ellory JC, Cossins AR. The relationship between hypotonically-induced taurine and K fluxes in trout red blood cells. *Pflügers Arch Eur J Physiol.* 2000;440(3):467–475.
824. Kirk K, Ellory JC, Young JD. Transport of organic substrates via a volume-activated channel. *J Biol Chem.* 1992;267(33):23475–23478.
825. Culliford SJ, Bernhardt I, Ellory JC. Activation of a novel organic solute transporter in mammalian red blood cells. *J Physiol.* 1995;489(Pt 3): 755–765.
826. Huang CC, Basavappa S, Ellory JC. Volume-activated taurine permeability in cells of the human erythroleukemic cell line K562. *J Cell Physiol.* 1996;167(2):354–358.
827. Koomoa DL, Musch MW, Goldstein L. Osmotic stress stimulates the organic osmolyte channel in Xenopus laevis oocytes expressing skate (Raja erinacea) AE1. *J Exp Zool A Comp Exp Biol.* 2005;303(4):319–322.
828. Puffer AB, Meschter EE, Musch MW, Goldstein L. Membrane trafficking factors are involved in the hypotonic activation of the taurine channel in the little skate (*Raja erinacea*) red blood cell. *J Exp Zool A Comp Exp Biol.* 2006;305(7):594–601.
829. Musch MW, Hubert EM, Goldstein L. Volume expansion stimulates p72(syk) and p56(lyn) in skate erythrocytes. *J Biol Chem.* 1999;274(12): 7923–7928.

830. Fiévet B, Perset F, Gabillat N, Guizouarn H, Borgese F, Ripoche P, et al. Transport of uncharged organic solutes in xenopus oocytes expressing red cell anion exchangers (AE1s). *Proc Natl Acad Sci USA*. 1998;95(18): 10996–11001.

831. Motais R, Guizouarn H, Borgese F. The Swelling-Sensitive Osmolyte Channel. In: Bernhardt I, Ellory CJ, editors. *Red Cell Membrane Transport in Health and Disease*. Berlin, Heidelberg, New York, Hong Kong, London, Milan, Paris, Tokyo: Springer; 2003, pp. 153–171.

832. Muller-Eberhard HJ. Molecular organization and function of the complement system. *Annu Rev Biochem*. 1988;57:321–347.

833. Tranum-Jensen J, Bhakdi S, Bhakdi-Lehnen B, Bjerrum OJ, Speth V. Complement lysis: the ultrastructure and orientation of the C5b-9 complex on target sheep erythrocyte membranes. *Scand J Immunol*. 1978;7(1):45–46.

834. Mayer MM. Mechanism of cytolysis by complement. *Proc Natl Acad Sci USA*. 1972;69(10):2954–2958.

835. Lauf PK. Immunological and physiological characteristics of the rapid immune hemolysis of neuraminidase-treated sheep red cells produced by fresh guinea pig serum. *J Exp Med*. 1975;142(4):974–988.

836. Sims PJ, Lauf PK. Steady-state analysis of tracer exchange across the C5b-9 complement lesion in a biological membrane. *Proc Natl Acad Sci USA*. 1978;75(11):5669–5673.

837. Sims P, Lauf P. Analysis of solute diffusion across the C5b-9 membrane lesion of complement: evidence that individual C5b-9 complexes do not function as discrete, uniform pores. *J Immunol*. 1980;125(6):2617–2625.

838. Halperin JA, Nicholson-Weller A, Brugnara C, Tosteson DC. Complement induces a transient increase in membrane permeability in unlysed erythrocytes. *J Clin Invest*. 1988;82(2):594–600.

839. Halperin JA, Brugnara C, Nicholson-Weller A. Ca^{2+}-activated K^+ efflux limits complement-mediated lysis of human erythrocytes. *J Clin Invest*. 1989;83(5):1466–1471.

840. Bruce LJ, Pan RJ, Cope DL, Uchikawa M, Gunn RB, Cherry RJ et al. Altered structure and anion transport properties of band 3 (AE1, SLC4A1) in human red cells lacking glycophorin A. *J Biol Chem*. 2004;279(4): 2414–2420.

841. Nicolas V, Le Van Kim C, Gane P, Birkenmeier C, Cartron JP, Colin Y et al. Rh-RhAG/ankyrin-R, a new interaction site between the membrane bilayer

and the red cell skeleton, is impaired by Rh(null)-associated mutation. *J Biol Chem.* 2003;278(28):25526–25533.
842. Bruce LJ, Beckmann R, Ribeiro ML, Peters LL, Chasis JA, Delaunay J et al. A band 3-based macrocomplex of integral and peripheral proteins in the RBC membrane. *Blood.* 2003;101(10):4180–4188.
843. Dahl KN, Parthasarathy R, Westhoff CM, Layton DM, Discher DE. Protein 4.2 is critical to CD47-membrane skeleton attachment in human red cells. *Blood.* 2004;103(3):1131–1136.
844. Peters LL, Jindel HK, Gwynn B, Korsgren C, John KM, Lux SE et al. Mild spherocytosis and altered red cell ion transport in protein 4. 2-null mice. *J Clin Invest.* 1999;103(11):1527–1537.
845. Hassoun H, Palek J. Hereditary spherocytosis: a review of the clinical and molecular aspects of the disease. *Blood Rev.* 1996;10(3):129–147.
846. Nunomura W, Takakuwa Y, Parra M, Conboy J, Mohandas N. Regulation of protein 4.1R, p55, and glycophorin C ternary complex in human erythrocyte membrane. *J Biol Chem.* 2000;275(32):24540–24546.
847. Lauf PK. Active and Passive Monovalent Ion Transport Association with Membrane Antigens in Sheep Red Blood Cells: A Molecular Riddle. In: Ellory C, Bernhardt I, editors. *Red Cell Membrane Transport in Health and Disease*: Berlin, Heidelberg, New York, Hong Kong, London, Milan, Paris, Tokyo. Springer-Verlag; 2003, pp. 27–42.
848. Gambhir KK, Archer JA, Bradley CJ. Characteristics of human erythrocyte insulin receptors. *Diabetes.* 1978;27(7):701–708.
849. Bennett BD, Solar GP, Yuan JQ, Mathias J, Thomas GR, Matthews W. A role for leptin and its cognate receptor in hematopoiesis. *Curr Biol.* 1996;6(9):1170–1180.
850. Konstantinou-Tegou A, Kaloyianni M, Bourikas D, Koliakos G. The effect of leptin on Na^+-H^+ antiport (NHE 1) activity of obese and normal subjects erythrocytes. *Mol Cell Endocr.* 2001;183(1–2):11–18.
851. Aurbach GD, Spiegel AM, Gardner JD. Beta-adrenergic receptors, cyclic AMP, and ion transport in the avian erythrocyte. *Adv Cyclic Nucleotide Res.* 1975;5:117–132.
852. Caron MG, Limbird LE, Lefkowitz RJ. Biochemical characterization of the BETA-adrenergic receptor of the frog erythrocyte. *Mol Cell Biochem.* 1979;28(1):45–66.

853. Lad PM, Nielsen TB, Preston MS, Rodbell M. The role of the guanine nucleotide exchange reaction in the regulation of the beta-adrenergic receptor and in the actions of catecholamines and cholera toxin on adenylate cyclase in turkey erythrocyte membranes. *J Biol Chem.* 1980;255(3):988–995.
854. Stadel JM, DeLean A, Lefkowitz RJ. A high affinity agonist. Beta-adrenergic receptor complex is an intermediate for catecholamine stimulation of adenylate cyclase in turkey and frog erythrocyte membranes. *J Biol Chem.* 1980;255(4):1436–1441.
855. Furukawa H, Bilezikian JP, Loeb JN. Effects of ouabain and isoproterenol on potassium influx in the turkey erythrocyte: quantitative relation to ligand binding and cyclic AMP generation. *Biochim Biophys Acta — Biomembranes.* 1980;598(2):345–356.
856. Jeffery DR, Charlton RR, Venter JC. Reconstitution of turkey erythrocyte beta-adrenergic receptors into human erythrocyte acceptor membranes. Demonstration of guanine nucleotide regulation of agonist affinity. *J Biol Chem.* 1980;255(11):5015–5018.
857. Vauquelin G, Bottari S, Andre C, Jacobsson B, Strosberg AD. Interaction between beta-adrenergic receptors and guanine nucleotide sites in turkey erythrocyte membranes. *Pro Natl Acad Sci USA.* 1980;77(7):3801–3805.
858. Rimon G, Hanski E, Levitzki A. Temperature dependence of beta receptor, adenosine receptor, and sodium fluoride stimulated adenylate cyclase from turkey erythrocytes. *Biochemistry.* 1980;19(19):4451–4460.
859. Briggs MM, Lefkowitz RJ. Parallel modulation of catecholamine activation of adenylate cyclase and formation of the high-affinity agonist-receptor complex in turkey erythrocyte membranes by temperature and cis-vaccenic acid. *Biochemistry.* 1980;19(19):4461–4466.
860. Braun S, Arad H, Levitzki A. The interaction of Mn^{2+} with turkey erythrocyte adenylate cyclase. *Biochim Biophys Acta — Protein Structure and Molecular Enzymology.* 1982;705(1):55–62.
861. Stadel JM, Nambi P, Lavin TN, Heald SL, Caron MG, Lefkowitz RJ. Catecholamine-induced desensitization of turkey erythrocyte adenylate cyclase. Structural alterations in the beta-adrenergic receptor revealed by photoaffinity labeling. *J Biol Chem.* 1982;257(16):9242–9245.
862. Stadel JM, Nambi P, Shorr RG, Sawyer DF, Caron MG, Lefkowitz RJ. Catecholamine-induced desensitization of turkey erythrocyte adenylate

cyclase is associated with phosphorylation of the beta-adrenergic receptor. *Proc Natl Acad of Sci USA.* 1983;80(11):3173–3177.
863. Nambi P, Sibley DR, Stadel JM, Michel T, Peters JR, Lefkowitz RJ. Cell-free desensitization of catecholamine-sensitive adenylate cyclase. Agonist- and cAMP-promoted alterations in turkey erythrocyte beta-adrenergic receptors. *J Biol Chem.* 1984;259(7):4629–4633.
864. Peters JR, Nambi P, Sibley DR, Lefkowitz RJ. Enhanced adenylate cyclase activity of turkey erythrocytes following treatment with [beta]-adrenergic receptor antagonists. *Eur J Pharmacol.* 1984;107(1):43–52.
865. Nambi P, Peters JR, Sibley DR, Lefkowitz RJ. Desensitization of the turkey erythrocyte beta-adrenergic receptor in a cell-free system. Evidence that multiple protein kinases can phosphorylate and desensitize the receptor. *J Biol Chem.* 1985;260(4):2165–2171.
866. Severne Y, Wemers C, Vauquelin G. Agonist/N-ethylmaleimide mediated inactivation of [beta]-adrenergic receptors: molecular aspects. *Biochem Pharmacol.* 1985;34(10):1611–1617.
867. Pedersen SE, Ross EM. Functional activation of beta-adrenergic receptors by thiols in the presence or absence of agonists. *J Biol Chem.* 1985; 260(26):14150–14157.
868. Stadel JM, Rebar R, Shorr RGL, Nambi P, Crooke ST. Biochemical characterization of phosphorylated beta-adrenergic receptors from catecholamine-desensitized turkey erythrocytes. *Biochemistry.* 1986; 25(12):3719–3724.
869. Stadel JM, Rebar R, Crooke ST. Catecholamine-induced desensitization of adenylate cyclase-coupled beta-adrenergic receptors in turkey erythrocytes: evidence for a two-step mechanism. *Biochemistry.* 1987; 26(18):5861–5866.
870. Stadel JM, Rebar R, Crooke ST. Alkaline phosphatase relieves desensitization of adenylate cyclase-coupled beta-adrenergic receptors in avian erythrocyte membranes. *Biochem J.* 1988;252(3):771–776.
871. Hertel C, Nunnally MH, Wong SK, Murphy EA, Ross EM, Perkins JP. A truncation mutation in the avian beta-adrenergic receptor causes agonist-induced internalization and GTP-sensitive agonist binding characteristic of mammalian receptors. *J Biol Chem.* 1990;265(29):17988–17994.
872. Rooney TA, Hager R, Thomas AP. Beta-adrenergic receptor-mediated phospholipase C activation independent of cAMP formation in turkey

erythrocyte membranes. *J Biological Chem.* 1991;266(23): 15068–15074.
873. James SR, Vaziri C, Walker TR, Milligan G, Downes CP. The turkey erythrocyte beta-adrenergic receptor couples to both adenylate cyclase and phospholipase C via distinct G-protein alpha subunits. *Biochem J.* 1994; 304(Pt 2):359–364.
874. Ugur O, Onaran HO. Allosteric equilibrium model explains steady-state coupling of beta-adrenergic receptors to adenylate cyclase in turkey erythrocyte membranes. *Biochem J.* 1997;323(Pt 3):765–776.
875. Simpson IA, Pfeuffer T. Functional desensitisation of beta-adrenergic receptors of avian erythrocytes by catecholamines and adenosine 3′,5′-phosphate. *Eur J Biochem.* 1980;111(1):111–116.
876. Popov KM, Bulargina TV, Severin ES. Factors essential for desensitization of pigeon erythrocyte adenylate cyclase responsiveness in a cell-free system. *Biochem Int.* 1985;10(5):723–731.
877. Lefkowitz RJ, Mullikin D, Caron MG. Regulation of beta-adrenergic receptors by guanyl-5′-yl imidodiphosphate and other purine nucleotides. *J Biol Chem.* 1976;251(15):4686–4692.
878. Strulovici B, Stadel JM, Lefkowitz RJ. Functional integrity of desensitized beta-adrenergic receptors. *J Biol Chem.* 1983;258(10):6410–6414.
879. Cherksey BD, Zadunaisky JA, Murphy RB. Cytoskeletal constraint of the beta-adrenergic receptor in frog erythrocyte membranes. *Proc Natl Acad Sci USA.* 1980;77(11):6401–6405.
880. Salama A, Nikinmaa M. Effect of oxygen tension on catecholamine-induced formation of cAMP and on swelling of carp red blood cells. *Am J Physiol Cell Physiol.* 1990;259(5):C723–C726.
881. Tetens V, Lykkeboe G, Christensen NJ. Potency of adrenaline and noradrenaline for beta-adrenergic proton extrusion from red cells of rainbow trout, Salmo gairdneri. *J Exp Biol.* 1988;134:267–280.
882. Perry SF, 2nd, Wood CM, Thomas S, Walsh PJ. Adrenergic inhibition of carbon dioxide excretion by trout red blood cells *in vitro* is mediated by activation of Na^+/H^+ exchange. *J Exp Biol.* 1991;157:367–380.
883. Caldwell S, Rummer JL, Brauner CJ. Blood sampling techniques and storage duration: effects on the presence and magnitude of the red blood cell [beta]-adrenergic response in rainbow trout (*Oncorhynchus mykiss*). *Comp Biochem Physiol — Part A.* 2006;144(2):188–195.

884. Thomas JB, Gilmour KM. The impact of social status on the erythrocyte [beta]-adrenergic response in rainbow trout, *Oncorhynchus mykiss*. *Comp Biochem Physiol — Part A*. 2006;143(2):162–172.
885. Setchenska MS, Arnstein HR. Characteristics of the beta-adrenergic adenylate cyclase system of developing rabbit bone-marrow erythroblasts. *Biochem J*. 1983;210(2):559–566.
886. Setchenska MS, Bonanou-Tzedaki SA, Arnstein HRV. Classification of [beta]-adrenergic subtypes in immature rabbit bone marrow erythroblasts. *Biochem Pharmacol*. 1986;35(21):3679–3684.
887. Limbird LE, Gill DM, Lefkowitz RJ. Agonist-promoted coupling of the beta-adrenergic receptor with the guanine nucleotide regulatory protein of the adenylate cyclase system. *Proc Natl Acad Sci USA*. 1980;77(2): 775–779.
888. Stiles GL, Stadel JM, De Lean A, Lefkowitz RJ. Hypothyroidism modulates beta adrenergic receptor adenylate cyclase interactions in rat reticulocytes. *J Clin Invest*. 1981;68(6):1450–1455.
889. Montandon JB, Porzig H. *In vitro* maturation of the rat reticulocyte beta-adrenoceptor adenylate cyclase system. *Biomed Biochim Acta*. 1983; 42(11–12):S197–S201.
890. Shane E, Yeh M, Feigin AS, Owens JM, Bilezikian JP. Cytosol activator protein from rat reticulocytes requires the stimulatory guanine nucleotide-binding protein for its actions on adeiiylate cyclase. *Endocrinol*. 1985; 117(1):255–263.
891. Yamashita A, Kurokawa T, Dan'ura T, Higashi K, Ishibashi S. Induction of desensitization by phorbol ester to [beta]-adrenergic agonist stimulation in adenylate cyclase system of rat reticulocytes. *Biochem Biophys Res Commun*. 1986;138(1):125–130.
892. Moudry R, Porzig H. Regulation of [beta]-adrenergic responses during *in vitro* differentiation of mouse erythroleukemia cells. *Exp Cell Res*. 1990; 191(2):278–285.
893. Cussac D, Schaak S, Denis C, Flordellis C, Calise D, Paris H. High level of α2-adrenoceptor in rat foetal liver and placenta is due to α2B-subtype expression in haematopoietic cells of the erythrocyte lineage. *Br J Pharmacol*. 2001;133(8):1387–1395.
894. Sager G. Receptor binding sites for beta-adrenergic ligands on human erythrocytes. *Biochem Pharmacol*. 1982;31(1):99–104.

895. Mazza P, Salvadori A, Baudo S, Fanari P, Fontana M, Ruga S et al. Catecholamine-stimulated potassium transport in erythrocytes from normal and obese śubjects. *Minerva Medica*. 1992;83(10):615–619.
896. Marques F, Bicho MP. Activation of a NADH dehydrogenase in the human erythrocyte by beta-adrenergic agonists: possible involvement of a G protein in enzyme activation. *Neurosignals*. 1997;6(2):52–61.
897. Horga JF, Gisbert J, De Agustín JC, Hernández M, Zapater P. A beta-2-adrenergic receptor activates adenylate cyclase in human erythrocyte membranes at physiological calcium plasma concentrations. *Blood Cells Mol Dis*. 2000;26(3):223–228.
898. Sohn DH, Kim HD. Effects of adenosine receptor agonists on volume-activated ion transport in pig red cells. *J Cell Physiol*. 1991;146(2):318–324.
899. von Lindern M, Zauner W, Mellitzer G, Steinlein P, Fritsch G, Huber K et al. The glucocorticoid receptor cooperates with the erythropoietin receptor and c-Kit to enhance and sustain proliferation of erythroid progenitors *in vitro*. *Blood*. 1999;94(2):550–559.
900. Schneider H, Chaovapong W, Matthews DJ, Karkaria C, Cass RT, Zhan H et al. Homodimerization of erythropoietin receptor by a bivalent monoclonal antibody triggers cell proliferation and differentiation of erythroid precursors. *Blood*. 1997;89(2):473–482.
901. Stellacci E, Di Noia A, Di Baldassarre A, Migliaccio G, Battistini A, Migliaccio AR. Interaction between the glucocorticoid and erythropoietin receptors in human erythroid cells. *Exp Hematol*. 2009;37(5):559–572.
902. Sluyter R, Shemon AN, Barden JA, Wiley JS. Extracellular ATP increases cation fluxes in human erythrocytes by activation of the P2X7 receptor. *J Biol Chem*. 2004;279(43):44749–44755.
903. Sluyter R, Shemon AN, Hughes WE, Stevenson RO, Georgiou JG, Eslick GD et al. Canine erythrocytes express the P2X7 receptor: greatly increased function compared with human erythrocytes. *Am J Physiol Reg Integ Comp Physiol*. 2007;293(5):R2090–2098.
904. Sluyter R, Shemon AN, Wiley JS. P2X7 receptor activation causes phosphatidylserine exposure in human erythrocytes. *Biochem Biophys Res Commun*. 2007;355(1):169–173.
905. Murphy SC, Harrison T, Hamm HE, Lomasney JW, Mohandas N, Haldar K. Erythrocyte G protein as a novel target for malarial chemotherapy. *PLoS Med*. 2006;3(12):e528.

906. Downes CP, Berrie CP, Hawkins PT, Stephens L, Boyer JL, Harden TK. Receptor and G-protein-dependent regulation of turkey erythrocyte phosphoinositidase C. *Philos Trans R Soc Lond B Biol Sci.* 1988;320(1199):267–280.
907. Berrie CP, Hawkins PT, Stephens LR, Harden TK, Downes CP. Phosphatidylinositol 4,5-bisphosphate hydrolysis in turkey erythrocytes is regulated by P2y purinoceptors. *Mol Pharmacol.* 1989;35(4):526–532.
908. Martin MW, Harden TK. Agonist-induced desensitization of a P2Y-purinergic receptor-regulated phospholipase C. *J Biol Chem.* 1989;264(33): 19535–19539.
909. Morris AJ, Waldo GL, Downes CP, Harden TK. A receptor and G-protein-regulated polyphosphoinositide-specific phospholipase C from turkey erythrocytes. II. P2Y-purinergic receptor and G-protein-mediated regulation of the purified enzyme reconstituted with turkey erythrocyte ghosts. *J Biol Chem.* 1990;265(23):13508–13514.
910. Vaziri C, Downes CP. G-protein-mediated activation of turkey erythrocyte phospholipase C by beta-adrenergic and P2y-purinergic receptors. *Biochem J.* 1992;284(Pt 3):917–922.
911. Stephens LR, Berrie CP, Irvine RF. Agonist-stimulated inositol phosphate metabolism in avian erythrocytes. *Biochem J.* 1990;269(1):65–72.
912. Sartorello R, Garcia CRS. Activation of a P2Y4-like purinoceptor triggers an increase in cytosolic Ca^{2+} in the red blood cells of the lizard *Ameiva ameiva* (Squamata, Teiidae). *Braz J Med Biol Res.* 2005;38:5–10.
913. Light DB, Dahlstrom PK, Gronau RT, Baumann NL. Extracellular ATP activates a P2 receptor in necturus erythrocytes during hypotonic swelling. *J Membr Biol.* 2001;182(3):193–202.
914. Geering K. The functional role of the [beta]-subunit in the maturation and intracellular transport of Na,K-ATPase. *FEBS Letters.* 1991;285(2): 189–193.
915. Dobretsov M, Stimers JR. Neuronal function and alpha3 isoform of Na/K ATPase. *Front Biosci.* 2005;10 2373–2396.
916. Wakabayashi S, Sardet C, Fafournoux P, Counillon L, Meloche S, Pagés G et al. Structure function of the growth factor-activatable Na^+/H^+ exchanger (NHE1). *Rev Physiol Biochem Pharmacol.* 1992;119:157–186.
917. Chen YC, Cadnapaphornchai MA, Schrier, RW. Clinical update on renal aquaporins. *Biol Cell* 2005;97:357–371.

3
Erythropoiesis

Maxim Pimkin[*] and Mitchell J Weiss[*,†,‡]

[*] *Department of Pediatrics, Division of Hematology, The Children's Hospital of Philadelphia, PA;* [†] *University of Pennsylvania School of Medicine, Philadelphia, PA, 19104 USA*

3.1 Introduction

Several unique features distinguish red blood cell (RBC) production (erythropoiesis). Firstly, synthetic demands and turnover are massive. The human body produces approximately 2.4×10^6 erythrocytes each second in a process that is dynamic and highly regulated to maintain circulating erythrocyte numbers within a narrow physiological range. Secondly, terminal erythropoiesis is highly focused on a singular specialized purpose: the production and maintenance of large amounts of hemoglobin to serve as the major blood oxygen carrier. Finally, the mature erythrocyte acquires streamlined membrane and internal structures that withstand repeated passages through small blood vessels. Erythropoiesis involves sequential proliferation and differentiation of the multipotent hematopoietic stem cell (HSC) into a mature erythrocyte (Fig. 3.1). Reduced proliferative capacity and restricted developmental potential accompany transitions through this hierarchy. A complex interplay of regulated hormones, cytokines and hematopoietic microenvironment drives the action of transcription factors, which establish erythroid specific programs of gene expression and simultaneously

---

[‡]Corresponding author: weissmi@email.chop.edu

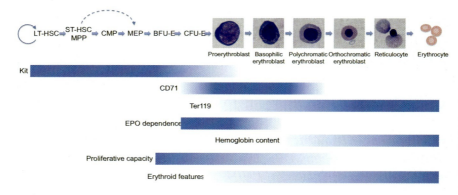

Figure 3.1 Developmental hierarchy of erythropoiesis. The dotted line indicates a possible early divergence of MEP. The bars indicate the timing of some developmental changes during cellular maturation. Erythroid features include characteristic morphologic changes illustrated in the figure, loss of various organelles and signature changes in gene expression (not shown). Micrographs courtesy of James Palis.

repress programs of alternate lineages. The final stages of maturation are characterized by high-level expression of erythroid-specific genes during several specialized cell divisions, culminating in ejection of the cell nucleus and degradation of numerous other organelles.

The rapid turnover of RBCs places extreme demands on the hematopoietic system, which must maintain homeostasis in the healthy state and rapidly adapt to various stresses, such as blood loss, enhanced erythrocyte destruction (hemolysis) or hypoxia. Understanding the mechanisms of erythropoiesis provides general insights into normal physiology and developmental biology. Indeed, numerous fundamental paradigms of gene expression, cellular development and structural biology arose from studies of erythropoiesis, globin gene expression and hemoglobin function. These studies have also led to effective treatments for numerous diseases of erythropoiesis including anemias (too few RBCs) and polycythemias (too many RBCs). Despite these advances, major gaps in our understanding of erythropoiesis exist and erythrocyte diseases remain a common cause of mortality and morbidity worldwide.

3.2 Ontogeny of Erythropoiesis

The hematopoietic system, one of the earliest tissues to develop in the conceptus, is derived from the mesodermal germ layer.[1,2] Unlike the adult erythropoietic system, fetal hematopoiesis is not at equilibrium. It must generate relatively large amounts of functional erythroid cells to accommodate the rapidly enlarging embryo prior to establishment of the adult bone marrow hematopoietic microenvironment. Several temporally overlapping waves of erythropoiesis occur during vertebrate ontogeny at specific anatomic locations to produce erythrocytes with distinct structural and functional properties. At least five anatomic sites participate in the formation of blood cells during ontogeny: yolk sac, placenta, aorta-gonad-mesonephros (AGM), fetal liver and bone marrow. Many aspects of developmental erythropoiesis are conserved among vertebrate species, which has led to the development of several valuable model organisms including zebrafish, chicken and mouse.[3]

3.2.1 *Primitive erythropoiesis: The yolk sac*

The first wave of erythropoiesis, termed primitive, occurs in the yolk sac of human and mouse conceptuses, beginning prior to the onset of circulation (reviewed in).[4–7] The first erythroid cells arise in yolk sac blood islands, which are enveloped by layers of endothelial cells and endodermal cells (Fig. 3.2). Blood islands become detectable at embryonic day (E) 7.5 of mice[8–10] and between E16–20 in human embryos.[11] Specialized yolk sac mesodermal-derived cells called hemangioblasts represent a bipotential hematopoietic-endothelial cell progenitor.[12–16]

Yolk sac-derived primitive erythrocytes are distinguished by their large size, presence of a nucleus at the time of release into the bloodstream and expression of embryonic-type hemoglobins. Primitive erythrocytes enter the bloodstream at the onset of circulation during the 3–4th week of human gestation and E8.5 in the mouse embryo. These cells differentiate synchronously as a single cohort undergoing progressive hemoglobinization, nuclear condensation and, eventually, enucleation.[17] Unlike definitive erythropoiesis, primitive erythroid cells are a transient population and not rooted in a self-renewing stem cell compartment.

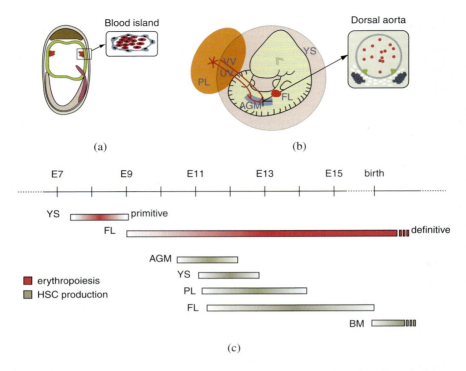

Figure 3.2 Ontogeny of mouse erythropoiesis. The anatomical sites of erythropoiesis are shown. (a) Blood islands in the yolk sac of a E7.5 mouse embryo. (b) Formation of HSCs in the AGM region of a E11 mouse embryo. (c) Timing of hematopoiesis during mouse development. Erythropoiesis is indicated in red and HSC production in grey. AGM, aorta-gonads-mesonephros; E, embryonic day; FL, fetal liver; HSC, hematopoietic stem cell; PL, placenta; UV, umbilical vessel; VV, vitelline vessel; YS; yolk sac. Adapted from Ref. (7).

3.2.2 Definitive erythropoiesis

The second wave of embryonic erythropoiesis, termed definitive, produces smaller cells that synthesize fetal and/or adult-type globins and expel their nuclei before entering the circulation.

3.2.2.1 Fetal liver

The switch to definitive erythropoiesis occurs when the fetal liver begins producing and releasing definitive erythroid cells at about 30 days of human gestation or E10 in mice embryos. The fetal liver is unable to produce hematopoietic progenitors *de novo,* but rather, is seeded by hematopoietic progenitors and HSCs derived from at least two distant sites. In mice, the first wave of definitive erythroid progenitors originates in the yolk sac at E8.25.[18–21] These cells, and possibly also placenta/amnion/chorion hematopoietic progenitors, colonize fetal liver, giving rise to early definitive fetal liver erythropoiesis.[21–28] However, these extraembryonically derived progenitors appear to be transient populations that probably do not contribute permanently to the adult HSC compartment.[29,30]

The major source of fetal liver HSCs is the AGM, specifically the dorsal aorta, where distinct intralumenal clusters of cells derived from hemogenic endothelium become apparent in mouse embryo at E10.5 and the human embryo at 4–5 weeks of gestation.[2,31–36] As dorsal aorta HSCs seed the fetal liver, the contribution of yolk sac-derived progenitors remits. Dorsal aorta-derived HSCs proliferate in the fetal liver and ultimately migrate to bone marrow to support long term hematopoiesis during late gestation and postnatal life.[37] Consequently, a gradual switch from primitive to definitive erythrocytes in the circulation occurs between 7–10 weeks of gestation in the human embryo and between E12–E16 in the mouse.[8,38,39] By the 5th month of human gestation, fetal liver becomes the predominant site of erythropoiesis. Compared to yolk sac erythropoiesis, fetal liver RBC differentiation is less synchronous, arising continuously from an expanding population of resident HSCs.[40] While fetal liver definitive erythrocytes are smaller than their primitive counterparts (120 μm *versus* 200 μm, respectively), they are macrocytic compared to mature adult erythrocytes, which measure 80 μm in diameter.

Adhesion molecules play important roles in colonization of fetal liver by hematopoietic stem and progenitor cells, as demonstrated by the observation that β1-integrin-deficient mouse embryos lack fetal liver erythropoiesis although yolk sac primitive erythropoiesis is intact.[41,42] Maturation and enucleation of definitive erythroblasts in the fetal liver occurs within erythroblastic islands consisting of erythroid precursors surrounding a central macrophage, similar to what occurs in the bone marrow (see Section 3.5.1).[43,44]

3.2.2.2 Bone marrow

Around month 6 of human gestation, HSCs migrate from fetal liver to bone marrow and hepatic erythropoiesis declines, nearly ceasing around birth. Bone marrow becomes the predominant source of erythropoiesis during late prenatal development and postnatally. The marrow cavities of most bones generate RBCs. In neonates, the predominant production sites are sinusoidal cavities of the long bones and the axial skeleton, while in adult life, erythropoiesis mainly occurs in the vertebral bodies, ribs and sternum.[45]

In addition to structural differences noted earlier, several important functional characteristics distinguish erythroid cells produced in different ontogenic phases. While definitive erythropoiesis is exquisitely dependent on erythropoietin (EPO), the role of this hormone in the primitive erythroid development appears to be less pronounced.[46,47] However, recent studies have uncovered novel roles of EPO in the survival and proliferation of primitive erythroblasts (James Palis, personal communication). *In vitro* cultures of fetal-derived definitive erythroid cells appear to have a higher proliferative potential and a greater sensitivity to stem cell factor (SCF) than erythroid progenitors from adults.

Mice are perhaps the most commonly used model system for the study of erythropoiesis. However, it is worth noting that several important features distinguish murine and human erythroid development. In particular, mice do not express fetal globins and immediately switch to adult type β-globin at the onset of definitive erythropoiesis. In mice, the spleen remains a functional hematopoietic organ maintaining a relatively minor fraction of normal erythropoietic output, particularly in early postnatal life. In addition, mouse spleen plays a dominant role during erythropoietic stress.[48] In healthy humans, the spleen plays virtually no role in steady-state erythrocyte production, although it remains a site for extramedullary erythropoiesis during human hemolytic disorders (see Section 3.4.2).

3.2.3 Recent paradigm shifts in developmental erythropoiesis

Several classical paradigms of primitive and definitive erythropoiesis have been reevaluated and revised recently. For example, it has long been believed that primitive erythrocytes retain their nuclei. However, recent studies demonstrate that they enucleate after entering the circulation.[17] In mice, primitive erythrocyte enucleation is thought to occur during transit through the fetal liver, in close association with resident macrophages.[49] In contrast, recent studies suggest that terminal maturation and enucleation of human primitive erythroblasts occurs in association with macrophages within first trimester placental villi.[50] Until recently, it has been difficult to trace the fate of primitive erythrocytes during embryogenesis because they are rapidly outnumbered by emerging definitive cells. However, in mice it is possible to tag primitive erythroid cells using transgenic fluorescent markers driven by lineage-specific regulatory sequences. Using this approach, studies demonstrated that primitive erythrocytes are not eliminated from the circulation during the onset of definitive erythropoiesis, as once believed, but instead form a stable population that persists at least until the end of gestation.[51]

3.2.4 *Hemoglobin switching*

Hemoglobin is a hetero-tetramer composed of two α-like and two β-like globin subunits, each bound to a heme prosthetic group. An iron molecule embedded within heme binds oxygen reversibly to facilitate its transfer from pulmonary capillary beds to all tissues. During ontogeny, different globin polypeptides are produced sequentially in a developmental stage-specific fashion (Fig. 3.3). Yolk sac erythrocytes express mainly β-like globins ε and γ and α-like globins ζ and α, resulting in four tetrameric combinations: Gower-1 ($\zeta_2\varepsilon_2$), Gower-2 ($\alpha_2\varepsilon_2$), Portland-1 ($\zeta_2\gamma_2$) and Portland-2 ($\zeta_2\beta_2$). Fetal liver erythropoiesis in humans is characterized by the production of adult type α-globin and the fetal β-like chain γ, resulting in the formation of fetal hemoglobin (HbF, $\alpha_2\gamma_2$). There are two fetal globin genes in humans, *HBG1* ($^A\gamma$) and *HBG2* ($^G\gamma$), which differ by only a single amino acid substitution. Compared with HbA, embryonic and fetal hemoglobins have higher oxygen affinities, presumably to facilitiate efficient extraction from maternal blood. As hematopoiesis shifts from liver to bone marrow, hemoglobin synthesis switches from F to adult-type HbA ($\alpha_2\beta_2$). At term, HbA to HbF ratio is about 1:1 and by 7 months of

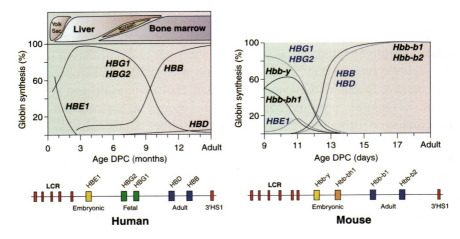

Figure 3.3 Developmental switching of β-like globin gene expression in human (left) and mouse (right). The graphs in the upper panels indicate timing of globin gene expression during embryonic development and after birth, with anatomic origins shown for humans. Bottom panels depict organization of the beta-globin locus. The human and mouse β-like gene clusters are shown with each gene indicated as a colored box. Upstream DNase hypersensitive sites (HS) and downstream 3'HS1 are shown as red boxes. In the right graph, the expression of endogenous mouse globin genes are indicated as black straight lines. Expression of human β-like globins in transgenic mice containing the entire human β-like globin cluster is indicated by blue dashed lines. Adapted from Ref. 55.

age, adult levels of HbF (<1%) are achieved. In normal adults, production of HbF is restricted to a few cells, termed F cells.[52,53] Clonal analysis demonstrated that F cells are generated from the same progenitors as the cells producing HbA.[54]

Of note, there is significant evolutionary diversity of hemoglobin switching patterns between mammalian species, which has complicated the use of mammalian models in the study of human hemoglobin switching.[55] Only Old World monkeys express the γ globin chain in a human-like temporal pattern whereas mice lack γ globin coding genes and switch from embryonic to adult globins during the primitive-to-definitive erythropoiesis transition.[56,57] Human β-globin locus transgenes in the mouse express γ globin in the early fetal liver, and these have proven to be useful instruments for hemoglobin switching research.[58,59] However, recent studies indicate that the regulation of human γ globin gene expression in these mice is more consistent with the murine embyonic globins, illustrating limitations to this experimental approach.[60]

Understanding and manipulating hemoglobin switching is of great clinical significance. Adults who maintain high HbF levels in a condition known as hereditary persistence of fetal hemoglobin (HPFH) have alleviated symptoms of concomitant β-hemoglobinopathies, such as β-thalassemia and sickle cell anemia.[53,61,62] Therefore, derepression of silenced γ globin represents an important strategy to treat these disorders. As proof of principle, hydroxyurea raises HbF levels in adults and alleviates symptoms of sickle cell anemia in many patients.[63–65] However, hydroxyurea is not fully effective. It is likely that more potent HbF-inducing agents will be identified through efforts to better understand the molecular mechanisms of fetal globin-switching. Major advancements in this area have been provided from recent genome-wide association studies to identify genetic determinants of HbF expression in adults.[53] Most cases of HPFH map to the β globin locus itself. However, two distant loci, MYB and BCL11A, also appear to regulate HbF expression.[66–72]

3.2.4.1 *BCL11A*

Originally described as a myeloid or B cell proto-oncogene, BCL11A is expressed in a stage-specific manner in human ontogeny, with fetal and embryonic erythroid cells producing shorter variants of this protein. A BCL11A polymorphism that reduces mRNA levels is associated with elevated HbF production in adult subjects and siRNA knockdown of BCL11A derepresses HbF production in cultured human erythroblasts. These experiments, along with biochemical studies, indicate that BCL11A forms a critical component of a multiprotein nuclear complex that represses γ globin synthesis by modulating local chromatin structure and the three dimensional configuration of the β globin gene cluster. BCL11A-associated proteins that may facilitate this function include the NuRD chromatin remodelling complex, SOX6, GATA1 and FOG1.[73] Interestingly, haploinsufficiency of KLF1, an erythroid transcription factor that represses BCL11A gene expression, causes HPFH.[74]

The timing of BCL11A expression may represent a key point of divergence in the mammalian evolution of hemoglobin switching. BCL11A-null mice fail to silence production of embryonic globins (murine βh1, human ζ and ε) during the transition from primitive to definitive erythropoiesis. Therefore, in the mouse, BCL11A appears to restrict expression of embryonic β-like globins to the primitive erythrocytes, whereas in humans it restricts expression of γ globin to fetal and early postnatal definitive erythropoiesis.[60]

A recent study demonstrated that two chromosome 13-encoded microRNAs stimulate γ globin synthesis by repressing MYB.[75] This explains the longstanding clinical observation that patients with full or partial trisomy 13 have a delayed fetal-to-adult globin switch and maintain elevated HbF levels throughout life. However, mechanisms by which MYB regulates γ globin production are unknown. In addition, several other nuclear factors, including NF-E4, DRED/TR2/TR4, COUP-TF and MBD2 have been implicated (reviewed in).[55] Compared to BCL11A, these factors appear to exert smaller effects on Hb switching and their roles are less well understood.

3.3 Erythroid Differentiation and Maturation

3.3.1 *Generation of committed erythroid progenitors from hematopoietic stem cells*

Long term (LT)-HSCs are a rare population capable of both self-renewal and differentiation into all mature hematopoietic lineages. By definition, a single LT-HSC can repopulate the entire hematopoietic system of a host animal. LT-HSCs give rise to intermediate and short term (ST)-HSCs with more limited self-renewal capacities. HSCs rarely divide and the primary level of homeostatic control is executed by the immature progeny of HSC which display more restricted developmental and self renewal potentials, yet still maintain high proliferative capacities.[76–78] The exact paths through which HSCs become committed erythroid cells and other types of committed progenitors is controversial (reviewed in).[79] The most widely accepted model postulates that hematopoietic committment involves sequential binary decisions. Thus, HSCs differentiate into multipotential progenitors (MPPs) that subsequently give rise to either a common myeloid progenitor (CMP) or a common lymphoid progenitor (CLP). The CMP differentiates further into bipotential granulocyte-macrophage progenitors (GMP) or megakaryocyte-erythroid progenitors (MEP).[80-82] Finally, committed erythroid progenitors arise from MEPs. Developmental plasticity appears to be maintained in multipotent progenitors partly through low level "promiscuous" expression of lineage-specific genes, a phenomenon termed "lineage priming". Subsequent cell fate commitment is accompanied by marked upregulation of appropriate lineage-specific genes with concommittant repression of alternate lineage programs (Fig. 3.1). This model is validated by numerous experimental studies and provides a useful conceptual framework for further understanding of the hematopoietic heirarchies. However, *in vivo* hematopoiesis is more likely to occur as a continuum rather than a set of distinct developmental stages. Moreover, alternative developmental pathways have recently been defined. For example, a lymphoid primed multipotent progenitor (LMPP) with both lymphoid and granulocyte-macrophage, but not erythroid-megakaryocyte lineage potential has been identified.[83] Later studies provided additional support for this model by showing similar gene expression profiles in early lymphoid and granulocyte-macrophage lineages, while MEP cells displayed a distinct transcriptional pattern.[84,85] Thus, it is possible that pluripotent bone marrow cells employ multiple pathways to terminal erythroid commitment.

3.3.2 Early committed erythroid progenitors

Terminal erythroid commitment of MEPs is likely a stochastic process influenced by transcription factors including KLF1, FLI1 and MYB,[86-89] and microRNAs, including miR-150, which represses MYB.[90,91] Erythroid progenitors are defined functionally by their ability to generate hemoglobinized colonies in semi-solid media.[45] Exclusion of lineage-positive cell types enriches primary bone marrow and fetal liver cultures for colony-forming erythroid progenitors.[92,93] Alternatively, early committed erythroid progenitors can be identified and purified through their distinguishing patterns of developmentally expressed surface antigens.[45,94]

3.3.2.1 BFU-E

The earliest committed erythroid progenitor is the burst-forming unit-erythroid (BFU-E) which gives rise to multiclustered colonies (bursts) containing several thousand hemoglobinized cells. *In vivo*, BFU-E cells represent a quiescent reserve with only 10–20% of cells in cycle at any given time. They occur at a frequency of 40–120 per 10^5 bone marrow cells from healthy humans.[95] BFU-E cells are also present in peripheral blood (10–40 per 10^5 mononuclear cells). The ability of progenitors to circulate is important for redistribution and regeneration of the bone marrow after local damage or transplantation. Thus, circulating BFU-E increase after chemotherapy or cytokine treatment.[96]

Growth and maturation of BFU-E is supported by several cytokines including SCF, thrombopoietin, interleukin (IL)-3, IL-6, IL-11, and Flt3 ligand.[45,97] *In vivo*, these cytokines are produced by bone marrow stromal and accessory cells and exert regulatory control over erythropoiesis. SCF and IL-3 exert proliferative effects on BFU-E. In contrast, tumor necrosis factor (TNF)-α, transforming growth factor (TGF)-β and interferon (IFN)-γ inhibit proliferation and survival of BFU-E.[98,99] Overexpression of these cytokines may contribute to inflammation-associated anemia.[100] While EPO is a major physiological regulator of erythropoiesis (see Chapter 6), BFU-E cells themselves are not dependent on this cytokine. Rather, their progeny, termed colony forming unit-erythroid (CFU-E) cells, require EPO for survival.

3.3.2.2 CFU-E

BFU-E cells give rise to CFU-E cells, which are defined by their ability to generate single small colonies containing 8–125 hemoglobinized cells when cultured for 7 days (humans) or 2–3 days (mice) *in vitro*. Most CFU-E divide and differentiate rapidly during *in vitro* culture. These cells are 5–10 times more abundant in the bone marrow than BFU-E and are exquisitely dependent on EPO for survival. A recent study showed that CFU-E maturation is functionally linked with cell division.[93] Specifically, upregulation of transferrin receptor (CD71) marks late CFU-E progenitors that are synchronized in S-phase. Progression of these cells through

S-phase is required for several differentiation events, including the onset of EPO dependance, switch of the β-globin locus into a more open chromatin conformation and activation of GATA-1 activity. Consequently, arresting S-phase progression in this population blocked induction of erythroid-specific genes. Furthermore, downregulation of the cell cycle inhibitor $p57^{KIP2}$ and the myeloid transcription factor PU.1 is necessary for this transition, and mutual inhibition between PU.1 and S-phase progression is proposed to serve as a mechanism ensuring synchronization between cell cycle and differentiation.[93]

3.3.3 *Maturation of late stage erythroid precursors (erythroblasts)*

CFU-E progenitors give rise to more mature erythroid precursors with distinct morphological features (Fig. 3.1). The immediate progeny of CFU-E is a proerythroblast, which differentiates sequentially into basophilic erythroblast, polychromatic erythroblast, orthochromatic erythroblast, reticulocyte and finally, mature erythrocyte (Fig. 3.1). The differentiation of CFU-E into a mature erythrocyte occurs in 2–3 days in the mouse and involves 3 to 5 cell divisions. Morphological changes during this progression include decreased cell size, accumulation of hemoglobin which imparts red color and finally, loss of the nucleus and other organelles. Before enucleation, the RNA content of maturing erythroblasts progressively decreases while nuclear chromatin becomes condensed. Reticulocytes detach from the bone marrow and enter the circulation where the final stages of maturation occur.

3.3.3.1 Developmental changes in membrane structure

Erythrocytes must be slippery and distensible in order to withstand repeated circulatory passages through small diameter capillary beds. This requires formation of a highly specialized membrane structure involving production and assembly of lineage specific proteins (reviewed in).[101] Most RBC-specific membrane proteins, including spectrin, glycophorins, ankyrin, band 4.1, and band 3, accumulate after the CFU-E stage. In contrast, the membrane glycoproteins p95, p105 and vimentin are downregulated or totally lost during maturation (also discussed below). Accumulation of erythroid-specific membrane proteins is regulated both at the transcriptional level and through tissue-specific splicing events.[102,103]

The expression of distinct membrane proteins at different stages of erythropoiesis is utilized to track maturation of defined progenitors. For example, the erythroid specific protein glycophorin (or Ter119 in mice), begins to be expressed at the proerythroblast stage and persists in mature RBCs. Transferrin receptors (CD71) are upregulated in CFU-E to polychromatophilic erythroblast stages to support iron acquisition for hemoglobin synthesis, and are subsequently reduced as hemoglobin synthesis winds down.[104] Reticulocytes express relatively small amounts of transferrin receptors that are shed within small lipid vesicles termed exosomes.[92,105] Thus, erythroid precursors at various stages can be identified or purified by flow cytometry based on expression of Ter119 and CD71.[92,106,107] Another strategy for erythroid progenitor purification utilizes Ter119 and the adhesion molecule CD44, which peaks at the CFU-E-proerythroblast stages and subsequently declines during terminal maturation.[108]

3.3.3.2 Erythroblast enucleation

Nuclear condensation occurs in all vertebrates during late erythropoiesis but only mammalian erythroblasts undergo enucleation during terminal maturation. Specifically, the vertebrate orthochromatic erythroblast expels its nucleus to become a reticulocyte. This process may help to optimize RBC deformability, decrease blood viscosity and reduce internal consumption of oxygen from metabolic demands of a nucleated cell. Enucleation is preceded by condensation and peripheralization of the nucleus during the basophilic and polychromatic stages. This process is associated with histone modifications. For example, histone deacetylation by HDAC2 is essential for chromatin condensation and subsequent enucleation.[109] Another study demonstrated that the proto-oncogene MYC, which drives proliferation of early erythroblasts, also impairs chromatin condensation and enucleation, supporting the longstanding observation that MYC downregulation is necessary for terminal maturation.[110]

Extrusion of the erythroblast nucleus occurs through mechanisms involving cytoskeletal remodeling.[111,112] During this process, actin filaments form a contractile ring between the cytoplasm and nucleus. Enucleation depends on the action of Rac GTPases — a family of proteins that regulate actin cytoskeleton structure and dynamics.[113] Rac GTPases bind the downstream effector mDia1, a formin protein required for nucleation of unbranched actin filaments to facilitate enucleation.[114] Non-apoptotic activation of caspases has also been implicated in enucleation of erythroblasts.[115,116] In addition, a recent study indicated that enucleation of human and mouse erythroblasts is dependent upon formation and movement of cytoplasmic endocytic vesicles adjacent to the extruding nucleus.[117] During enucleation, the extruded nucleus is surrounded by a thin layer of cytoplasm and plasma membrane with assymetric distribution of resident proteins. For example, cell surface transferrin receptor and glycophorin A are preferentially redistributed to the reticulocyte.[114] Upon separation from the incipient reticulocyte, the extruded nucleus exposes phosphatidylserine on its plasma membrane, providing a recognition signal for engulfment and digestion by nearby macrophages (discussed below).[118]

Overall, several molecular pathways including remodeling of nuclear architecture, cytoskeletal reorganization, vesicle trafficking and features of apoptosis regulate nuclear expulsion during terminal erythroid maturation. However, a comprehensive understanding of how these features are coordinated and interrelated remains lacking.

3.3.4 *Maturation of the reticulocyte*

Maturation of human reticulocytes occurs over approximately 3 days in a process of dramatic remodeling that begins in the bone marrow and continues after release into the bloodstream.[119] This process includes complete loss of mitochondria, endocytic vesicles, ribosomes, Golgi cisternae and rough endoplasmic reticulum, as well as significant portions of the cell membrane and resident proteins.[120,121] Loss of these organelles occurs through numerous mechanisms including autophagy, exocytosis and proteolytic degradation.

3.3.4.1 Autophagy-mediated organelle degradation

Autophagy is a fundamental catabolic process in which cellular proteins and organelles are sequestered in a double-membrane vesicle called an autophagosome, which fuses with lysosomes where the enclosed matter is degraded by hydrolases.[122] Autophagy is important for the removal of damaged cellular components and also for normal development. Mitochondria within autophagosomes were observed in developing RBCs more than 40 years ago.[123,124] More recent studies provide substantial proof that mitochondrial clearance is partially dependent on autophagy and reveal erythroid-specific mechanisms of autophagy regulation (reviewed in).[125,126]

During late stage erythropoiesis, the BCL-2 family member Nix accumulates on the outer mitochondrial membrane where it promotes fusion to autophagosomes by binding two resident proteins LC3 and GABARAP.[127–130] $Nix^{-/-}$ erythrocytes retain their mitochondria but still form autophagosomes. Consequently, $Nix^{-/-}$ mice exhibit hemolytic anemia attributed to the production of reactive oxygen species by persistent mitochondria.[131,132]

Ulk1, a ubiquitously expressed serine-threonine kinase and member of the ATG-related family of autophagosomal proteins also functions in mitochondrial clearance.[133] $Ulk1^{-/-}$ mice display mild anemia with retained mitochondria in circulating reticulocytes. Interestingly, $Ulk1^{-/-}$ reticulocytes also exhibit impaired clearance of RNA-bound ribosomes. Both Nix and Ulk1 are dispensable for starvation-induced autophagy, suggesting more specific roles for these proteins in developmental clearance of erythroid organelles.

Two recent studies examined hematopoietic cells lacking Atg7, an essential autophagy protein. Lethally irradiated mice reconstituted with $Atg7^{-/-}$ fetal liver HSCs displayed anemia and impaired (but not completely blocked) erythroid mitochondrial clearance.[134] Similarly, loss of Atg7 through hematopoietic-specific gene deletion caused severe anemia.[134,135] $Atg7^{-/-}$ erythroid cells retained damaged mitochondria with altered membrane potential. Overall, it is now clear that autophagy facilitates erythroid mitochondrial clearance although additional studies are required to further elucidate the mechanisms. It will be particularly interesting to further characterize erythroid-specific adaptations of autophagy *versus* erythroid manifestations of universally conserved autophagic processes.

3.3.4.2 Autophagy-independent mitochondrial clearance

Erythroid mitochondria are also degraded by autophagy-independent proteolytic mechanisms. For example, the enzyme 15-lipooxygenase (15-LOX) integrates specifically into mitochondrial membranes, resulting in organelle permeabilization to components of the proteolytic system.[136] Inhibition of 15-LOX impairs mitochondrial clearance in reticulocytes.[137] Interestingly, 15-LOX mRNA is present at relatively high levels in early erythroid precursors, but is translationally silenced through interactions with RNA binding proteins that inhibit ribosome assembly.[138,139]

3.3.4.3 Reticulocyte remodeling by the exosome pathway

The characteristic biconcave shape and plasticity of the RBC are determined by the molecular makeup of the plasma membrane, which undergoes significant remodeling during terminal erythroid maturation.[101,140] This process results in approximately 20% loss of the cell surface area and a significant increase in the overall membrane stability.[112,141–143] Indeed, immediately after enucleation, the reticulocyte is still larger than an erythrocyte and has an irregular, multilobular shape (Fig. 3.1). Its remarkable transformation into the smaller, fully mature biconcave RBC appears to depend, at least in part, on the formation and subsequent ejection of vesicles, called exosomes, which contain obsolete membrane components and membrane-bound organelles. For example, this pathway facilitates the characteristic loss of transferrin receptors (CD71), which are no longer needed for iron acquisition once hemoglobin synthesis is complete.[144–147] Thus, transferrin receptors on the reticulocyte plasma membrane are internalized by clathrin coated pits, forming an endosome. Subsequently, small intralumenal vesicles, exosomes, are created by inward budding of the endosomal membrane. The resulting multivesicular endosome (MVE) fuses with the reticulocyte plasma membrane, ejecting the small exosomes containing the transferrin receptor. Other proteins, both transmembrane and cytosolic, are also removed from the maturing reticulocyte *via* the exosome pathway. These include β1 integrin, Na$^+$, K$^+$-ATPase, aquaporin-1, Glut-4 glucose transporter and Hsc70 molecular chaperone (148 and references therein). Exosome-mediated loss of integrins probably facilitates release of reticulocytes from erythroblastic islands and helps to prevent adhesion of mature RBCs to the vascular endothelium. Interestingly, exosome-mediated disposal of the aquaporin-1 water channel is regulated by the tonicity of the surrounding medium, suggesting a developmental response to osmotic stress.[149]

The mechanisms of protein sorting into exosomes are complex and poorly understood. In one pathway, ubiquitination provides a signal for unwanted proteins to be directed to the exosomes after the ubiquitin moiety is recognized by the components of the ESCRT protein complex in the endosomal membrane.[150,151] Other mechanisms include partitioning of proteins destined for disposal into lipid rafts on the endosomal membrane and transport of intralumenal endosomal proteins into exosomes (reviewed in).[148] The sphingolipid ceramide is believed to facilitate the formation of some exosomes.[152]

Of note, exosome formation was first discovered in reticulocytes[144] and is now believed to represent a major mechanism for cell remodeling and communication in many tissues. Interestingly, exosomes produced by a variety of non-erythroid cell types contain protein coding mRNAs and microRNAs.[153–158] These vesicles can be engulfed by recipient cells, where the donor RNAs modulate gene expression. In this fashion, reticulocyte exosomes could mediate molecular crosstalk with nearby cells in the hematopoietic niche.

3.3.5 The erythropoietic niche

3.3.5.1 Erythroblastic islands

Committed erythroid progenitors cultured *in vitro* with various cytokines (most importantly EPO) proliferate and mature, suggesting that this process is largely cell-autonomous. However, the efficiency of erythropoiesis *in vitro* is significantly lower than what occurs *in vivo*, indicating an important role for the bone marrow microenvironment or "niche" (reviewed in).[44] *In vivo*, erythrocytes form in the context of "erythroblastic islands" consisting of 5–30 erythroblasts surrounding a central macrophage, often termed the "nurse cell",[159] Erythroblastic islands are fragile and usually cannot be observed on bone marrow smears, although they reform *in vitro* in bone marrow cells cultured with extracellular matrix proteins.[160] Erythroblastic islands can also be observed directly in fresh preparations of fetal liver tissue.[43] The structure of erythroblastic islands is dynamic and highly regulated. As erythroblasts mature, they move toward the periphery of the island, and the expelled nuclei are engulfed by the macrophage (Fig 3.4). Although erythroblastic islands are detected throughout the bone marrow, a recent study suggested that the islands migrate toward sinusoids as their erythroblasts mature.[161] The density of erythroblastic islands correlates with erythropoietic output and is negatively regulated by the hematocrit. This regulatory loop was first demostrated in hallmark experiments showing that RBC hypertransfusion decreased dramatically the number of bone marrow erythroblastic islands.[162] A variety of positive and negative regulatory feedback mechanisms operate within the island. These include soluble factors, cell-cell and cell-extracellular matrix interactions that define the role of erythroblastic islands as the erythropoietic niche.

3.3.5.2 *Cell-cell interactions in the niche*

Numerous adhesion molecules mediate cell-cell interactions within the erythroblastic island, including Emp/Emp, $\alpha_4\beta_1$ integrin/VCAM-1 and ICAM-4/α_v.[163,164] Erythroblast macrophage protein (Emp) is present on both the erythroblast and the macrophage plasma membranes, facilitating erythroblast/macrophage association *via* a homophilic interaction.[165] Emp-null embryos die perinatally of severe anemia and Emp-deficient erythroblasts fail to enucleate even in the presence of wild-type macrophages.[166] Additionally, culturing erythroblasts in the presence of anti-Emp antibody causes a dramatic decrease in erythroid proliferation and maturation accompanied by an increase in apoptosis.[167] Similarly, an α_4 blocking antibody decreases the number of islands and the number of erythroblasts per island in fetal liver.[49] Interestingly, a secreted isoform of ICAM-4, termed ICAM-4S, is upregulated during terminal erythroid differentiation and has been shown to displace ICAM-4 from the α_v integrin

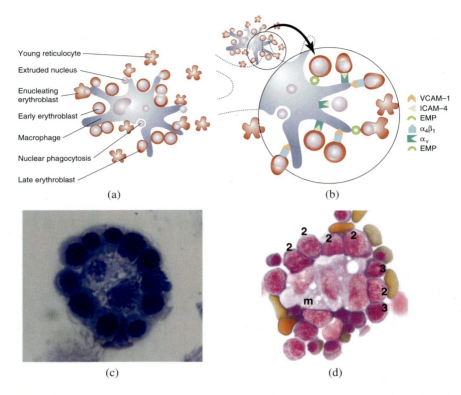

Figure 3.4 The erythroblastic island. (a) Cartoon diagramming the structure of the island with relevant cell types indicated in the left panel. (b) Cartoon showing some of the intermolecular interactions between macrophage and developing erythroid cells. (Panels A and B adapted from Ref. 44. (c) Micrograph of a bone marrow erythroblastic island with developing erythroid cells surrounding two central macrophages (courtesy of James Palis). (d) Micrograph of a mouse fetal liver (E15.5) erythroblastic island containing erythroblasts at a synchronous differentiation stage. m, central microphage; numbers indicate erythroblast developmental stages. Image reproduced from Ref. 43.

binding sites on macrophages.[168] Such competitive binding suggests a possible mechanism for the detachment of incipient RBCs from the erythroblastic islands.

During enucleation, Emp and β_1 integrin partition asymmetrically to the departing nucleus facilitating attachment to nearby macrophage.[169] In addition, the expelled nuclei express phosphatidylserine on the outer leaflet of the plasma membrane likely due to loss of ATP-dependent phospholipid transport. Together, these events mediate recognition and phagocytosis of erythroid nuclei by the central macrophages of the erythroblastic island.[118,170] Defects in erythroblastic island structure and function have been demonstrated in retinoblastoma tumor supressor protein (Rb)-null mice.[112–115] The anemia observed in Rb-null animals has been attributed to defects in both the central macrophage and erythroblasts, indicating autonomous and nonautonomous roles for Rb in erythroid maturation. However, this topic is controversial and complex, as other studies indicate that the role for Rb in erythropoiesis is largely cell autonomous.[171,172]

Direct contact with macrophages sustains high rates of erythroblast proliferation through EPO-independent mechanisms.[173] This interaction is clinically important, as illustrated by malignancy-induced anemias which are often unresponsive to EPO, perhaps due to perturbations in macrophage function, which may include secretion of inhibitory cytokines. In a mouse model, chemical depletion of macrophages disables erythropoietic response to bleeding.[174] Two macrophage proteins, macrophage membrane protein Ephrin-2 and SCF, interact with receptors on erythroid progenitors, EphB4 and c-Kit, respectively, to induce erythroid proliferation.[175–177] Interestingly, mice that are unable to synthesize the membrane-restricted form of SCF are anemic, despite relatively high levels of circulating SCF.[178] This underscores the importance of cytokine-mediated regulation through erythroblast-niche interactions. At the same time, erythroblasts engage in functionally important homotypic interactions within the niche. Thus, late stage orthochromatic erythroblasts express Fas ligand on their surface, which interacts with Fas death receptor on less mature erythroblasts, inducing their apoptosis *via* caspase-mediated cleavage of the critical erythroid transcription factor GATA-1.[179] This process is modulated by EPO which protects early erythroblasts from Fas ligand-induced toxicity. During erythropoiesis, this negative feedback loop may serve as one physiological mechanism to limit steady-state erythrocyte production which normally occurs at low EPO levels, while accelerating erythroid output during stress (Section 3.4.2 of this Chapter and Merav Socolovsky, personal communication). Similarly, Fas/Fas ligand interaction limits fetal liver erythroblast production so that it does not exceed the developmentally programmed rate.[43]

3.3.5.3 Extracellular matrix

Niche extracellular matrix proteins also modulate various aspects of erythropoiesis. Erythroblast surface integrins $\alpha_4\beta_1$ and $\alpha_5\beta_1$ bind several sites on fibronectin.[180,181] *In vitro*, attachment to fibronectin promotes proliferation of erythroblasts through EPO independent mechanisms.[182] In addition, fibronectin facilitates the maturation of cultured erythroid progenitors into reticulocytes.[183]

3.3.6 Balancing hemoglobin synthesis through protein quality control

The transcriptional/translational activity of the maturing red cell is dominated by the production of hemoglobin, which constitutes ~ 95% of total protein in mature erythrocytes.[184–186] Moreover, individual hemoglobin components including α and β globins, heme and iron are cytotoxic in their free forms. Consequently, the developing erythroblast must balance heme and globin production with acquisition of iron. As mentioned earlier, transferrin receptors are upregulated in late CFU-E to facilitiate iron uptake. The availability of heme is an important pre-requisite for the correct folding of globin chains, and some evidence suggests that heme binding occurs co-translationally.[187,188] Heme biosynthetic enzymes are induced during erythroid maturation, and heme synthesis is coordinated with the production of globin proteins. Specifically, nuclear factor Bach I and cytoplasmic heme regulated inhibitor of translation (HRI) repress globin transcription and translation, respectively, when the supply of heme in the erythroid cells is low.[189–192] In addition, erythroid cells must balance the production of individual hemoglobin subunits (reviewed in).[193] How this is achieved remains unknown and is particularly vexing, given the fact that α and β hemoglobin genes are located on different chromosomes in different chromatin environments.[61] The importance of balanced globin synthesis is illustrated by thalassemias, a group of genetic disorders characterized by deficiencies of globin production. In β-thalassemia, one or more β-globin genes are defective, causing accumulation of excess α-globin. In the absence of its binding partner, α-globin initiates an auto-oxidative cascade and ultimately precipitates, damaging erythroid precursors in a process termed ineffective erythropoiesis and reducing the lifespan of mature erythrocytes.[194,195]

Erythroid cells are able to mitigate harmful effects of globin chain imbalance. Indeed, 50% reduction of β-globin synthesis in β-globin gene haploinsufficiency (β-thalassemia trait) is reasonably well tolerated in humans. These protective mechanisms probably involve general protein quality control systems used by most cells to stabilize and/or degrade unstable toxic proteins.[193] Components of protein quality control include molecular chaperones, regulated proteolysis and autophagy. Both "public" (non-substrate-specific) and "private" (substrate-specific) molecular chaperones are implicated in the control of erythropoiesis (reviewed in).[196] Interestingly, some of these chaperones do not participate in hemoglobin quality control but instead perform other regulatory functions in erythroid development. For example, the ubiquitously expressed molecular chaperone Hsp70 mediates EPO-dependent erythroid survival by preventing caspase-mediated cleavage of GATA-1.[197] In addition, Hsp70 binds the pro-apoptotic mitochondrial protein AIF and prevents its transport to the nucleus where it induces DNA fragmentation and, ultimately, apoptotic cell death.[198]

3.3.6.1 α-hemoglobin-stabilizing protein

α-hemoglobin-stabilizing protein (AHSP) is an erythroid-specific molecular chaperone whose expression is induced by GATA-1 during late erythropoiesis.[199] AHSP is a private chaperone that specifically binds and stabilizes free α-globin *in vivo* and *in vitro*.[199–203] AHSP and β-globin occupy roughly the same binding interface on the surface of α-globin; thus, their binding is mutually exclusive.[202] In addition, AHSP induces a unique change in the heme binding pocket of α-globin, whereby heme iron is oxidized and coordinated at both axial positions by histidine side chains (bis-histidyl coordination). This conformation inhibits the heme-catalyzed production of reactive oxygen species, abating the oxidative damage caused by accumulation of free α-hemoglobin.[204] Kinetic measurements suggest that AHSP binding to α-hemoglobin is 20-fold faster but 10,000-fold weaker than binding of β-hemoglobin.[205,206] Thus, in an equimolar mix of α-hemoglobin, β-hemoglobin and AHSP the α-hemoglobin-AHSP complex forms initially, but α-hemoglobin is eventually displaced by β-hemoglobin to form HbA tetramers.[201,207] This suggests a role for AHSP as a transient stabilizer of α-hemoglobin, aiding in HbA assembly *in vivo*. Indeed, a role for AHSP in normal erythropoiesis is evident from studies of *Ahsp*[-/-] mice, which exhibit mild hemolytic anemia with hemoglobin precipitation.[199,208] In addition, loss of AHSP exacerbates β-thalassemia in mice.[208] In selected cases, gene polymorphisms or mutations that alter AHSP expression or function may influence the human β-thalassemia phenotype.[206,209,210] However, it is unlikely that the AHSP gene itself represents a major genetic modifier of β-thalassemia.[211,212]

3.3.6.2 *The ubiquitin/proteasome system*

Early experiments demonstrated that reticulocyte lysates have the capacity to degrade free α-hemoglobin while intact HbA tetramers are protected from degradation.[213] This activity was increased in β-thalassemia reticulocytes and blocked in the presence of protease inhibitors.[214,215] Interestingly, Nobel Prize winning experiments characterizing the ubiquitin/proteasome system in reticulocytes used modified or denatured hemoglobin as a convenient substrate.[216–218] These studies suggested that excess hemoglobin can be removed by proteolytic degradation. Reports by Shaeffer and colleagues[219–223] demonstrated that reticulocytes degrade excess α-hemoglobin in an ATP-dependent manner and that conjugation of α-hemoglobin chains to ubiquitin is essential for this process. Free β and γ globin were later shown to be degraded similarly.[224] Furthermore, intracellular aggregates of precipitated α-globin colocalize with ubiquitin in β-thalassemic erythroblasts.[225,226] Proteosome- and ATP-independent degradation of oxidant-damaged hemoglobin occurs in reticulocytes and mature erythrocytes, but the molecular identity of the relevant enzyme(s) remains unknown.[227,228] These studies strongly implicate ubiquitin-dependent and independent proteosome systems as essential components of erythroid protein quality control for hemoglobin synthesis. However, further studies are required to elucidate the associated mechanisms and their regulation during normal and pathological erythropoiesis. Long term manipulation of these systems may represent a therapeutic strategy for treating anemias associated with accumulation of unstable hemoglobins.

3.4 Regulation of Erythropoiesis

3.4.1 *Regulatory action of hormones and cytokines*

Increased erythropoietic output in anemia or hypoxia correlates with circulating EPO levels (see Chapter 6). In addition, early erythroid progenitors are responsive to multiple cytokines that influence proliferation and/or survival both positively (TPO, GM-CSF, IL-3, IL-11, SCF) and negatively (IL-6, TGF-β, TNF-α, IFN-γ, SCF). Many of these cytokines are produced by the stromal cells and macrophages in the erythropoietic niche. It is thought that high concentrations of the cytokines are present within the niche and that this may be due at least partly to binding of the cytokines to stromal cell membranes and/or extracellular matrix. It is likely that erythroid progenitors respond to the sum of cytokines present at the niche, imparting erythropoiesis with significant adaptive plasticity.

3.4.1.1 *Erythropoietin (see also Chapter 6)*

EPO is a 34 KDa glycoprotein containing approximately 40% sialic acid-rich carbohydrate by mass. It has a half-life of 7–8 hours in plasma, and its concentration can increase more than 1000-fold during severe hypoxia. Binding of EPO to its receptor (EPO-R) induces dimerization of the intracellular receptor domains and activation of the JAK2 tyrosine kinase, which phosphorylates itself and EPO-R initiating recruitement of multiple downstream signaling molecules, including STAT5, phosphoinositide 3-kinase, phospholipase Cγ, protein kinase C, Grb2, Shc and Ras. EPO stimulates erythropoiesis by promoting proliferation and survival of committed erythroid progenitors.[229,230] For example, Stat5, a signalling molecule in the EPO cascade, promotes survival of erythroblasts by inducing the antiapoptotic gene bcl-xl[106,231,232] and other effectors.[233,234]

During steady state erythropoiesis EPO dependence is restricted within the late CFU-E to late basophilic erythroblast stages. This is reflected in part by temporal expression of EPO-R. Indeed, BFU-E express only about 20–50 molecules of EPO-R on their cell surface. This number increases to 300–500 by the late CFU-E stage. Subsequently, EPO-R expression decreases precipitously as proerythroblasts mature, becoming undetectable in reticulocytes.[47,235–238] Currently, there is no convincing evidence that EPO influences lineage commitment.[93] Thus, $Epo^{-/-}$ and $EpoR^{-/-}$ embryos die by E13.5 with normal numbers of BFU-E and CFU-E that can be identified in colony assays after restoration of EpoR by retroviral transfer.[47] Maturation of erythroid precursors beyond the basophilic erythroblast can proceed without the support of EPO but is enhanced by adhesion to a fibronectin matrix, as well as other components of the erythropoietic niche[182] (see Section 3.5.3).

3.4.1.2 Stem cell factor (SCF)

Naturally occuring mutations in the genes encoding SCF and its tyrosine kinase receptor, c-Kit (CD117), cause severe anemia resulting from defects in both HSC maintenance and erythropoiesis (reviewed in).[239] Stem cell factor (SCF, Kit ligand) is produced by stromal cells in hematopoietic niches, including erythropoietic, whereas c-Kit is expressed on HSCs, multipotential progenitors, BFU-E and CFU-E. Binding of SCF to c-Kit induces receptor dimerization, autophosphorylation and activation of numerous signaling pathways, including phosphoinositide-3 kinase, Src kinases and phospholipase Cγ. SCF acts synergistically with EPO to promote expansion of erythroid progenitors, apparently *via* an interaction between c-Kit and EPO-R.[240] Expression of c-Kit declines as erythroid cells mature, in part due to repressive actions of the transcription factor GATA-1 on the Kit gene (see Section 3.4.3.1 below).

3.4.1.3 Other positively acting cytokines in the erythropoietic niche

Soluble cytokines secreted by bone marrow macrophages include burst-promoting activity (BPA) and insulin-like growth factor 1 (IGF-1), which induce proliferation of both BFU-E and CFU-E.[241,242] The recently discovered soluble factor Gas6 is secreted by erythroblasts in the erythroid niche in response to EPO. Binding of Gas6 to its receptors enhances EPO signaling in erythroblasts and prevents niche macrophages from secreting erythroid inhibitory factors.[243]

3.4.1.4 *Cytokines that inhibit erythropoiesis*

Chronic inflammation and malignancy are characterized by accumulation of many cytokines that inhibit erythropoiesis, including IL-6, TGF-β, TNF-α and IFN-γ.[44] For example, TNF-α may decrease proliferaton of erythroid progenitors or elicit an apoptotic response by inducing caspase-mediated cleavage of GATA-1.[179,244] IL-6 induces secretion of hepcidin, which impedes erythropoiesis indirectly by inhibiting enterocyte iron absorption and iron export from macrophages.[100,245]

3.4.2 Stress erythropoiesis

Under steady-state conditions, about 1% of circulating RBCs are replaced daily. This basal rate of erythropoiesis can increase up to 10-fold during insufficient tissue oxygen delivery, for example with blood loss or increased oxygen demands. Under these circumstances, reticulocytosis reflects increased erythropoietic output. This response, termed "stress erythropoiesis", is regulated differently from the steady state (reviewed in)[246] and is primarily driven by high concentrations of EPO with secondary input by glucocorticoids, SCF and bone morphogenetic protein 4 (Bmp4).[106,247–251] In mice, a unique set of BFU-E and CFU-E "stress progenitors" arise in the spleen.[249,250,252] These progenitors uniquely require Bmp4/Smad5 signaling. Flexed tail mice, which harbor a mutation in the Smad5 gene, lack splenic stress BFU-E and fail to mount adequate erythropoietic stress responses.[250]

3.4.2.1 *Unique actions of EPO in stress erythropoiesis*

Under erythropoietic stress, expression of EPO-R extends to virtually all stages of the hematopoietic hierarchy, including HSCs and BFU-E.[246,253,254] In contrast to bone marrow BFU-E, splenic stress BFU-E are

EPO-responsive, although proerythroblasts and basophilic erythroblasts remain the prominent sites of EPO action.[107] Mice with genetic defects that partially inhibit EPO-R signaling exhibit normal steady-state erythropoiesis, but impaired erythropoietic stress responses.[255,256] Indeed, it is estimated that less than 10% of EPO receptors are occupied during steady-state erythropoiesis, and this number increases markedly during stress.[257]

3.4.2.2 *EPO-dependent survival of stress progenitors*

As mentioned earlier, apoptosis of erythroid precursors is regulated by Fas/Fas ligand interactions within erythroblastic islands. Interestingly, the number of apoptotic, as well as Fas- or Fas ligand-expressing erythroblasts is significantly higher in the mouse spleen during steady-state erythropoiesis than in the bone marrow.[107] Roughly 60% of spleen erythroblasts appear apoptotic and die before reaching maturity. During stress, high levels of EPO downregulate Fas/Fas ligand on spleen erythroblasts, which promotes their survival and increases erythropoietic output. Thus, the mouse spleen appears to be a reserve erythropoietic organ containing a special population of Fas^+ erythroblasts that is continually replaced without reaching maturity. Recruiting these cells during stress by reducing the fraction undergoing apoptosis may provide a faster way of increasing erythropoietic output compared to generating additional cells *de novo* from HSCs. Of note, many aspects of stress erythropoiesis have been ascertained through studies of mice. The extent to which these mechanisms, for example, the roles of Bmp4/Smad5 and Fas signaling, apply to human stress erythropoiesis *in vivo* is unknown.

3.4.2.3 *Glucocorticoid hormones in stress erythropoiesis*

The glucocorticoid receptor (GR) represents an additional important regulator of stress erythropoiesis.[248,249,258,259] Activation of GR inhibits differentiation and promotes proliferation of erythroid progenitors *in vitro*.[94] This property has been exploited to expand erythroid progenitors *in vitro* (Section 3.5). *In vivo*, many stress conditions, including anemia, stimulate production of glucocorticoid hormones, which is predicted to enhance erythropoiesis. In support, targeted disruption of GR in mice inhibits stress erythropoiesis.[248]

3.4.3 *Control of erythropoiesis by nuclear factors*

Nuclear control of erythropoiesis conforms to a handful of fundamental general principles (reviewed in)[87]:

1. Transcription factors are modular proteins with distinct domains mediating DNA binding and interactions with other proteins;
2. Transcription factors function within multiprotein complexes and are critically dependent on chromatin environment;
3. Transcription factor function is modulated by various posttranslational modifications, such as phosphorylation, acetylation and ubiquitination;
4. Most transcription factors can both activate and repress gene expression, depending on the availability of binding partners and local chromatin context. Recent genome-wide studies of transcription factor chromatin occupancy and induced global transcription patterns highlight the complexity and combinatorial nature of transcription regulation in erythropoiesis.[260–266]

The important role of higher-order chromatin structures and chromatin modifications in transcription regulation, as well as an intimate association of the differentiation events with the cell cycle clock are increasingly recognized.[267–269] For example, a recent study demonstrated the presence of "transcription factories" in erythroid nuclei. These factories contain high local concentrations of the transcription factor KLF1, which is believed to coordinate recruitment and expression of numerous erythroid genes that are physically dispersed throughout the genome.[359] It is likely that other erythroid transcription factors also function in the highly organized context of transcription factories, providing for more efficient and coordinated control of gene expression (reviewed in).[269] Three major nuclear regulators of erythropoiesis: GATA-1, SCL and KLF1 (EKLF) are reviewed here. Other relevant erythroid transcription factors, including NF-E2 and Gfi-1, are described in recent reviews.[3,55,87]

3.4.3.1 *GATA-1 and GATA-2*

Six related vertebrate GATA family member proteins facilitate the development of various tissues. GATA-1 and GATA-2 primarily function in hematopoiesis and act sequentially during erythroid development. These proteins contain two highly conserved tandemly arranged zinc finger motifs that recognize DNA through a (T/A)GATA(A/G) consensus motif that is present in virtually all erythroid-specific genes.[270,271] The carboxyl (C) finger is necessary and sufficient for DNA binding while the amino (N) finger stabilizes binding to extended sequences consisting of repeated GATA consensus motifs.[272–275] Both zinc fingers facilitate protein–protein interactions with other nuclear factors (reviewed in).[86] Expression of GATA-2 predominates during the early stages of hematopoiesis and is essential for maintaining hematopoietic stem and progenitor cells.[276–278] In the later stages of erythroid development, GATA-2 expression declines and GATA-1 expression increases.[276,279] It is believed that GATA-2 initiates the erythroid program and becomes replaced by GATA-1 in terminal erythroid differentiation; this process is usually referred to as the "GATA" switch (reviewed in).[87,280,281] Replacement of GATA-1 for GATA-2 causes gene specific, sometimes opposing effects. For example, the Kit gene, which encodes the SCF receptor c-Kit, is actively expressed during early erythropoiesis and downregulated upon cell maturation. Kit expression is activated by GATA-2 and repressed by GATA-1 *via* alternate regulation of three dimensional chromatin loop conformations.[282] GATA-2 transcription itself is also inhibited by GATA-1,[283] while GATA-2 activates both its own expression and GATA-1 expression. In contrast, numerous other GATA factor-dependent gene loci, including globins and other erythroid-specific proteins, are strongly activated during the GATA factor switch. In these cases, GATA-2 occupancy may facilitate the initiation and/or maintenence of a permissive, open chromatin state with low level gene expression that is significantly enhanced upon replacement with GATA-1.

The GATA1 gene is X-linked and germline missense mutations cause dyserythropoietic anemia and thrombocytopenia in male patients. These mutations alter the N-zinc finger by inhibiting its ability to bind DNA or interact with partner proteins such as FOG1 (reviewed in).[281]

GATA1 null mouse embryos die between E10.5 and E11.5 of development due to severe anemia with a block to proerythroblast maturation in the primitive erythroid lineage.[284–286] GATA-1 is also important for the formation of platelets/megakaryocytes, eosinophils, mast cells[287] and dendritic cells.[288,289]

GATA-1 mediated gene activation and repression depends on physical interactions with numerous nuclear proteins, as well as with other genetic loci, by way of DNA looping. Both target specificity and repression/activation efficiency are regulated in this way. Recent genome-wide studies of GATA-1 chromatin occupancy provide important insights into the relative contribution of *cis*- and *trans*-regulatory elements in GATA-dependent regulation of gene expression and how gene activation *versus* repression is distinguished.[262,263,290,291] Most or all functions of GATA-1 (and GATA-2) in erythroid development require interaction with FOG-1.[292–294] Indeed, $Fog1^{-/-}$ mice display similar, albeit less severe, erythroid defects as those observed in GATA1 null embryos. FOG-1 facilitates the assembly of multi-protein complexes around DNA-bound GATA-1. For example, GATA-1 cooperates with FOG-1 to bring a distant enhancer (the locus control region) close to promoters at the β-globin locus resulting in gene activation.[295] FOG-1 can also facilitate gene repression, at least in part by recruiting co-repressors such as the NuRD complex and/or CtBP.[296,297] However, NuRD has also been found at GATA-1-activated genes and mechanisms that distinguish gene activation *versus* repression by GATA-FOG complexes are not fully understood.[297]

GATA-1 also interacts with the master myeloid transcription factor PU.1, which stimulates generation of monocytic, granulocytic, and lymphoid lineages.[298] Forced expression of PU.1 blocks erythroid differentiation and it is postulated that downregulation of PU.1 is necessary for erythroid lineage commitment.[299–301] Indeed, PU.1 and GATA-1 cross antagonize each other *via* direct protein interaction.[302,303] Additionally, GATA-1 inhibits expression of the PU.1/Sfpi1 gene.[304] It appears that the relative stoichiometries of PU.1 and GATA-1 in multipotent progenitors influence myeloid *versus* erythroid lineage commitment decisions. Thus, GATA-1 enhances erythropoiesis not only by inducing lineage-specific genes, but also by repressing the expression of alternate lineage genes, at least in part by inhibiting PU.1.

3.4.3.2 Stem cell leukemia (SCL, TAL1, TCL5)

SCL/TAL1, a basic helix-loop-helix type transcription factor, functions in association with a wide variety of partner proteins (reviewed in).[305] SCL/TAL1 was discovered as a rearranged gene in acute T-cell leukemia associated with the 11p13 chromosomal translocation.[306] It is expressed at high level in erythroid precursors and binds the E box consensus motif CANNTG.[307–309] E box motifs are found at numerous erythroid genes, including, KLF1, β-globin and GATA1.[310–313] SCL complexes can activate or repress transcription by recruiting coactivators (p300, CBP) or corepressors (Sin3A, Eto-2).[314–317] In erythroid cells, SCL is associated with a transcription factor complex containing GATA-1, LMO2, LDB1 and several other proteins.[318,319] This complex can assemble on DNA containing juxtaposed E-box and GATA consensus binding motifs. However, this transcription factor complex can also occupy GATA motifs lacking adjacent E-box sites. Moreover, gene complementation studies indicate that many functions of SCL are independent of its DNA binding activity.[260,261,263,290,320]

SCL is regarded to establish both primitive and definitive blood lineages in the embryo but is largely dispensable for maintenance of adult HSCs.[321–327] Specifically, SCL appears to induce hematopoietic committment of hemangioblasts.[328–330] Forced overexpression of SCL promotes erythroid differentiation in erythroleukemia cell lines and multipotential progenitor cells.[331,332] The erythroid defects observed upon ablation of SCL in adult HSCs resemble the phenotype caused by GATA-1 or FOG-1 loss, consistent with functional and physical interactions between these proteins. Recent studies indicate that co-occupancy with SCL generally distinguishes GATA-1 activated from GATA-1-repressed genes.[262,333]

3.4.3.3 KLF1 and related proteins

Erythroid Krüppel-Like Factor (EKLF, KLF1) is a member of a transcription factor family with homology to *Drosophila melanogaster* Krüppel embryonic pattern regulator (reviewed in).[334,335] Krüppel-like factors contain three zinc finger domains which allow binding to extended "CACCC box" motifs in numerous erythroid expressed genes. KLF1 is mainly an erythroid-specific transcription factor, with relatively low level expression in mast cells.[336,337] In humans, CACCC box mutations at the β-globin promoter abrogate binding of KLF1, causing β+ thalassemia.[338–341] Homozygous loss of KLF1 in mice causes severe anemia and embryonic lethality between E14 and E16 due to defects in definitive erythropoiesis.[342,343] Although KLF1 is expressed in all ontogenic stages of erythropoiesis and can be detected at embryonic globin gene loci, it appears to be dispensable for yolk sac erythropoiesis.[344] $Klf1^{-/-}$ erythrocytes display typical features of β thalassemia, including hypochromia, poikilocytosis, elevated α/β-globin ratio and Heinz bodies consisting of α-globin precipitates. Thus, it was initially believed that the predominant function of KLF1 is to drive the high-level expression of β-globin, in part *via* formation of a chromatin loop at the β-globin locus.[345] However, subsequent studies demonstrated that KLF1 plays a broader role in erythropoiesis.[342,346–348] Thus, rescue of β-globin expression failed to prevent embryonic lethality of $Klf1^{-/-}$ mice.[349] Subsequently, an extensive number of KLF1 target genes were identified by chromatin immunoprecipitation and gene expression profiling. These KLF1 target genes include α-globin, Ahsp and genes encoding erythroid membrane proteins, heme biosynthetic enzymes and cell cycle regulators.[265,346,350–354] The mouse hemolytic anemia mutation Nan has recently been traced to a single amino acid substitution in KLF1.[355] KLF1 appears to activate gene expression by recruiting the general transcription factors CBP/p300 and the SWI/SNF chromatin remodelling complex.[356–358]

Embryonic globins are expressed normally in $Klf1^{-/-}$ animals (reviewed in).[55,87] Moreover, γ globin synthesis is de-repressed in $Klf1^{-/-}$ mice containing a human β-globin locus transgene. These early findings, which suggested a role for KLF1 in hemoglobin switching, were verified by more recent studies showing that heterozygosity for a KLF1-null mutation causes HPFH in humans (with or without dyserythropoietic anemia) and that KLF1 regulates the expression of the γ globin repressor BCL11A[74,360–362] (discussed in Section 3.2.4.1). Interestingly, another study reported that only compound heterozygotes for two co-existing KLF1 mutations (S270X nonsense and K332Q missense) lead to HPFH, while patients with isolated KLF1 mutations have normal levels of HbF.[361]

Several additional Krüppel family members that are expressed more broadly than KLF1 also function in erythroid cells. For example, Fetal Krüppel-Like Factor (FKLF, KLF11) activates the ε-, γ- and β-globin gene promoters, but not other erythroid promoters that contain the consensus CACCC box.[354,363] Basic Krüppel-Like Factor (BKLF, KLF3), whose role in erythropoiesis is unclear, appears to be activated by KLF1.[364,365] Sp1 and Sp3 are expressed in multiple tissues and targeted disruption of these genes causes anemia along with multiple other developmental defects.[366,367] Thus, there is functional overlap and interplay between multiple KLF factors that are expressed during erythropoiesis. However, KLF1 appears to play an essential, predominant role.

3.4.4 *MicroRNAs and erythropoiesis*

MicroRNAs (miRNAs) are small (~22 nucleotide) RNAs that repress gene expression post-transcriptionally by binding to specific target mRNAs to induce their nucleolytic cleavage and/or inhibit translation (reviewed in).[368,369] Roughly 500-1000 different miRNAs encoded throughout mammalian genomes are believed to regulate virtually all aspects of tissue development. Mutations in the *Ago2* gene, which encodes a protein that mediates miRNA biogenesis and effector functions, produce anemia with impaired erythroid maturation in mice[370,371] and zebrafish.[372] Roles for specific miRNAs in erythropoiesis are now beginning to be defined (reviewed in).[91,373] For example, miR-451 facilitates terminal erythroid maturation and protects against oxidant stress.[90,374,375] Erythroblast enucleation is inhibited by miR-191, which is normally downregulated during terminal maturation.[376] However, hundreds of additional miRNAs are expressed dynamically during erythropoiesis.[376-382] How these miRNAs collectively orchestrate erythroid gene expression and terminal maturation is not understood.

MicroRNAs are also expressed in mature erythroblasts, which are transcriptionally and translationally inert.[383-385] These miRNAs may simply remain as remnants from erythroblasts, or more interestingly, could exert atypical functions in erythrocyte physiology. MicroRNA expression is dysregulated in numerous erythroid diseases including sickle cell anemia[384,386] and polycythemia vera,[377,387–389] although the mechanistic basis of these findings and their implications for disease pathophysiology are poorly understood.

3.5 Generation of RBCs *in vitro*

Detailed studies of erythropoiesis are greatly facilitated by the ability to recapitulate this process under defined conditions *in vitro*. This is usually achieved by culturing multipotent hematopoietic progenitors and/or committed erythroid progenitors in a mixture of nutrients, cytokines and hormones, typically EPO, SCF, dexamethasone and IGF-1. In this fashion, erythroid precursors can be expanded and differentiated relatively synchronously. Moreover, precursors are amenable to genetic manipulation *via* retroviruses or siRNAs in order to assess the effects of enforced overexpression or reduction of proteins or RNAs of interest. The most common species studied are human and mice, largely because appropriate cytokines and culture methods are readily available. Sources for mammalian primary erythroid progenitors include peripheral blood, fetal liver, bone marrow and spleen.[92,93,390–392] Fetal liver is highly enriched for committed erythroid progenitors and therefore provides a robust tissue source for studies of erythroid maturation. Typically, early murine fetal liver erythroid precursors are purified by lineage depletion, then cultured in various cytokines, most importantly EPO. Differentiation ensues over about three days, as evidenced by induction of the erythroid lineage surface marker Ter119, downregulation of transferrin receptor (CD71) and characteristic changes in cell morphology.[92] This approach has been used to elucidate various aspects of erythropoiesis including enucleation,[110,114] transcriptional regulation,[94,393] cell cycle control and chromatin modification,[93] and microRNAs.[374,394]

Purified peripheral blood CD34$^+$ cells also provide a source of hematopoietic progenitors for studies of erythropoiesis (reviewed in).[395,396] This approach produced large amounts of erythroblasts without the need for bone marrow sampling. Indeed, up to 5.8×10^8 erythroblasts at 90% purity can be produced from the buffy coat obtained from 1 unit of blood. A recent study indicates that the CD34⁻ fraction of buffy coat cells has an even higher erythroid expansion potential than CD34⁺ cells.[397] Human umbilical cord blood is another convenient source of expandable HSC/progenitors for studies of erythropoiesis.[398–400]

The expansion potential of erythroid cultures depends on the starting material and the growth conditions. SCF, IL-3 and Flt3 ligand induce proliferation of early progenitors and EPO can often be withheld for several days until terminal differentiation is desired. Perhaps the greatest expansion can be obtained when the stress erythropoiesis signaling is activated by glucocorticoids. In combination with SCF and EPO, dexamethasone delays differentiation and induces a dramatic proliferation of early committed erythroid progenitors.[401] Subsequent removal of dexamethasone and adjustment of cytokines induces rapid terminal differentiation. The use of feeder cells *in vitro* to recapitulate the erythropoietic niche enhances terminal differentiation of erythroblasts including enucleation.[402] However, it may be possible to generate mature, anucleate human erythrocytes efficiently without stromal cells.[398]

It is also possible to generate erythroid progeny from *in vitro* differentiation of human and mouse embryonic stem (ES) and induced pluripotent stem (iPS) cells[403–408] (reviewed in).[408–411] Use of this experimental system provides insights into the ontogeny of erythropoiesis, since embryonic stem cell differentiation recapitulates some early events of embryogenesis[412] (reviewed in).[4,413] Moreover, ES/iPS cells are capable of extended self-renewal, in contrast to primary tissues which exhibit more restricted expansion in culture. Since ES and iPS cells are particularly amenable to genetic modification, their use for blood production can potentially lead to the development of customized blood products, or "designer blood".[409] However, it is important to keep in mind that 1 unit of donor blood contains approximately 2×10^{12} erythrocytes. At this time, new technologies are required to efficiently and economically generate such quantities of cells *ex vivo* from any source of progenitors.[414]

3.6 Summary

Erythropoiesis is accompanied by dramatic change in cellular function, behaviour and appearance during the transition from HSC to RBC.

Despite the highly specialized nature of this process, studies of erythropoiesis have provided groundbreaking insights into major aspects of general biology and medicine. These include developmental control of gene expression, transcription factor function, chromatin structure, exosome formation, ubiquitin-mediated proteolysis, cytokine signaling and the genetic basis for human disease. Despite years of highly productive research, the field remains rich with new opportunities for scientific discovery and medical applications. Most of the cellular and genetic processes described within this chaper are not fully understood and the mechanisms of many worldwide anemias, including anemias in developing countries and anemias of the elderly, remain unknown.[415-418] Two practical problems of intense current interest are to develop new treatments for hemoglobinopathies and to optimize the *in vitro* production of erythrocytes for transfusion medicine. Certainly, achieving these clinical goals will be facilitated by a combination of translational studies and ongoing laboratory efforts to better understand the basic biology of erythropoiesis.

Acknowledgements

We thank Gerd Blobel, Alan Cantor, James Palis, Vijay Sankaran, Merav Socolovsky, and Nancy Speck for critical reading of this chapter and for helpful suggestions. We also thank James Palis and Merav Socolovsky for providing microphotographs for figures. We thank the many scientists who have contributed to the study of erythropoiesis and apologize to those whose work we did not cite due to space limitations or oversight.

Mitchell J. Weiss's studies on erythropoiesis are supported by the National Institutes of Health, The Lukemia and Lymphoma Society, The Jane Fisherman Grinberg Endowed Chair at The Children's Hospital of Philadelphia Research Institute and the DiGaetano family.

Bibliography

1. Dzierzak A. A developmental approach to erythropoiesis. In: Steinberg MH, Forget BJ, Higgs DR, Weatherall DJ, editors. Disorders of Hemoglobin: *Genetics, Pathophysiology and Clinical Management.* New York, NY: Cambridge University Press; 2009, pp. 3–23.

2. Dzierzak E, Speck NA. Of lineage and legacy: the development of mammalian hematopoietic stem cells. *Nat Immunol.* 2008;9(2):129–136.
3. Tsiftsoglou AS, Vizirianakis IS, Strouboulis J. Erythropoiesis: model systems, molecular regulators, and developmental programs. *IUBMB Life.* 2009;61(8):800–830.
4. Palis J. Ontogeny of erythropoiesis. *Curr Opin Hematol.* 2008; 15(3): 155–161.
5. Palis J, Malik J, McGrath KE, Kingsley PD. Primitive erythropoiesis in the mammalian embryo. *Int J Dev Biol.* 2010;54(6–7):1011–1018.
6. McGrath K, Palis J. Ontogeny of erythropoiesis in the mammalian embryo. *Curr Top Dev Biol.* 2008;82:1–22.
7. Ottersbach K, Smith A, Wood A, Göttgens B. Ontogeny of haematopoiesis: recent advances and open questions. *Br J Haematol.* 2010;148(3):343–355.
8. Russel ES, Bernstein SE. Blood and blood formation. In: Green EL, editor. *Biology of Laboratory Mouse.* 2nd ed. New York: McGraw-Hill; 1966, pp. 351–372.
9. Haar JL, Ackerman GA. A phase and electron microscopic study of vasculogenesis and erythropoiesis in the yolk sac of the mouse. *Anat Rec.* 1971;170(2):199–223.
10. Silver L, Palis J. Initiation of murine embryonic erythropoiesis: a spatial analysis. *Blood.* 1997;89(4):1154–1164.
11. Tavian M, Peault B. Embryonic development of the human hematopoietic system. *Int J Dev Biol.* 2005;49(2–3):243–250.
12. Park C, Ma YD, Choi K. Evidence for the hemangioblast. *Exp Hematol.* 2005;33(9):965–970.
13. Shalaby F, Rossant J, Yamaguchi TP, Gertsenstein M, Wu XF, Breitman ML, *et al.* Failure of blood-island formation and vasculogenesis in Flk-1-deficient mice. *Nature.* 1995;376(6535):62–66.
14. Fehling HJ, Lacaud G, Kubo A, Kennedy M, Robertson S, Keller G, *et al.* Tracking mesoderm induction and its specification to the hemangioblast during embryonic stem cell differentiation. *Development.* 2003;130(17):4217–4227.
15. Ferkowicz MJ, Yoder MC. Blood island formation: longstanding observations and modern interpretations. *Exp Hematol.* 2005;33(9):1041–1047.
16. Huber TL, Kouskoff V, Fehling HJ, Palis J, Keller G. Haemangioblast commitment is initiated in the primitive streak of the mouse embryo. *Nature.* 2004;432(7017):625–630.

17. Kingsley PD, Malik J, Fantauzzo KA, Palis J. Yolk sac-derived primitive erythroblasts enucleate during mammalian embryogenesis. *Blood.* 2004;104(1):19–25.
18. Lux CT, Yoshimoto M, McGrath K, Conway SJ, Palis J, Yoder MC. All primitive and definitive hematopoietic progenitor cells emerging before E10 in the mouse embryo are products of the yolk sac. *Blood.* 2008;111(7):3435–3438.
19. England SJ, McGrath KE, Frame JM, Palis J. Immature erythroblasts with extensive *ex vivo* self-renewal capacity emerge from the early mammalian fetus. *Blood.* 2011;117(9):2708–2717.
20. Palis J, Chan RJ, Koniski A, Patel R, Starr M, Yoder MC. Spatial and temporal emergence of high proliferative potential hematopoietic precursors during murine embryogenesis. *Proc Natl Acad Sci USA.* 2001;98(8):4528–4533.
21. Palis J, Robertson S, Kennedy M, Wall C, Keller G. Development of erythroid and myeloid progenitors in the yolk sac and embryo proper of the mouse. *Development.* 1999;126(22):5073–5084.
22. Moore MA, Metcalf D. Ontogeny of the haemopoietic system: yolk sac origin of *in vivo* and *in vitro* colony forming cells in the developing mouse embryo. *Br J Haematol.* 1970;18(3):279–296.
23. Toles JF, Chui DH, Belbeck LW, Starr E, Barker JE. Hemopoietic stem cells in murine embryonic yolk sac and peripheral blood. *Proc Natl Acad Sci USA.* 1989;86(19):7456–7459.
24. Huang H, Auerbach R. Identification and characterization of hematopoietic stem cells from the yolk sac of the early mouse embryo. *Proc Natl Acad Sci USA.* 1993;90(21):10110–10114.
25. Gekas C, Dieterlen-Lievre F, Orkin SH, Mikkola HK. The placenta is a niche for hematopoietic stem cells. *Dev Cell.* 2005;8(3):365–375.
26. Gekas C, Rhodes KE, Van Handel B, Chhabra A, Ueno M, Mikkola HK. Hematopoietic stem cell development in the placenta. *Int J Dev Biol.* 2010;54(6–7):1089–1098.
27. Ottersbach K, Dzierzak E. The murine placenta contains hematopoietic stem cells within the vascular labyrinth region. *Dev Cell.* 2005 Mar;8(3):377–387.
28. Ottersbach K, Dzierzak E. The placenta as a haematopoietic organ. *Int J Dev Biol.* 2010;54(6–7):1099–1106.

29. Ottersbach K, Smith A, Wood A, Gottgens B. Ontogeny of haematopoiesis: recent advances and open questions. *Br J Haematol.* 2009;148(3): 343–355.
30. Samokhvalov IM, Samokhvalova NI, Nishikawa S. Cell tracing shows the contribution of the yolk sac to adult haematopoiesis. *Nature.* 2007; 446(7139):1056–1061.
31. Medvinsky A, Dzierzak E. Definitive hematopoiesis is autonomously initiated by the AGM region. *Cell.* 1996;86(6):897–906.
32. Muller AM, Medvinsky A, Strouboulis J, Grosveld F, Dzierzak E. Development of hematopoietic stem cell activity in the mouse embryo. *Immunity.* 1994;1(4):291–301.
33. de Bruijn MF, Ma X, Robin C, Ottersbach K, Sanchez MJ, Dzierzak E. Hematopoietic stem cells localize to the endothelial cell layer in the midgestation mouse aorta. *Immunity.* 2002;16(5):673–683.
34. Labastie MC, Cortes F, Romeo PH, Dulac C, Peault B. Molecular identity of hematopoietic precursor cells emerging in the human embryo. *Blood.* 1998;92(10):3624–3635.
35. Garcia-Porrero JA, Godin IE, Dieterlen-Lievre F. Potential intraembryonic hemogenic sites at pre-liver stages in the mouse. *Anat Embryol (Berl).* 1995;192(5):425–435.
36. Yokomizo T, Dzierzak E. Three-dimensional cartography of hematopoietic clusters in the vasculature of whole mouse embryos. *Development.* 2010; 137(21):3651–3661.
37. Sheng G. Primitive and definitive erythropoiesis in the yolk sac: a birds eye view. *Int J Dev Biol.* 2010;54(6–7):1033–1043.
38. Kovach JS, Marks PA, Russell ES, Epler H. Erythroid cell development in fetal mice: ultrastructural characteristics and hemoglobin synthesis. *J Mol Biol.* 1967;25(1):131–142.
39. Rifkind RA, Chui D, Epler H. An ultrastructural study of early morphogenetic events during the establishment of fetal hepatic erythropoiesis. *J Cell Biol.* 1969;40(2):343–365.
40. Orkin SH, Zon LI. Hematopoiesis: an evolving paradigm for stem cell biology. *Cell.* 2008;132(4):631–644.
41. Fassler R, Meyer M. Consequences of lack of beta 1 integrin gene expression in mice. *Genes Dev.* 1995;9(15):1896–1908.

42. Hirsch E, Iglesias A, Potocnik AJ, Hartmann U, Fassler R. Impaired migration but not differentiation of haematopoietic stem cells in the absence of beta1 integrins. *Nature*. 1996;380(6570):171–175.
43. Socolovsky M, Murrell M, Liu Y, Pop R, Porpiglia E, Levchenko A. Negative autoregulation by FAS mediates robust fetal erythropoiesis. *PLoS Biol*. 2007;5(10):e252.
44. Chasis JA, Mohandas N. Erythroblastic islands: niches for erythropoiesis. *Blood*. 2008;112(3):470–478.
45. Papayannopoulou T, D'Andrea AD, Abkowitz JL, Migliaccio AR. Biology of erythropoiesis, erythroid differentiation, and maturation. In: Hoffman R, Benz EJ, Jr., Shattil SJ, Furie B, Cohen HJ, Silberstein LE, *et al*., editors. Hematology: *Basic Principles and Practice*. Philadelphia, PA: Churchill Livingstone Elsevier; 2009.
46. Lin CS, Lim SK, D'Agati V, Costantini F. Differential effects of an erythropoietin receptor gene disruption on primitive and definitive erythropoiesis. *Genes Dev*. 1996;10(2):154–164.
47. Wu H, Liu X, Jaenisch R, Lodish HF. Generation of committed erythroid BFU-E and CFU-E progenitors does not require erythropoietin or the erythropoietin receptor. *Cell*. 1995;83(1):59–67.
48. Papayannopoulou T, Finch CA. On the *in vivo* action of erythropoietin: a quantitative analysis. *J Clin Invest*. 1972;51(5):1179–1185.
49. McGrath KE, Kingsley PD, Koniski AD, Porter RL, Bushnell TP, Palis J. Enucleation of primitive erythroid cells generates a transient population of "pyrenocytes" in the mammalian fetus. *Blood*. 2008;111(4):2409–2417.
50. Van Handel B, Prashad SL, Hassanzadeh-Kiabi N, Huang A, Magnusson M, Atanassova B, *et al*. The first trimester human placenta is a site for terminal maturation of primitive erythroid cells. *Blood*. 2010;116(17):3321–3330.
51. Fraser ST, Isern J, Baron MH. Maturation and enucleation of primitive erythroblasts during mouse embryogenesis is accompanied by changes in cell-surface antigen expression. *Blood*. 2007;109(1):343–352.
52. Boyer SH, Belding TK, Margolet L, Noyes AN. Fetal hemoglobin restriction to a few erythrocytes (F cells) in normal human adults. *Science*. 1975;188(4186):361–363.
53. Thein SL, Menzel S. Discovering the genetics underlying foetal haemoglobin production in adults. *Br J Haematol*. 2009;145(4):455–467.

54. Papayannopoulou T, Brice M, Stamatoyannopoulos G. Hemoglobin F synthesis *in vitro*: evidence for control at the level of primitive erythroid stem cells. *Proc Natl Acad Sci USA*. 1977;74(7):2923–2927.
55. Sankaran VG, Xu J, Orkin SH. Advances in the understanding of haemoglobin switching. *Br J Haematol*. 2010;149(2):181–194.
56. Johnson RM, Buck S, Chiu CH, Gage DA, Shen TL, Hendrickx AG, et al. Humans and old world monkeys have similar patterns of fetal globin expression. *J Exp Zool*. 2000;288(4):318–326.
57. Johnson RM, Gumucio D, Goodman M. Globin gene switching in primates. *Comp Biochem Physiol A Mol Integr Physiol*. 2002;133(3):877–883.
58. Dillon N, Grosveld F. Human gamma-globin genes silenced independently of other genes in the beta-globin locus. *Nature*. 1991;350(6315):252–254.
59. Strouboulis J, Dillon N, Grosveld F. Developmental regulation of a complete 70-kb human beta-globin locus in transgenic mice. *Genes Dev*. 1992;6(10):1857–1864.
60. Sankaran VG, Xu J, Ragoczy T, Ippolito GC, Walkley CR, Maika SD, et al. Developmental and species-divergent globin switching are driven by BCL11A. *Nature*. 2009;460(7259):1093–1097.
61. Stamatoyannopoulos G. Control of globin gene expression during development and erythroid differentiation. *Exp Hematol*. 2005;33(3):259–271.
62. Weatherall DJ. Phenotype-genotype relationships in monogenic disease: lessons from the thalassaemias. *Nat Rev Genet*. 2001;2(4):245–255.
63. Platt OS. Hydroxyurea for the treatment of sickle cell anemia. *N Engl J Med*. 2008;358(13):1362–1369.
64. Platt OS, Orkin SH, Dover G, Beardsley GP, Miller B, Nathan DG. Hydroxyurea enhances fetal hemoglobin production in sickle cell anemia. *J Clin Invest*. 1984;74(2):652–656.
65. Trompeter S, Roberts I. Haemoglobin F modulation in childhood sickle cell disease. *Br J Haematol*. 2009;144(3):308–316.
66. Thein SL, Menzel S, Peng X, Best S, Jiang J, Close J, et al. Intergenic variants of HBS1L-MYB are responsible for a major quantitative trait locus on chromosome 6q23 influencing fetal hemoglobin levels in adults. *Proc Natl Acad Sci USA*. 2007;104(27):11346–11351.
67. Lettre G, Sankaran VG, Bezerra MA, Araujo AS, Uda M, Sanna S, et al. DNA polymorphisms at the BCL11A, HBS1L-MYB, and beta-globin loci

associate with fetal hemoglobin levels and pain crises in sickle cell disease. *Proc Natl Acad Sci USA*. 2008 Aug 19;105(33):11869–11874.
68. So CC, Song YQ, Tsang ST, Tang LF, Chan AY, Ma ES, *et al*. The HBS1L-MYB intergenic region on chromosome 6q23 is a quantitative trait locus controlling fetal haemoglobin level in carriers of beta-thalassaemia. *J Med Genet*. 2008;45(11):745–751.
69. Uda M, Galanello R, Sanna S, Lettre G, Sankaran VG, Chen W, *et al*. Genome-wide association study shows BCL11A associated with persistent fetal hemoglobin and amelioration of the phenotype of beta-thalassemia. *Proc Natl Acad Sci USA*. 2008;105(5):1620–1625.
70. Menzel S, Garner C, Gut I, Matsuda F, Yamaguchi M, Heath S, *et al*. A QTL influencing F cell production maps to a gene encoding a zinc-finger protein on chromosome 2p15. *Nat Genet*. 2007;39(10):1197–1199.
71. Nuinoon M, Makarasara W, Mushiroda T, Setianingsih I, Wahidiyat PA, Sripichai O, *et al*. A genome-wide association identified the common genetic variants influence disease severity in beta0-thalassemia/hemoglobin E. *Hum Genet*. 2009;127(3):303–314.
72. Galanello R, Sanna S, Perseu L, Sollaino MC, Satta S, Lai ME, *et al*. Amelioration of Sardinian beta thalassemia by genetic modifiers. *Blood*. 2009;114(18):3935–3937.
73. Sankaran VG, Menne TF, Xu J, Akie TE, Lettre G, Van Handel B, *et al*. Human fetal hemoglobin expression is regulated by the developmental stage-specific repressor BCL11A. *Science*. 2008;322(5909):1839–1842.
74. Borg J, Papadopoulos P, Georgitsi M, Gutierrez L, Grech G, Fanis P, *et al*. Haploinsufficiency for the erythroid transcription factor KLF1 causes hereditary persistence of fetal hemoglobin. *Nat Genet*. 2010;42(9): 801–805.
75. Sankaran VG, Menne TF, Scepanovic D, Vergilio JA, Ji P, Kim J, *et al*. MicroRNA-15a and -16-1 act *via* MYB to elevate fetal hemoglobin expression in human trisomy 13. *Proc Natl Acad Sci USA*. 2011;108(4):1519–1524.
76. Bradford GB, Williams B, Rossi R, Bertoncello I. Quiescence, cycling, and turnover in the primitive hematopoietic stem cell compartment. *Exp Hematol*. 1997;25(5):445–453.
77. Cheshier SH, Morrison SJ, Liao X, Weissman IL. *In vivo* proliferation and cell cycle kinetics of long-term self-renewing hematopoietic stem cells. *Proc Natl Acad Sci USA*. 1999;96(6):3120–3125.

78. Passegue E, Wagers AJ, Giuriato S, Anderson WC, Weissman IL. Global analysis of proliferation and cell cycle gene expression in the regulation of hematopoietic stem and progenitor cell fates. *J Exp Med.* 2005;202(11): 1599–1611.
79. Murre C. Defining the pathways of early adult hematopoiesis. *Cell Stem Cell.* 2007;1(4):357–8.
80. Akashi K, Traver D, Miyamoto T, Weissman IL. A clonogenic common myeloid progenitor that gives rise to all myeloid lineages. *Nature.* 2000;404(6774):193–197.
81. Kondo M, Weissman IL, Akashi K. Identification of clonogenic common lymphoid progenitors in mouse bone marrow. *Cell.* 1997;91(5):661–672.
82. Manz MG, Miyamoto T, Akashi K, Weissman IL. Prospective isolation of human clonogenic common myeloid progenitors. *Proc Natl Acad Sci USA.* 2002;99(18):11872–11877.
83. Adolfsson J, Mansson R, Buza-Vidas N, Hultquist A, Liuba K, Jensen CT, *et al.* Identification of Flt3$^+$ lympho-myeloid stem cells lacking erythro-megakaryocytic potential a revised road map for adult blood lineage commitment. *Cell.* 2005;121(2):295–306.
84. Mansson R, Hultquist A, Luc S, Yang L, Anderson K, Kharazi S, *et al.* Molecular evidence for hierarchical transcriptional lineage priming in fetal and adult stem cells and multipotent progenitors. *Immunity.* 2007;26(4):407–419.
85. Pronk CJ, Rossi DJ, Mansson R, Attema JL, Norddahl GL, Chan CK, *et al.* Elucidation of the phenotypic, functional, and molecular topography of a myeloerythroid progenitor cell hierarchy. *Cell Stem Cell.* 2007;1(4): 428–442.
86. Cantor AB, Orkin SH. Transcriptional regulation of erythropoiesis: an affair involving multiple partners. *Oncogene.* 2002;21(21):3368–3376.
87. Blobel GA, Weiss MJ. Nuclear factors that regulate erythropoiesis. In: Steinberg MH, Forget BJ, Higgs DR, Weatherall DJ, editors. Disorders of Hemoglobin: Genetics, *Pathophysiology and Clinical Management.* New York, NY: Cambridge University Press; 2009, pp. 62–85.
88. Emambokus N, Vegiopoulos A, Harman B, Jenkinson E, Anderson G, Frampton J. Progression through key stages of haemopoiesis is dependent on distinct threshold levels of c-Myb. *EMBO J.* 2003;22(17):4478–4488.
89. Carpinelli MR, Hilton DJ, Metcalf D, Antonchuk JL, Hyland CD, Mifsud SL, *et al.* Suppressor screen in Mpl–/– mice: c-Myb mutation causes

supraphysiological production of platelets in the absence of thrombopoietin signaling. *Proc Natl Acad Sci USA*. 2004;101(17):6553–6558.
90. Yu D, dos Santos CO, Zhao G, Jiang J, Amigo JD, Khandros E, et al. miR-451 protects against erythroid oxidant stress by repressing 14-3-3zeta. *Genes Dev*. 2010;24(15):1620–1633.
91. Zhao G, Yu D, Weiss MJ. MicroRNAs in erythropoiesis. *Curr Opin Hematol*. 2010;17(3):155–162.
92. Zhang J, Socolovsky M, Gross AW, Lodish HF. Role of ras signaling in erythroid differentiation of mouse fetal liver cells: functional analysis by a flow cytometry-based novel culture system. *Blood*. 2003;102(12):3938–3946.
93. Pop R, Shearstone JR, Shen Q, Liu Y, Hallstrom K, Koulnis M, et al. A key commitment step in erythropoiesis is synchronized with the cell cycle clock through mutual inhibition between PU.1 and S-phase progression. *PLoS Biol*. 2010;8(9):e1000484.
94. Flygare J, Rayon Estrada V, Shin C, Gupta S, Lodish HF. HIF-1 {alpha} synergizes with glucocorticoids to promote BFU-E progenitor self-renewal. *Blood*. 2010;117(12):3435–3444.
95. Migliaccio A, Papayannopoulou T. Erythropoiesis. In: Steinberg MH, Forget BG, Higgs DR, Nagel RL, editors. Disorders of Hemoglobin: Genetics, *Pathophysiology and Clinical Management*. Cambridge, UK: Cambridge University Press; 2001, pp. 52–71.
96. To LB, Haylock DN, Simmons PJ, Juttner CA. The biology and clinical uses of blood stem cells. *Blood*. 1997;89(7):2233–2258.
97. Testa NG. Structure and regulation of the erythroid system at the level of progenitor cells. *Crit Rev Oncol Hematol*. 1989;9(1):17–35.
98. Broxmeyer HE, Williams DE. The production of myeloid blood cells and their regulation during health and disease. *Crit Rev Oncol Hematol*. 1988;8(3):173–226.
99. Keller JR, Mantel C, Sing GK, Ellingsworth LR, Ruscetti SK, Ruscetti FW. Transforming growth factor beta 1 selectively regulates early murine hematopoietic progenitors and inhibits the growth of IL-3-dependent myeloid leukemia cell lines. *J Exp Med*. 1988;168(2):737–750.
100. Means RT. Hepcidin and cytokines in anaemia. *Hematology*. 2004;9(5–6): 357–362.
101. Mohandas N, Gallagher PG. Red cell membrane: past, present, and future. *Blood*. 2008;112(10):3939–3948.

102. Yamamoto ML, Clark TA, Gee SL, Kang JA, Schweitzer AC, Wickrema A, et al. Alternative pre-mRNA splicing switches modulate gene expression in late erythropoiesis. *Blood*. 2009;113(14):3363–3370.
103. Hou VC, Conboy JG. Regulation of alternative pre-mRNA splicing during erythroid differentiation. *Curr Opin Hematol*. 2001;8(2):74–79.
104. Iacopetta BJ, Morgan EH, Yeoh GC. Transferrin receptors and iron uptake during erythroid cell development. *Biochim Biophys Acta*. 1982;687(2):204–210.
105. Huebers HA, Finch CA. The physiology of transferrin and transferrin receptors. *Physiol Rev*. 1987;67(2):520–582.
106. Socolovsky M, Nam H, Fleming MD, Haase VH, Brugnara C, Lodish HF. Ineffective erythropoiesis in Stat5a(−/−)5b(−/−) mice due to decreased survival of early erythroblasts. *Blood*. 2001;98(12):3261–3273.
107. Liu Y, Pop R, Sadegh C, Brugnara C, Haase VH, Socolovsky M. Suppression of Fas-FasL coexpression by erythropoietin mediates erythroblast expansion during the erythropoietic stress response *in vivo*. *Blood*. 2006;108(1):123–133.
108. Chen K, Liu J, Heck S, Chasis JA, An X, Mohandas N. Resolving the distinct stages in erythroid differentiation based on dynamic changes in membrane protein expression during erythropoiesis. *Proc Natl Acad Sci USA*. 2009;106(41):17413–17418.
109. Ji P, Yeh V, Ramirez T, Murata-Hori M, Lodish HF. HDAC2 is required for chromatin condensation and subsequent enucleation of cultured mouse fetal erythroblasts. *Haematologica*. 2010;95(12):2013–2021.
110. Jayapal SR, Lee KL, Ji P, Kaldis P, Lim B, Lodish HF. Downregulation of MYC is essential for terminal erythroid maturation. *J Biol Chem*. 2010;285(51):40252–40265.
111. Koury ST, Koury MJ, Bondurant MC. Cytoskeletal distribution and function during the maturation and enucleation of mammalian erythroblasts. *J Cell Biol*. 1989;109(6 Pt 1):3005–3013.
112. Chasis JA, Prenant M, Leung A, Mohandas N. Membrane assembly and remodeling during reticulocyte maturation. *Blood*. 1989;74(3):1112–1120.
113. Ji P, Lodish HF. Rac GTPases play multiple roles in erythropoiesis. *Haematologica*. 2010;95(1):2–4.
114. Ji P, Jayapal SR, Lodish HF. Enucleation of cultured mouse fetal erythroblasts requires Rac GTPases and mDia2. *Nat Cell Biol*. 2008;10(3):314–321.

115. Carlile GW, Smith DH, Wiedmann M. Caspase-3 has a nonapoptotic function in erythroid maturation. *Blood*. 2004;103(11):4310–4316.
116. Zermati Y, Garrido C, Amsellem S, Fishelson S, Bouscary D, Valensi F, et al. Caspase activation is required for terminal erythroid differentiation. *J Exp Med*. 2001;193(2):247–254.
117. Keerthivasan G, Small S, Liu H, Wickrema A, Crispino JD. Vesicle trafficking plays a novel role in erythroblast enucleation. *Blood*. 2010;116(17):3331–3340.
118. Yoshida H, Kawane K, Koike M, Mori Y, Uchiyama Y, Nagata S. Phosphatidylserine-dependent engulfment by macrophages of nuclei from erythroid precursor cells. *Nature*. 2005;437(7059):754–758.
119. Mel HC, Prenant M, Mohandas N. Reticulocyte motility and form: studies on maturation and classification. *Blood*. 1977;49(6):1001–1009.
120. Gronowicz G, Swift H, Steck TL. Maturation of the reticulocyte *in vitro*. *J Cell Sci*. 1984 Oct;71:177–197.
121. Koury MJ, Koury ST, Kopsombut P, Bondurant MC. *In vitro* maturation of nascent reticulocytes to erythrocytes. *Blood*. 2005;105(5):2168–2174.
122. Yang Z, Klionsky DJ. Mammalian autophagy: core molecular machinery and signaling regulation. *Curr Opin Cell Biol*. 2010;22(2):124–131.
123. Kent G, Minick OT, Volini FI, Orfei E. Autophagic vacuoles in human red cells. *Am J Pathol*. 1966;48(5):831–857.
124. Heynen MJ, Tricot G, Verwilghen RL. Autophagy of mitochondria in rat bone marrow erythroid cells. Relation to nuclear extrusion. *Cell Tissue Res*. 1985;239(1):235–239.
125. Mortensen M, Ferguson DJ, Simon AK. Mitochondrial clearance by autophagy in developing erythrocytes: clearly important, but just how much so? *Cell Cycle*. 2010;9(10):1901–1906.
126. Mizushima N, Levine B. Autophagy in mammalian development and differentiation. *Nat Cell Biol*. 2010;12(9):823–830.
127. Novak I, Kirkin V, McEwan DG, Zhang J, Wild P, Rozenknop A, et al. Nix is a selective autophagy receptor for mitochondrial clearance. *EMBO Rep*. 2010;11(1):45–51.
128. Kanki T. Nix, a receptor protein for mitophagy in mammals. *Autophagy*. 2010;6(3):433–435.

129. Sandoval H, Thiagarajan P, Dasgupta SK, Schumacher A, Prchal JT, Chen M, et al. Essential role for Nix in autophagic maturation of erythroid cells. *Nature*. 2008;454(7201):232–235.
130. Zhang J, Ney PA. NIX induces mitochondrial autophagy in reticulocytes. *Autophagy*. 2008;4(3):354–356.
131. Diwan A, Koesters AG, Odley AM, Pushkaran S, Baines CP, Spike BT, et al. Unrestrained erythroblast development in Nix–/– mice reveals a mechanism for apoptotic modulation of erythropoiesis. *Proc Natl Acad Sci USA*. 2007;104(16):6794–6799.
132. Schweers RL, Zhang J, Randall MS, Loyd MR, Li W, Dorsey FC, et al. NIX is required for programmed mitochondrial clearance during reticulocyte maturation. *Proc Natl Acad Sci USA*. 2007;104(49):19500–19505.
133. Kundu M, Lindsten T, Yang CY, Wu J, Zhao F, Zhang J, et al. Ulk1 plays a critical role in the autophagic clearance of mitochondria and ribosomes during reticulocyte maturation. *Blood*. 2008;112(4):1493–1502.
134. Zhang J, Randall MS, Loyd MR, Dorsey FC, Kundu M, Cleveland JL, et al. Mitochondrial clearance is regulated by Atg7-dependent and -independent mechanisms during reticulocyte maturation. *Blood*. 2009;114(1):157–164.
135. Mortensen M, Ferguson DJ, Edelmann M, Kessler B, Morten KJ, Komatsu M, et al. Loss of autophagy in erythroid cells leads to defective removal of mitochondria and severe anemia *in vivo*. *Proc Natl Acad Sci USA*. 2010;107(2):832–837.
136. van Leyen K, Duvoisin RM, Engelhardt H, Wiedmann M. A function for lipoxygenase in programmed organelle degradation. *Nature*. 1998;395(6700):392–395.
137. Grullich C, Duvoisin RM, Wiedmann M, van Leyen K. Inhibition of 15-lipoxygenase leads to delayed organelle degradation in the reticulocyte. *FEBS Lett*. 2001;489(1):51–54.
138. Naarmann IS, Harnisch C, Muller-Newen G, Urlaub H, Ostareck-Lederer A, Ostareck DH. DDX6 recruits translational silenced human reticulocyte 15-lipoxygenase mRNA to RNP granules. *RNA*. 2010;16(11):2189–2204.
139. Ostareck DH, Ostareck-Lederer A, Shatsky IN, Hentze MW. Lipoxygenase mRNA silencing in erythroid differentiation: the 3'UTR regulatory complex controls 60S ribosomal subunit joining. *Cell*. 2001;104(2):281–290.

140. Liu J, Guo X, Mohandas N, Chasis JA, An X. Membrane remodeling during reticulocyte maturation. *Blood*. 2010;115(10):2021–2027.
141. Waugh RE, McKenney JB, Bauserman RG, Brooks DM, Valeri CR, Snyder LM. Surface area and volume changes during maturation of reticulocytes in the circulation of the baboon. *J Lab Clin Med*. 1997;129(5):527–535.
142. Come SE, Shohet SB, Robinson SH. Surface remodelling of reticulocytes produced in response to erythroid stress. *Nat New Biol*. 1972;236(66): 157–158.
143. Waugh RE, Mantalaris A, Bauserman RG, Hwang WC, Wu JH. Membrane instability in late-stage erythropoiesis. *Blood*. 2001;97(6):1869–1875.
144. Pan BT, Johnstone RM. Fate of the transferrin receptor during maturation of sheep reticulocytes *in vitro*: selective externalization of the receptor. *Cell*. 1983;33(3):967–978.
145. Johnstone RM, Adam M, Hammond JR, Orr L, Turbide C. Vesicle formation during reticulocyte maturation. Association of plasma membrane activities with released vesicles (exosomes). *J Biol Chem*. 1987;262(19): 9412–9420.
146. Harding C, Heuser J, Stahl P. Receptor-mediated endocytosis of transferrin and recycling of the transferrin receptor in rat reticulocytes. *J Cell Biol*. 1983;97(2):329–339.
147. Pan BT, Teng K, Wu C, Adam M, Johnstone RM. Electron microscopic evidence for externalization of the transferrin receptor in vesicular form in sheep reticulocytes. *J Cell Biol*. 1985;101(3):942–948.
148. Blanc L, Vidal M. Reticulocyte membrane remodeling: contribution of the exosome pathway. *Curr Opin Hematol*. 2010;17(3):177–183.
149. Blanc L, Liu J, Vidal M, Chasis JA, An X, Mohandas N. The water channel aquaporin-1 partitions into exosomes during reticulocyte maturation: implication for the regulation of cell volume. *Blood*. 2009;114(18): 3928–3934.
150. Raiborg C, Rusten TE, Stenmark H. Protein sorting into multivesicular endosomes. *Curr Opin Cell Biol*. 2003;15(4):446–455.
151. Katzmann DJ, Odorizzi G, Emr SD. Receptor downregulation and multivesicular-body sorting. *Nat Rev Mol Cell Biol*. 2002;3(12):893–905.
152. Trajkovic K, Hsu C, Chiantia S, Rajendran L, Wenzel D, Wieland F, *et al.* Ceramide triggers budding of exosome vesicles into multivesicular endosomes. *Science*. 2008;319(5867):1244–1247.

153. Valadi H, Ekstrom K, Bossios A, Sjostrand M, Lee JJ, Lotvall JO. Exosome-mediated transfer of mRNAs and microRNAs is a novel mechanism of genetic exchange between cells. *Nat Cell Biol.* 2007;9(6):654–659.
154. Meckes DG, Jr., Shair KH, Marquitz AR, Kung CP, Edwards RH, Raab-Traub N. Human tumor virus utilizes exosomes for intercellular communication. *Proc Natl Acad Sci USA.* 2010;107(47):20370–20375.
155. Katakowski M, Buller B, Wang X, Rogers T, Chopp M. Functional microRNA is transferred between glioma cells. *Cancer Res.* 2010;70(21): 8259–8263.
156. Zhang Y, Liu D, Chen X, Li J, Li L, Bian Z, et al. Secreted monocytic miR-150 enhances targeted endothelial cell migration. *Mol Cell.* 2010;39(1): 133–144.
157. Kosaka N, Iguchi H, Yoshioka Y, Takeshita F, Matsuki Y, Ochiya T. Secretory mechanisms and intercellular transfer of microRNAs in living cells. *J Biol Chem.* 2010;285(23):17442–17452.
158. Pegtel DM, Cosmopoulos K, Thorley-Lawson DA, van Eijndhoven MA, Hopmans ES, Lindenberg JL, et al. Functional delivery of viral miRNAs via exosomes. *Proc Natl Acad Sci USA.* 2010;107(14):6328–6333.
159. Bessis M. Erythroblastic island, functional unity of bone marrow. *Rev Hematol.* 1958;13(1):8–11.
160. Allen TD, Dexter TM. Ultrastructural aspects of erythropoietic differentiation in long-term bone marrow culture. *Differentiation.* 1982;21(2):86–94.
161. Yokoyama T, Etoh T, Kitagawa H, Tsukahara S, Kannan Y. Migration of erythroblastic islands toward the sinusoid as erythroid maturation proceeds in rat bone marrow. *J Vet Med Sci.* 2003;65(4):449–452.
162. Mohandas N, Prenant M. Three-dimensional model of bone marrow. *Blood.* 1978;51(4):633–643.
163. Sadahira Y, Yoshino T, Monobe Y. Very late activation antigen 4-vascular cell adhesion molecule 1 interaction is involved in the formation of erythroblastic islands. *J Exp Med.* 1995;181(1):411–415.
164. Lee G, Lo A, Short SA, Mankelow TJ, Spring F, Parsons SF, et al. Targeted gene deletion demonstrates that the cell adhesion molecule ICAM-4 is critical for erythroblastic island formation. *Blood.* 2006;108(6):2064–2071.
165. Hanspal M, Hanspal JS. The association of erythroblasts with macrophages promotes erythroid proliferation and maturation: a 30-kD heparin-binding protein is involved in this contact. *Blood.* 1994;84(10):3494–3504.

166. Soni S, Bala S, Gwynn B, Sahr KE, Peters LL, Hanspal M. Absence of erythroblast macrophage protein (Emp) leads to failure of erythroblast nuclear extrusion. *J Biol Chem*. 2006;281(29):20181–20189.
167. Hanspal M, Smockova Y, Uong Q. Molecular identification and functional characterization of a novel protein that mediates the attachment of erythroblasts to macrophages. *Blood*. 1998;92(8):2940–2950.
168. Lee G, Spring FA, Parsons SF, Mankelow TJ, Peters LL, Koury MJ, et al. Novel secreted isoform of adhesion molecule ICAM-4: potential regulator of membrane-associated ICAM-4 interactions. *Blood*. 2003;101(5): 1790–1797.
169. Lee JC, Gimm JA, Lo AJ, Koury MJ, Krauss SW, Mohandas N, et al. Mechanism of protein sorting during erythroblast enucleation: role of cytoskeletal connectivity. *Blood*. 2004;103(5):1912–1919.
170. Allen TD, Testa NG. Cellular interactions in erythroblastic islands in long-term bone marrow cultures, as studied by time-lapse video. *Blood Cells*. 1991;17(1):29–38; discussion 9–43.
171. Sankaran VG, Orkin SH, Walkley CR. Rb intrinsically promotes erythropoiesis by coupling cell cycle exit with mitochondrial biogenesis. *Genes Dev*. 2008;22(4):463–475.
172. Spike BT, Dibling BC, Macleod KF. Hypoxic stress underlies defects in erythroblast islands in the Rb-null mouse. *Blood*. 2007;110(6):2173–2181.
173. Rhodes MM, Kopsombut P, Bondurant MC, Price JO, Koury MJ. Adherence to macrophages in erythroblastic islands enhances erythroblast proliferation and increases erythrocyte production by a different mechanism than erythropoietin. *Blood*. 2008;111(3):1700–1708.
174. Sadahira Y, Yasuda T, Yoshino T, Manabe T, Takeishi T, Kobayashi Y, et al. Impaired splenic erythropoiesis in phlebotomized mice injected with CL2MDP-liposome: an experimental model for studying the role of stromal macrophages in erythropoiesis. *J Leukoc Biol*. 2000;68(4): 464–470.
175. Inada T, Iwama A, Sakano S, Ohno M, Sawada K, Suda T. Selective expression of the receptor tyrosine kinase, HTK, on human erythroid progenitor cells. *Blood*. 1997;89(8):2757–2765.
176. Suenobu S, Takakura N, Inada T, Yamada Y, Yuasa H, Zhang XQ, et al. A role of EphB4 receptor and its ligand, ephrin-B2, in erythropoiesis. *Biochem Biophys Res Commun*. 2002;293(3):1124–1131.

177. Muta K, Krantz SB, Bondurant MC, Dai CH. Stem cell factor retards differentiation of normal human erythroid progenitor cells while stimulating proliferation. *Blood.* 1995;86(2):572–580.
178. Flanagan JG, Chan DC, Leder P. Transmembrane form of the kit ligand growth factor is determined by alternative splicing and is missing in the Sld mutant. *Cell.* 1991;64(5):1025–1035.
179. De Maria R, Zeuner A, Eramo A, Domenichelli C, Bonci D, Grignani F, et al. Negative regulation of erythropoiesis by caspase-mediated cleavage of GATA-1. *Nature.* 1999;401(6752):489–493.
180. Rosemblatt M, Vuillet-Gaugler MH, Leroy C, Coulombel L. Coexpression of two fibronectin receptors, VLA-4 and VLA-5, by immature human erythroblastic precursor cells. *J Clin Invest.* 1991;87(1):6–11.
181. Vuillet-Gaugler MH, Breton-Gorius J, Vainchenker W, Guichard J, Leroy C, Tchernia G, et al. Loss of attachment to fibronectin with terminal human erythroid differentiation. *Blood.* 1990;75(4):865–873.
182. Eshghi S, Vogelezang MG, Hynes RO, Griffith LG, Lodish HF. Alpha4beta1 integrin and erythropoietin mediate temporally distinct steps in erythropoiesis: integrins in red cell development. *J Cell Biol.* 2007; 177(5):871–880.
183. Patel VP, Lodish HF. A fibronectin matrix is required for differentiation of murine erythroleukemia cells into reticulocytes. *J Cell Biol.* 1987;105 (6 Pt 2):3105–3118.
184. Nienhuis AW, Benz EJ, Jr. Regulation of hemoglobin synthesis during the development of the red cell (third of three parts). *N Engl J Med.* 1977; 297(26):1430–1436.
185. Nienhuis AW, Benz EJ, Jr. Regulation of hemoglobin synthesis during the development of the red cell. (second of three parts). *N Engl J Med.* 1977;297(25):1371–1381.
186. Nienhuis AW, Benz EJ, Jr. Regulation of hemoglobin synthesis during the development of the red cell (first of three parts). *N Engl J Med.* 1977; 297(24):1318–1328.
187. Komar AA, Kommer A, Krasheninnikov IA, Spirin AS. Cotranslational folding of globin. *J Biol Chem.* 1997;272(16):10646–10651.
188. Komar AA, Kommer A, Krasheninnikov IA, Spirin AS. Cotranslational heme binding to nascent globin chains. *FEBS Lett.* 1993;326(1–3): 261–263.
189. Tahara T, Sun J, Nakanishi K, Yamamoto M, Mori H, Saito T, et al. Heme positively regulates the expression of beta-globin at the locus control

region *via* the transcriptional factor Bach1 in erythroid cells. *J Biol Chem.* 2004;279(7):5480–5487.

190. Tahara T, Sun J, Igarashi K, Taketani S. Heme-dependent up-regulation of the alpha-globin gene expression by transcriptional repressor Bach1 in erythroid cells. *Biochem Biophys Res Commun.* 2004;324(1): 77–85.

191. Sun J, Brand M, Zenke Y, Tashiro S, Groudine M, Igarashi K. Heme regulates the dynamic exchange of Bach1 and NF-E2-related factors in the Maf transcription factor network. *Proc Natl Acad Sci USA.* 2004;101(6): 1461–1466.

192. Chen JJ. Regulation of protein synthesis by the heme-regulated eIF2alpha kinase: relevance to anemias. *Blood.* 2007;109(7):2693–2699.

193. Khandros E, Weiss MJ. Protein quality control during erythropoiesis and hemoglobin synthesis. *Hematol Oncol Clin North Am.* 2010;24(6): 1071–1088.

194. Rivella S. Ineffective erythropoiesis and thalassemias. *Curr Opin Hematol.* 2009;16(3):187–194.

195. Fessas P. Inclusions of hemoglobin erythroblasts and erythrocytes of thalassemia. *Blood.* 1963;21:21–32.

196. Weiss MJ, dos Santos CO. Chaperoning erythropoiesis. *Blood.* 2009; 113(10):2136–2144.

197. Ribeil JA, Zermati Y, Vandekerckhove J, Cathelin S, Kersual J, Dussiot M, *et al.* Hsp70 regulates erythropoiesis by preventing caspase-3-mediated cleavage of GATA-1. *Nature.* 2007;445(7123):102–105.

198. Lui JC, Kong SK. Heat shock protein 70 inhibits the nuclear import of apoptosis-inducing factor to avoid DNA fragmentation in TF-1 cells during erythropoiesis. *FEBS Lett.* 2007;581(1):109–117.

199. Kihm AJ, Kong Y, Hong W, Russell JE, Rouda S, Adachi K, *et al.* An abundant erythroid protein that stabilizes free alpha-haemoglobin. *Nature.* 2002;417(6890):758–763.

200. Yu X, Kong Y, Dore LC, Abdulmalik O, Katein AM, Zhou S, *et al.* An erythroid chaperone that facilitates folding of alpha-globin subunits for hemoglobin synthesis. *J Clin Invest.* 2007;117(7):1856–1865.

201. Zhou S, Olson JS, Fabian M, Weiss MJ, Gow AJ. Biochemical fates of alpha hemoglobin bound to alpha hemoglobin-stabilizing protein AHSP. *J Biol Chem.* 2006;281(43):32611–32618.

202. Feng L, Gell DA, Zhou S, Gu L, Kong Y, Li J, et al. Molecular mechanism of AHSP-mediated stabilization of alpha-hemoglobin. Cell. 2004;119(5): 629–640.
203. Gell D, Kong Y, Eaton SA, Weiss MJ, Mackay JP. Biophysical characterization of the alpha-globin binding protein alpha-hemoglobin stabilizing protein. J Biol Chem. 2002;277(43):40602–40609.
204. Feng L, Zhou S, Gu L, Gell DA, Mackay JP, Weiss MJ, et al. Structure of oxidized alpha-haemoglobin bound to AHSP reveals a protective mechanism for haem. Nature. 2005;435(7042):697–701.
205. Mollan TL, Yu X, Weiss MJ, Olson JS. The role of alpha-hemoglobin stabilizing protein in redox chemistry, denaturation, and hemoglobin assembly. Antioxid Redox Signal. 2009;12(2):219–231.
206. Brillet T, Baudin-Creuza V, Vasseur C, Domingues-Hamdi E, Kiger L, Wajcman H, et al. Alpha-hemoglobin stabilizing protein (AHSP), a kinetic scheme of the action of a human mutant, AHSPV56G. J Biol Chem. 2010;285(23):17986–17992.
207. Baudin-Creuza V, Vasseur-Godbillon C, Pato C, Prehu C, Wajcman H, Marden MC. Transfer of human alpha- to beta-hemoglobin via its chaperone protein: evidence for a new state. J Biol Chem. 2004;279(35):36530–36533.
208. Kong Y, Zhou S, Kihm AJ, Katein AM, Yu X, Gell DA, et al. Loss of alpha-hemoglobin-stabilizing protein impairs erythropoiesis and exacerbates beta-thalassemia. J Clin Invest. 2004;114(10):1457–1466.
209. Lai MI, Jiang J, Silver N, Best S, Menzel S, Mijovic A, et al. Alpha-haemoglobin stabilising protein is a quantitative trait gene that modifies the phenotype of beta-thalassaemia. Br J Haematol. 2006;133(6):675–682.
210. dos Santos CO, Costa FF. AHSP and beta-thalassemia: a possible genetic modifier. Hematology. 2005;10(2):157–161.
211. Wang Z, Yu W, Li Y, Shang X, Zhang X, Xiong F, et al. Analysis of alpha-hemoglobin-stabilizing protein (AHSP) gene as a genetic modifier to the phenotype of beta-thalassemia in Southern China. Blood Cells Mol Dis. 2010;45(2):128–132.
212. Viprakasit V, Tanphaichitr VS, Chinchang W, Sangkla P, Weiss MJ, Higgs DR. Evaluation of alpha hemoglobin stabilizing protein (AHSP) as a genetic modifier in patients with beta thalassemia. Blood. 2004;103(9):3296–3299.
213. Hanash SM, Rucknagel DL. Proteolytic activity in erythrocyte precursors. Proc Natl Acad Sci USA. 1978;75(7):3427–3431.

214. Braverman AS, Lester D. Evidence for increased proteolysis in intact beta thalassemia erythroid cells. *Hemoglobin.* 1981;5(6):549–564.
215. Loukopoulos D, Karoulias A, Fessas P. Proteolysis in thalassemia: studies with protease inhibitors. *Ann N Y Acad Sci.* 1980;344:323–335.
216. Ciehanover A, Hod Y, Hershko A. A heat-stable polypeptide component of an ATP-dependent proteolytic system from reticulocytes. *Biochem Biophys Res Commun.* 1978;81(4):1100–1105.
217. Hershko A, Ciechanover A, Rose IA. Resolution of the ATP-dependent proteolytic system from reticulocytes: a component that interacts with ATP. *Proc Natl Acad Sci USA.* 1979;76(7):3107–3110.
218. Hershko A, Heller H, Elias S, Ciechanover A. Components of ubiquitin-protein ligase system. Resolution, affinity purification, and role in protein breakdown. *J Biol Chem.* 1983;258(13):8206–8214.
219. Shaeffer JR. Turnover of excess hemoglobin alpha chains in beta-thalassemic cells is ATP-dependent. *J Biol Chem.* 1983;258(21): 13172–13177.
220. Shaeffer JR. ATP-dependent proteolysis of hemoglobin alpha chains in beta-thalassemic hemolysates is ubiquitin-dependent. *J Biol Chem.* 1988;263(27):13663–13669.
221. Shaeffer JR. Monoubiquitinated alpha globin is an intermediate in the ATP-dependent proteolysis of alpha globin. *J Biol Chem.* 1994;269(35): 22205–22210.
222. Shaeffer JR, Cohen RE. Ubiquitin aldehyde increases adenosine triphosphate-dependent proteolysis of hemoglobin alpha-subunits in beta-thalassemic hemolysates. *Blood.* 1997;90(3):1300–1308.
223. Shaeffer JR, Kania MA. Degradation of monoubiquitinated alpha-globin by 26S proteasomes. *Biochemistry.* 1995;34(12):4015–4021.
224. Adachi K, Lakka V, Zhao Y, Surrey S. Ubiquitylation of nascent globin chains in a cell-free system. *J Biol Chem.* 2004;279(40): 41767–41774.
225. Wickramasinghe SN, Lee MJ. Evidence that the ubiquitin proteolytic pathway is involved in the degradation of precipitated globin chains in thalassaemia. *Br J Haematol.* 1998;101(2):245–250.
226. Wickramasinghe SN, Lee MJ, Furukawa T, Eguchi M, Reid CD. Composition of the intra-erythroblastic precipitates in thalassaemia and congenital dyserythropoietic anaemia (CDA): identification of a new type of CDA with intra-erythroblastic precipitates not reacting with monoclonal

antibodies to alpha- and beta-globin chains. *Br J Haematol.* 1996;93(3):576–585.
227. Fagan JM, Waxman L. The ATP-independent pathway in red blood cells that degrades oxidant-damaged hemoglobin. *J Biol Chem.* 1992;267(32): 23015–23022.
228. Fagan JM, Waxman L, Goldberg AL. Red blood cells contain a pathway for the degradation of oxidant-damaged hemoglobin that does not require ATP or ubiquitin. *J Biol Chem.* 1986;261(13):5705–5713.
229. Koury MJ, Bondurant MC. Erythropoietin retards DNA breakdown and prevents programmed death in erythroid progenitor cells. *Science.* 1990;248(4953):378–381.
230. Kelley LL, Koury MJ, Bondurant MC, Koury ST, Sawyer ST, Wickrema A. Survival or death of individual proerythroblasts results from differing erythropoietin sensitivities: a mechanism for controlled rates of erythrocyte production. *Blood.* 1993;82(8):2340–2352.
231. Socolovsky M, Fallon AE, Wang S, Brugnara C, Lodish HF. Fetal anemia and apoptosis of red cell progenitors in Stat5a–/–5b–/– mice: a direct role for Stat5 in Bcl-X(L) induction. *Cell.* 1999;98(2):181–191.
232. Dolznig H, Grebien F, Deiner EM, Stangl K, Kolbus A, Habermann B, et al. Erythroid progenitor renewal *versus* differentiation: genetic evidence for cell autonomous, essential functions of EpoR, Stat5 and the GR. *Oncogene.* 2006;25(20):2890–2900.
233. Wojchowski DM, Sathyanarayana P, Dev A. Erythropoietin receptor response circuits. *Curr Opin Hematol.* 2010;17(3):169–176.
234. Fang J, Menon M, Zhang D, Torbett B, Oxburgh L, Tschan M, et al. Attenuation of EPO-dependent erythroblast formation by death-associated protein kinase-2. *Blood.* 2008;112(3):886–890.
235. Sawada K, Krantz SB, Sawyer ST, Civin CI. Quantitation of specific binding of erythropoietin to human erythroid colony-forming cells. *J Cell Physiol.* 1988;137(2):337–345.
236. Sawada K, Krantz SB, Kans JS, Dessypris EN, Sawyer S, Glick AD, et al. Purification of human erythroid colony-forming units and demonstration of specific binding of erythropoietin. *J Clin Invest.* 1987;80(2):357–366.
237. Wickrema A, Bondurant MC, Krantz SB. Abundance and stability of erythropoietin receptor mRNA in mouse erythroid progenitor cells. *Blood.* 1991;78(9):2269–22675.

238. Broudy VC, Lin N, Brice M, Nakamoto B, Papayannopoulou T. Erythropoietin receptor characteristics on primary human erythroid cells. *Blood.* 1991;77(12):2583–2590.
239. Munugalavadla V, Kapur R. Role of c-Kit and erythropoietin receptor in erythropoiesis. *Crit Rev Oncol Hematol.* 2005;54(1):63–75.
240. Wu H, Klingmuller U, Besmer P, Lodish HF. Interaction of the erythropoietin and stem-cell-factor receptors. *Nature.* 1995;377(6546): 242–246.
241. Kurtz A, Hartl W, Jelkmann W, Zapf J, Bauer C. Activity in fetal bovine serum that stimulates erythroid colony formation in fetal mouse livers is insulinlike growth factor I. *J Clin Invest.* 1985;76(4):1643–1648.
242. Sawada K, Krantz SB, Dessypris EN, Koury ST, Sawyer ST. Human colony-forming units-erythroid do not require accessory cells, but do require direct interaction with insulin-like growth factor I and/or insulin for erythroid development. *J Clin Invest.* 1989;83(5):1701–1709.
243. Angelillo-Scherrer A, Burnier L, Lambrechts D, Fish RJ, Tjwa M, Plaisance S, et al. Role of Gas6 in erythropoiesis and anemia in mice. *J Clin Invest.* 2008;118(2):583–596.
244. Dai C, Chung IJ, Jiang S, Price JO, Krantz SB. Reduction of cell cycle progression in human erythroid progenitor cells treated with tumour necrosis factor alpha occurs with reduced CDK6 and is partially reversed by CDK6 transduction. *Br J Haematol.* 2003;121(6):919–927.
245. Nemeth E, Ganz T. Regulation of iron metabolism by hepcidin. *Annu Rev Nutr.* 2006;26:323–342.
246. Socolovsky M. Molecular insights into stress erythropoiesis. *Curr Opin Hematol.* 2007;14(3):215–224.
247. von Lindern M, Zauner W, Mellitzer G, Steinlein P, Fritsch G, Huber K, et al. The glucocorticoid receptor cooperates with the erythropoietin receptor and c-Kit to enhance and sustain proliferation of erythroid progenitors in vitro. *Blood.* 1999;94(2):550–559.
248. Bauer A, Tronche F, Wessely O, Kellendonk C, Reichardt HM, Steinlein P, et al. The glucocorticoid receptor is required for stress erythropoiesis. *Genes Dev.* 1999;13(22):2996–3002.
249. Broudy VC, Lin NL, Priestley GV, Nocka K, Wolf NS. Interaction of stem cell factor and its receptor c-kit mediates lodgment and acute expansion of hematopoietic cells in the murine spleen. *Blood.* 1996;88(1):75–81.

250. Lenox LE, Perry JM, Paulson RF. BMP4 and Madh5 regulate the erythroid response to acute anemia. *Blood.* 2005;105(7):2741–2748.
251. Menon MP, Karur V, Bogacheva O, Bogachev O, Cuetara B, Wojchowski DM. Signals for stress erythropoiesis are integrated *via* an erythropoietin receptor-phosphotyrosine-343-Stat5 axis. *J Clin Invest.* 2006;116(3): 683–694.
252. Hara H, Ogawa M. Erthropoietic precursors in mice with phenylhydrazine-induced anemia. *Am J Hematol.* 1976;1(4):453–458.
253. Miyamoto T, Iwasaki H, Reizis B, Ye M, Graf T, Weissman IL, *et al.* Myeloid or lymphoid promiscuity as a critical step in hematopoietic lineage commitment. *Dev Cell.* 2002;3(1):137–147.
254. Forsberg EC, Serwold T, Kogan S, Weissman IL, Passegue E. New evidence supporting megakaryocyte-erythrocyte potential of flk2/flt3+ multipotent hematopoietic progenitors. *Cell.* 2006;126(2):415–426.
255. Zang H, Sato K, Nakajima H, McKay C, Ney PA, Ihle JN. The distal region and receptor tyrosines of the Epo receptor are non-essential for *in vivo* erythropoiesis. *EMBO J.* 2001;20(12):3156–3166.
256. Jegalian AG, Acurio A, Dranoff G, Wu H. Erythropoietin receptor haploinsufficiency and *in vivo* interplay with granulocyte-macrophage colony-stimulating factor and interleukin 3. *Blood.* 2002;99(7):2603–2605.
257. Syed RS, Reid SW, Li C, Cheetham JC, Aoki KH, Liu B, *et al.* Efficiency of signalling through cytokine receptors depends critically on receptor orientation. *Nature.* 1998;395(6701):511–516.
258. Wessely O, Bauer A, Quang CT, Deiner EM, von Lindern M, Mellitzer G, *et al.* A novel way to induce erythroid progenitor self renewal: cooperation of c-Kit with the erythropoietin receptor. *Biol Chem.* 1999;380(2):187–202.
259. Wessely O, Deiner EM, Beug H, von Lindern M. The glucocorticoid receptor is a key regulator of the decision between self-renewal and differentiation in erythroid progenitors. *EMBO J.* 1997;16(2):267–280.
260. Kassouf MT, Hughes JR, Taylor S, McGowan SJ, Soneji S, Green AL, *et al.* Genome-wide identification of TAL1's functional targets: insights into its mechanisms of action in primary erythroid cells. *Genome Res.* 2010;20(8):1064–1083.
261. Soler E, Andrieu-Soler C, de Boer E, Bryne JC, Thongjuea S, Stadhouders R, *et al.* The genome-wide dynamics of the binding of Ldb1 complexes during erythroid differentiation. *Genes Dev.* 2010;24(3):277–289.

262. Cheng Y, Wu W, Kumar SA, Yu D, Deng W, Tripic T, et al. Erythroid GATA1 function revealed by genome-wide analysis of transcription factor occupancy, histone modifications, and mRNA expression. *Genome Res.* 2009;19(12):2172–2184.
263. Yu M, Riva L, Xie H, Schindler Y, Moran TB, Cheng Y, et al. Insights into GATA-1-mediated gene activation *versus* repression *via* genome-wide chromatin occupancy analysis. *Mol Cell.* 2009;36(4):682–695.
264. Fujiwara T, O'Geen H, Keles S, Blahnik K, Linnemann AK, Kang Y-A, et al. Discovering hematopoietic mechanisms through genome-wide analysis of GATA factor chromatin occupancy. *Mol Cell.* 2009;36(4):667–681.
265. Tallack MR, Whitington T, Yuen WS, Wainwright EN, Keys JR, Gardiner BB, et al. A global role for KLF1 in erythropoiesis revealed by ChIP-seq in primary erythroid cells. *Genome Res.* 2010;20(8):1052–1063.
266. Novershtern N, Subramanian A, Lawton LN, Mak RH, Haining WN, McConkey ME, et al. Densely interconnected transcriptional circuits control cell states in human hematopoiesis. *Cell.* 2011;144(2):296–309.
267. Wozniak RJ, Bresnick EH. Epigenetic control of complex loci during erythropoiesis. *Curr Top Dev Biol.* 2008;82:55–83.
268. de Laat W, Klous P, Kooren J, Noordermeer D, Palstra RJ, Simonis M, et al. Three-dimensional organization of gene expression in erythroid cells. *Curr Top Dev Biol.* 2008;82:117–139.
269. Schoenfelder S, Clay I, Fraser P. The transcriptional interactome: gene expression in 3D. *Curr Opin Genet Dev.* 2010;20(2):127–133.
270. Evans T, Felsenfeld G. The erythroid-specific transcription factor Eryf1: a new finger protein. *Cell.* 1989;58(5):877–885.
271. Tsai SF, Martin DI, Zon LI, D'Andrea AD, Wong GG, Orkin SH. Cloning of cDNA for the major DNA-binding protein of the erythroid lineage through expression in mammalian cells. *Nature.* 1989;339(6224):446–451.
272. Martin DI, Orkin SH. Transcriptional activation and DNA binding by the erythroid factor GF-1/NF-E1/Eryf 1. *Genes Dev.* 1990;4(11):1886–1898.
273. Yang HY, Evans T. Distinct roles for the two cGATA-1 finger domains. *Mol Cell Biol.* 1992;12(10):4562–4570.
274. Whyatt DJ, deBoer E, Grosveld F. The two zinc finger-like domains of GATA-1 have different DNA binding specificities. *EMBO J.* 1993;12(13):4993–5005.

275. Trainor CD, Omichinski JG, Vandergon TL, Gronenborn AM, Clore GM, Felsenfeld G. A palindromic regulatory site within vertebrate GATA-1 promoters requires both zinc fingers of the GATA-1 DNA-binding domain for high-affinity interaction. *Mol Cell Biol.* 1996;16(5):2238–2247.
276. Leonard M, Brice M, Engel JD, Papayannopoulou T. Dynamics of GATA transcription factor expression during erythroid differentiation. *Blood.* 1993; 82(4):1071–1079.
277. Tsai FY, Keller G, Kuo FC, Weiss M, Chen J, Rosenblatt M, *et al.* An early haematopoietic defect in mice lacking the transcription factor GATA-2. *Nature.* 1994;371(6494):221–226.
278. Rodrigues NP, Janzen V, Forkert R, Dombkowski DM, Boyd AS, Orkin SH, *et al.* Haploinsufficiency of GATA-2 perturbs adult hematopoietic stem-cell homeostasis. *Blood.* 2005;106(2):477–484.
279. Shimizu R, Yamamoto M. Gene expression regulation and domain function of hematopoietic GATA factors. *Semin Cell Dev Biol.* 2005;16(1):129–136.
280. Kaneko H, Shimizu R, Yamamoto M. GATA factor switching during erythroid differentiation. *Curr Opin Hematol.* 2010;17(3):163–168.
281. Bresnick EH, Lee HY, Fujiwara T, Johnson KD, Keles S. GATA switches as developmental drivers. *J Biol Chem.* 2010;285(41):31087–31093.
282. Jing H, Vakoc CR, Ying L, Mandat S, Wang H, Zheng X, *et al.* Exchange of GATA factors mediates transitions in looped chromatin organization at a developmentally regulated gene locus. *Mol Cell.* 2008;29(2):232–242.
283. Grass JA, Boyer ME, Pal S, Wu J, Weiss MJ, Bresnick EH. GATA-1-dependent transcriptional repression of GATA-2 *via* disruption of positive autoregulation and domain-wide chromatin remodeling. *Proc Natl Acad Sci USA.* 2003;100(15):8811–8816.
284. Fujiwara Y, Browne CP, Cunniff K, Goff SC, Orkin SH. Arrested development of embryonic red cell precursors in mouse embryos lacking transcription factor GATA-1. *Proc Natl Acad Sci USA.* 1996;93(22): 12355–12358.
285. Weiss MJ, Keller G, Orkin SH. Novel insights into erythroid development revealed through *in vitro* differentiation of GATA-1 embryonic stem cells. *Genes Dev.* 1994;8(10):1184–1197.
286. Weiss MJ, Orkin SH. Transcription factor GATA-1 permits survival and maturation of erythroid precursors by preventing apoptosis. *Proc Natl Acad Sci USA.* 1995;92(21):9623–9627.

287. Crispino JD. GATA1 in normal and malignant hematopoiesis. *Semin Cell Dev Biol*. 2005 Feb;16(1):137–147.
288. Gutierrez L, Nikolic T, van Dijk TB, Hammad H, Vos N, Willart M, et al. Gata1 regulates dendritic-cell development and survival. *Blood*. 2007;110(6):1933–1941.
289. Gobel F, Taschner S, Jurkin J, Konradi S, Vaculik C, Richter S, et al. Reciprocal role of GATA-1 and vitamin D receptor in human myeloid dendritic cell differentiation. *Blood*. 2009;114(18):3813–3821.
290. Cheng Y, Wu W, Kumar SA, Yu D, Deng W, Tripic T, et al. Erythroid GATA1 function revealed by genome-wide analysis of transcription factor occupancy, histone modifications, and mRNA expression. *Genome Res*. 2009;19(12):2172–2184.
291. Fujiwara T, O'Geen H, Keles S, Blahnik K, Linnemann AK, Kang YA, et al. Discovering hematopoietic mechanisms through genome-wide analysis of GATA factor chromatin occupancy. *Mol Cell*. 2009;36(4):667–681.
292. Tsang AP, Visvader JE, Turner CA, Fujiwara Y, Yu C, Weiss MJ, et al. FOG, a multitype zinc finger protein, acts as a cofactor for transcription factor GATA-1 in erythroid and megakaryocytic differentiation. *Cell*. 1997;90(1): 109–119.
293. Tsang AP, Fujiwara Y, Hom DB, Orkin SH. Failure of megakaryopoiesis and arrested erythropoiesis in mice lacking the GATA-1 transcriptional cofactor FOG. *Genes Dev*. 1998;12(8):1176–1188.
294. Pevny L, Lin CS, D'Agati V, Simon MC, Orkin SH, Costantini F. Development of hematopoietic cells lacking transcription factor GATA-1. *Development*. 1995;121(1):163–172.
295. Vakoc CR, Letting DL, Gheldof N, Sawado T, Bender MA, Groudine M, et al. Proximity among distant regulatory elements at the beta-globin locus requires GATA-1 and FOG-1. *Mol Cell*. 2005;17(3):453–462.
296. Miccio A, Blobel GA. Role of the GATA-1/FOG-1/NuRD pathway in the expression of human beta-like globin genes. *Mol Cell Biol*. 2010;30(14): 3460–3470.
297. Miccio A, Wang Y, Hong W, Gregory GD, Wang H, Yu X, et al. NuRD mediates activating and repressive functions of GATA-1 and FOG-1 during blood development. *EMBO J*. 2010;29(2):442–456.
298. Koschmieder S, Rosenbauer F, Steidl U, Owens BM, Tenen DG. Role of transcription factors C/EBPalpha and PU.1 in normal hematopoiesis and leukemia. *Int J Hematol*. 2005;81(5):368–377.

299. Rao G, Rekhtman N, Cheng G, Krasikov T, Skoultchi AI. Deregulated expression of the PU.1 transcription factor blocks murine erythroleukemia cell terminal differentiation. *Oncogene.* 1997;14(1):123–131.
300. Yamada T, Kondoh N, Matsumoto M, Yoshida M, Maekawa A, Oikawa T. Overexpression of PU.1 induces growth and differentiation inhibition and apoptotic cell death in murine erythroleukemia cells. *Blood.* 1997;89(4): 1383–1393.
301. Delgado MD, Gutierrez P, Richard C, Cuadrado MA, Moreau-Gachelin F, Leon J. Spi-1/PU.1 proto-oncogene induces opposite effects on monocytic and erythroid differentiation of K562 cells. *Biochem Biophys Res Commun.* 1998;252(2):383–391.
302. Yamada T, Kihara-Negishi F, Yamamoto H, Yamamoto M, Hashimoto Y, Oikawa T. Reduction of DNA binding activity of the GATA-1 transcription factor in the apoptotic process induced by overexpression of PU.1 in murine erythroleukemia cells. *Exp Cell Res.* 1998;245(1):186–194.
303. Rekhtman N, Radparvar F, Evans T, Skoultchi AI. Direct interaction of hematopoietic transcription factors PU.1 and GATA-1: functional antagonism in erythroid cells. *Genes Dev.* 1999;13(11):1398–1411.
304. Chou ST, Khandros E, Bailey LC, Nichols KE, Vakoc CR, Yao Y, et al. Graded repression of PU.1/Sfpi1 gene transcription by GATA factors regulates hematopoietic cell fate. *Blood.* 2009;114(5):983–994.
305. Lecuyer E, Hoang T. SCL: from the origin of hematopoiesis to stem cells and leukemia. *Exp Hematol.* 2004;32(1):11–24.
306. Begley CG, Green AR. The SCL gene: from case report to critical hematopoietic regulator. *Blood.* 1999 May 1;93(9):2760–2770.
307. Green AR, Salvaris E, Begley CG. Erythroid expression of the 'helix-loop-helix' gene, SCL. *Oncogene.* 1991;6(3):475–479.
308. Visvader J, Begley CG, Adams JM. Differential expression of the LYL, SCL and E2A helix-loop-helix genes within the hemopoietic system. *Oncogene.* 1991;6(2):187–194.
309. Mouthon MA, Bernard O, Mitjavila MT, Romeo PH, Vainchenker W, Mathieu-Mahul D. Expression of tal-1 and GATA-binding proteins during human hematopoiesis. *Blood.* 1993;81(3):647–655.
310. Elnitski L, Miller W, Hardison R. Conserved E boxes function as part of the enhancer in hypersensitive site 2 of the beta-globin locus control region. Role of basic helix-loop-helix proteins. *J Biol Chem.* 1997;272(1):369–378.

311. Anderson KP, Crable SC, Lingrel JB. Multiple proteins binding to a GATA-E box-GATA motif regulate the erythroid Kruppel-like factor (EKLF) gene. *J Biol Chem.* 1998;273(23):14347–14354.
312. Vyas P, McDevitt MA, Cantor AB, Katz SG, Fujiwara Y, Orkin SH. Different sequence requirements for expression in erythroid and megakaryocytic cells within a regulatory element upstream of the GATA-1 gene. *Development.* 1999;126(12):2799–2811.
313. Anderson KP, Crable SC, Lingrel JB. The GATA-E box-GATA motif in the EKLF promoter is required for *in vivo* expression. *Blood.* 2000;95(5):1652–1655.
314. Huang S, Qiu Y, Stein RW, Brandt SJ. p300 functions as a transcriptional coactivator for the TAL1/SCL oncoprotein. *Oncogene.* 1999;18(35):4958–4967.
315. Huang S, Brandt SJ. mSin3A regulates murine erythroleukemia cell differentiation through association with the TAL1 (or SCL) transcription factor. *Mol Cell Biol.* 2000;20(6):2248–2259.
316. Schuh AH, Tipping AJ, Clark AJ, Hamlett I, Guyot B, Iborra FJ, *et al.* ETO-2 associates with SCL in erythroid cells and megakaryocytes and provides repressor functions in erythropoiesis. *Mol Cell Biol.* 2005;25(23):10235–10250.
317. Meier N, Krpic S, Rodriguez P, Strouboulis J, Monti M, Krijgsveld J, *et al.* Novel binding partners of Ldb1 are required for haematopoietic development. *Development.* 2006;133(24):4913–4923.
318. Cohen-Kaminsky S, Maouche-Chretien L, Vitelli L, Vinit MA, Blanchard I, Yamamoto M, *et al.* Chromatin immunoselection defines a TAL-1 target gene. *EMBO J.* 1998;17(17):5151–5160.
319. Wadman IA, Osada H, Grutz GG, Agulnick AD, Westphal H, Forster A, *et al.* The LIM-only protein Lmo2 is a bridging molecule assembling an erythroid, DNA-binding complex which includes the TAL1, E47, GATA-1 and Ldb1/NLI proteins. *EMBO J.* 1997;16(11):3145–3157.
320. Kassouf MT, Chagraoui H, Vyas P, Porcher C. Differential use of SCL/TAL-1 DNA-binding domain in developmental hematopoiesis. *Blood.* 2008;112(4):1056–1067.
321. Robb L, Lyons I, Li R, Hartley L, Kontgen F, Harvey RP, *et al.* Absence of yolk sac hematopoiesis from mice with a targeted disruption of the scl gene. *Proc Natl Acad Sci USA.* 1995;92(15):7075–7079.

322. Shivdasani RA, Mayer EL, Orkin SH. Absence of blood formation in mice lacking the T-cell leukaemia oncoprotein tal-1/SCL. *Nature*. 1995;373(6513):432–434.
323. Porcher C, Swat W, Rockwell K, Fujiwara Y, Alt FW, Orkin SH. The T cell leukemia oncoprotein SCL/tal-1 is essential for development of all hematopoietic lineages. *Cell*. 1996;86(1):47–57.
324. Hall MA, Curtis DJ, Metcalf D, Elefanty AG, Sourris K, Robb L, et al. The critical regulator of embryonic hematopoiesis, SCL, is vital in the adult for megakaryopoiesis, erythropoiesis, and lineage choice in CFU-S12. *Proc Natl Acad Sci USA*. 2003;100(3):992–997.
325. Visvader JE, Crossley M, Hill J, Orkin SH, Adams JM. The C-terminal zinc finger of GATA-1 or GATA-2 is sufficient to induce megakaryocytic differentiation of an early myeloid cell line. *Mol Cell Biol*. 1995;15(2):634–641.
326. Mikkola HK, Klintman J, Yang H, Hock H, Schlaeger TM, Fujiwara Y, et al. Haematopoietic stem cells retain long-term repopulating activity and multipotency in the absence of stem-cell leukaemia SCL/tal-1 gene. *Nature*. 2003;421(6922):547–551.
327. Curtis DJ, Hall MA, Van Stekelenburg LJ, Robb L, Jane SM, Begley CG. SCL is required for normal function of short-term repopulating hematopoietic stem cells. *Blood*. 2004;103(9):3342–3348.
328. D'Souza SL, Elefanty AG, Keller G. SCL/Tal-1 is essential for hematopoietic commitment of the hemangioblast but not for its development. *Blood*. 2005;105(10):3862–3870.
329. Dooley KA, Davidson AJ, Zon LI. Zebrafish scl functions independently in hematopoietic and endothelial development. *Dev Biol*. 2005;277(2): 522–536.
330. Patterson LJ, Gering M, Patient R. Scl is required for dorsal aorta as well as blood formation in zebrafish embryos. *Blood*. 2005;105(9):3502–3511.
331. Aplan PD, Nakahara K, Orkin SH, Kirsch IR. The SCL gene product: a positive regulator of erythroid differentiation. *EMBO J*. 1992;11(11):4073–4081.
332. Elwood NJ, Zogos H, Pereira DS, Dick JE, Begley CG. Enhanced megakaryocyte and erythroid development from normal human CD34(+) cells: consequence of enforced expression of SCL. *Blood*. 1998;91(10):3756–3765.
333. Tripic T, Deng W, Cheng Y, Zhang Y, Vakoc CR, Gregory GD, et al. SCL and associated proteins distinguish active from repressive GATA transcription factor complexes. *Blood*. 2009;113(10):2191–2201.

334. Pearson R, Fleetwood J, Eaton S, Crossley M, Bao S. Kruppel-like transcription factors: a functional family. *Int J Biochem Cell Biol*. 2008;40(10): 1996–2001.
335. Swamynathan SK. Kruppel-like factors: three fingers in control. *Hum Genomics*. 2010;4(4):263–270.
336. Miller IJ, Bieker JJ. A novel, erythroid cell-specific murine transcription factor that binds to the CACCC element and is related to the Kruppel family of nuclear proteins. *Mol Cell Biol*. 1993;13(5):2776–2786.
337. Southwood CM, Downs KM, Bieker JJ. Erythroid Kruppel-like factor exhibits an early and sequentially localized pattern of expression during mammalian erythroid ontogeny. *Dev Dyn*. 1996;206(3):248–259.
338. Orkin SH, Kazazian HH, Jr., Antonarakis SE, Goff SC, Boehm CD, Sexton JP, *et al*. Linkage of beta-thalassaemia mutations and beta-globin gene polymorphisms with DNA polymorphisms in human beta-globin gene cluster. *Nature*. 1982;296(5858):627–631.
339. Orkin SH, Antonarakis SE, Kazazian HH, Jr. Base substitution at position -88 in a beta-thalassemic globin gene. Further evidence for the role of distal promoter element ACACCC. *J Biol Chem*. 1984;259(14): 8679–8681.
340. Kulozik AE, Bellan-Koch A, Bail S, Kohne E, Kleihauer E. Thalassemia intermedia: moderate reduction of beta globin gene transcriptional activity by a novel mutation of the proximal CACCC promoter element. *Blood*. 1991;77(9):2054–2058.
341. Feng WC, Southwood CM, Bieker JJ. Analyses of beta-thalassemia mutant DNA interactions with erythroid Kruppel-like factor (EKLF), an erythroid cell-specific transcription factor. *J Biol Chem*. 1994;269(2): 1493–1500.
342. Nuez B, Michalovich D, Bygrave A, Ploemacher R, Grosveld F. Defective haematopoiesis in fetal liver resulting from inactivation of the EKLF gene. *Nature*. 1995;375(6529):316–318.
343. Perkins AC, Sharpe AH, Orkin SH. Lethal beta-thalassaemia in mice lacking the erythroid CACCC-transcription factor EKLF. *Nature*. 1995;375(6529): 318–322.
344. Zhou D, Pawlik KM, Ren J, Sun CW, Townes TM. Differential binding of erythroid Krupple-like factor to embryonic/fetal globin gene promoters during development. *J Biol Chem*. 2006;281(23):16052–16057.

345. Drissen R, Palstra RJ, Gillemans N, Splinter E, Grosveld F, Philipsen S, et al. The active spatial organization of the beta-globin locus requires the transcription factor EKLF. *Genes Dev.* 2004;18(20):2485–2490.
346. Hodge D, Coghill E, Keys J, Maguire T, Hartmann B, McDowall A, et al. A global role for EKLF in definitive and primitive erythropoiesis. *Blood.* 2006;107(8):3359–3370.
347. Lim SK, Bieker JJ, Lin CS, Costantini F. A shortened life span of EKLF–/– adult erythrocytes, due to a deficiency of beta-globin chains, is ameliorated by human gamma-globin chains. *Blood.* 1997;90(3):1291–1299.
348. Pilon AM, Arcasoy MO, Dressman HK, Vayda SE, Maksimova YD, Sangerman JI, et al. Failure of terminal erythroid differentiation in EKLF-deficient mice is associated with cell cycle perturbation and reduced expression of E2F2. *Mol Cell Biol.* 2008;28(24):7394–7401.
349. Perkins AC, Peterson KR, Stamatoyannopoulos G, Witkowska HE, Orkin SH. Fetal expression of a human Agamma globin transgene rescues globin chain imbalance but not hemolysis in EKLF null mouse embryos. *Blood.* 2000;95(5):1827–1833.
350. Shyu YC, Wen SC, Lee TL, Chen X, Hsu CT, Chen H, et al. Chromatin-binding *in vivo* of the erythroid kruppel-like factor, EKLF, in the murine globin loci. *Cell Res.* 2006;16(4):347–355.
351. Drissen R, von Lindern M, Kolbus A, Driegen S, Steinlein P, Beug H, et al. The erythroid phenotype of EKLF-null mice: defects in hemoglobin metabolism and membrane stability. *Mol Cell Biol.* 2005;25(12):5205–5214.
352. Pilon AM, Nilson DG, Zhou D, Sangerman J, Townes TM, Bodine DM, et al. Alterations in expression and chromatin configuration of the alpha hemoglobin-stabilizing protein gene in erythroid Kruppel-like factor-deficient mice. *Mol Cell Biol.* 2006;26(11):4368–4377.
353. Nilson DG, Sabatino DE, Bodine DM, Gallagher PG. Major erythrocyte membrane protein genes in EKLF-deficient mice. *Exp Hematol.* 2006;34(6):705–712.
354. Keys JR, Tallack MR, Hodge DJ, Cridland SO, David R, Perkins AC. Genomic organisation and regulation of murine alpha haemoglobin stabilising protein by erythroid Kruppel-like factor. *Br J Haematol.* 2007;136(1):150–157.
355. Siatecka M, Sahr KE, Andersen SG, Mezei M, Bieker JJ, Peters LL. Severe anemia in the Nan mutant mouse caused by sequence-selective disruption

of erythroid Kruppel-like factor. *Proc Natl Acad Sci USA.* 2010;107(34):15151–15156.
356. Armstrong JA, Bieker JJ, Emerson BM. A SWI/SNF-related chromatin remodeling complex, E-RC1, is required for tissue-specific transcriptional regulation by EKLF *in vitro*. *Cell.* 1998;95(1):93–104.
357. Tewari R, Gillemans N, Wijgerde M, Nuez B, von Lindern M, Grosveld F, *et al.* Erythroid Kruppel-like factor (EKLF) is active in primitive and definitive erythroid cells and is required for the function of 5'HS3 of the beta-globin locus control region. *EMBO J.* 1998;17(8):2334–2341.
358. Brown RC, Pattison S, van Ree J, Coghill E, Perkins A, Jane SM, *et al.* Distinct domains of erythroid Kruppel-like factor modulate chromatin remodeling and transactivation at the endogenous beta-globin gene promoter. *Mol Cell Biol.* 2002;22(1):161–170.
359. Schoenfelder S, Sexton T, Chakalova L, Cope NF, Horton A, Andrews S, *et al.* Preferential associations between co-regulated genes reveal a transcriptional interactome in erythroid cells. *Nat Genet.* 2010;42(1):53–61.
360. Zhou D, Liu K, Sun CW, Pawlik KM, Townes TM. KLF1 regulates BCL11A expression and gamma- to beta-globin gene switching. *Nat Genet.* 2010;42(9):742–744.
361. Satta S, Perseu L, Moi P, Asunis I, Cabriolu A, Maccioni L, *et al.* Compound heterozygosity for KLF1 mutations associated with remarkable increase of fetal hemoglobin and red cell protoporphyrin. *Haematologica.* 2011;96(5):767–770.
362. Arnaud L, Saison C, Helias V, Lucien N, Steschenko D, Giarratana MC, *et al.* A dominant mutation in the gene encoding the erythroid transcription factor KLF1 causes a congenital dyserythropoietic anemia. *Am J Hum Genet.* 2010;87(5):721–727.
363. Asano H, Li XS, Stamatoyannopoulos G. FKLF, a novel Kruppel-like factor that activates human embryonic and fetal beta-like globin genes. *Mol Cell Biol.* 1999;19(5):3571–3579.
364. Asano H, Li XS, Stamatoyannopoulos G. FKLF-2: a novel Kruppel-like transcriptional factor that activates globin and other erythroid lineage genes. *Blood.* 2000;95(11):3578–3584.
365. Funnell AP, Maloney CA, Thompson LJ, Keys J, Tallack M, Perkins AC, *et al.* Erythroid Kruppel-like factor directly activates the basic Kruppel-like factor gene in erythroid cells. *Mol Cell Biol.* 2007;27(7):2777–2790.

366. Marin M, Karis A, Visser P, Grosveld F, Philipsen S. Transcription factor Sp1 is essential for early embryonic development but dispensable for cell growth and differentiation. *Cell.* 1997;89(4):619–628.
367. Kruger I, Vollmer M, Simmons DG, Elsasser HP, Philipsen S, Suske G. Sp1/Sp3 compound heterozygous mice are not viable: impaired erythropoiesis and severe placental defects. *Dev Dyn.* 2007;236(8):2235–2244.
368. Bartel DP. MicroRNAs: genomics, biogenesis, mechanism, and function. *Cell.* 2004;116(2):281–297.
369. Carthew RW, Sontheimer EJ. Origins and mechanisms of miRNAs and siRNAs. *Cell.* 2009;136(4):642–655.
370. O'Carroll D, Mecklenbrauker I, Das PP, Santana A, Koenig U, Enright AJ, *et al.* A slicer-independent role for Argonaute 2 in hematopoiesis and the microRNA pathway. *Genes Dev.* 2007;21(16):1999–2004.
371. Cheloufi S, Dos Santos CO, Chong MM, Hannon GJ. A dicer-independent miRNA biogenesis pathway that requires ago catalysis. *Nature.* 2010; 465(7298):584–589.
372. Cifuentes D, Xue H, Taylor DW, Patnode H, Mishima Y, Cheloufi S, *et al.* A novel miRNA processing pathway independent of dicer requires Argonaute2 catalytic activity. *Science.* 2010;328(5986):1694–1698.
373. Lawrie CH. microRNA expression in erythropoiesis and erythroid disorders. *Br J Haematol.* 2010;150(2):144–151.
374. Patrick DM, Zhang CC, Tao Y, Yao H, Qi X, Schwartz RJ, *et al.* Defective erythroid differentiation in miR-451 mutant mice mediated by 14-3-3zeta. *Genes Dev.* 2010;24(15):1614–1619.
375. Rasmussen KD, Simmini S, Abreu-Goodger C, Bartonicek N, Di Giacomo M, Bilbao-Cortes D, *et al.* The miR-144/451 locus is required for erythroid homeostasis. *J Exp Med.* 2010;207(7):1351–1358.
376. Zhang L, Flygare J, Wong P, Lim B, Lodish HF. miR-191 regulates mouse erythroblast enucleation by down-regulating Riok3 and Mxi1. *Genes Dev.* 2011;25(2):119–124.
377. Bruchova H, Yoon D, Agarwal AM, Mendell J, Prchal JT. Regulated expression of microRNAs in normal and polycythemia vera erythropoiesis. *Exp Hematol.* 2007;35(11):1657–1667.
378. Choong ML, Yang HH, McNiece I. MicroRNA expression profiling during human cord blood-derived CD34 cell erythropoiesis. *Exp Hematol.* 2007; 35(4):551–564.

379. Dore LC, Amigo JD, Dos Santos CO, Zhang Z, Gai X, Tobias JW, et al. A GATA-1-regulated microRNA locus essential for erythropoiesis. *Proc Natl Acad Sci USA*. 2008;105(9):3333–3338.
380. Du TT, Fu YF, Dong M, Wang L, Fan HB, Chen Y, et al. Experimental validation and complexity of miRNA-mRNA target interaction during zebrafish primitive erythropoiesis. *Biochem Biophys Res Commun*. 2009; 381(4):688–693.
381. Merkerova M, Belickova M, Bruchova H. Differential expression of microRNAs in hematopoietic cell lineages. *Eur J Haematol*. 2008;81(4): 304–310.
382. Zhan M, Miller CP, Papayannopoulou T, Stamatoyannopoulos G, Song CZ. MicroRNA expression dynamics during murine and human erythroid differentiation. *Exp Hematol*. 2007;35(7):1015–1025.
383. Rathjen T, Nicol C, McConkey G, Dalmay T. Analysis of short RNAs in the malaria parasite and its red blood cell host. *FEBS Lett*. 2006;580(22): 5185–5188.
384. Chen SY, Wang Y, Telen MJ, Chi JT. The genomic analysis of erythrocyte microRNA expression in sickle cell diseases. *PLoS One*. 2008;3(6): e2360.
385. Kloosterman WP, Steiner FA, Berezikov E, de Bruijn E, van de Belt J, Verheul M, et al. Cloning and expression of new microRNAs from zebrafish. *Nucleic Acids Res*. 2006;34(9):2558–2569.
386. Sangokoya C, Telen MJ, Chi JT. MicroRNA miR-144 modulates oxidative stress tolerance and associates with anemia severity in sickle cell disease. *Blood*. 2010;116(20):4338–4348.
387. Slezak S, Jin P, Caruccio L, Ren J, Bennett M, Zia N, et al. Gene and microRNA analysis of neutrophils from patients with polycythemia vera and essential thrombocytosis: down-regulation of micro RNA-1 and -133a. *J Transl Med*. 2009;7:39.
388. Bruchova H, Merkerova M, Prchal JT. Aberrant expression of microRNA in polycythemia vera. *Haematologica*. 2008;93(7):1009–1016.
389. Bruchova H, Yoon D, Agarwal AM, Swierczek S, Prchal JT. Erythropoiesis in polycythemia vera is hyper-proliferative and has accelerated maturation. *Blood Cells Mol Dis*. 2009;43(1):81–87.
390. Dolznig H, Boulme F, Stangl K, Deiner EM, Mikulits W, Beug H, et al. Establishment of normal, terminally differentiating mouse erythroid

progenitors: molecular characterization by cDNA arrays. *FASEB J.* 2001; 15(8):1442–1444.
391. Fibach E, Manor D, Oppenheim A, Rachmilewitz EA. Proliferation and maturation of human erythroid progenitors in liquid culture. *Blood.* 1989;73(1):100–103.
392. Pope SH, Fibach E, Sun J, Chin K, Rodgers GP. Two-phase liquid culture system models normal human adult erythropoiesis at the molecular level. *Eur J Haematol.* 2000;64(5):292–303.
393. Hattangadi SM, Burke KA, Lodish HF. Homeodomain-interacting protein kinase 2 plays an important role in normal terminal erythroid differentiation. *Blood.* 2010;115(23):4853–4861.
394. Zhang L, Flygare J, Wong P, Lim B, Lodish HF. miR-191 regulates mouse erythroblast enucleation by down-regulating Riok3 and Mxi1. *Genes Dev.* 2011 Jan 15;25(2):119–124.
395. Migliaccio AR, Whitsett C, Migliaccio G. Erythroid cells *in vitro*: from developmental biology to blood transfusion products. *Curr Opin Hematol.* 2009;16(4):259–268.
396. Douay L, Giarratana MC. *Ex vivo* generation of human red blood cells: a new advance in stem cell engineering. *Methods Mol Biol.* 2009;482:127–140.
397. van den Akker E, Satchwell TJ, Pellegrin S, Daniels G, Toye AM. The majority of the *in vitro* erythroid expansion potential resides in CD34(–) cells, outweighing the contribution of CD34(+) cells and significantly increasing the erythroblast yield from peripheral blood samples. *Haematologica.* 2010;95(9):1594–1598.
398. Miharada K, Hiroyama T, Sudo K, Nagasawa T, Nakamura Y. Efficient enucleation of erythroblasts differentiated *in vitro* from hematopoietic stem and progenitor cells. *Nat Biotechnol.* 2006;24(10):1255–1256.
399. Neildez-Nguyen TM, Wajcman H, Marden MC, Bensidhoum M, Moncollin V, Giarratana MC, *et al.* Human erythroid cells produced *ex vivo* at large scale differentiate into red blood cells *in vivo*. *Nat Biotechnol.* 2002;20(5):467–472.
400. Fujimi A, Matsunaga T, Kobune M, Kawano Y, Nagaya T, Tanaka I, *et al. Ex vivo* large-scale generation of human red blood cells from cord blood CD34+ cells by co-culturing with macrophages. *Int J Hematol.* 2008;87(4):339–350.
401. von Lindern M, Zauner W, Mellitzer G, Steinlein P, Fritsch G, Huber K, *et al.* The glucocorticoid receptor cooperates with the erythropoietin

receptor and c-Kit to enhance and sustain proliferation of erythroid progenitors *in vitro*. *Blood*. 1999;94(2):550–559.
402. Giarratana MC, Kobari L, Lapillonne H, Chalmers D, Kiger L, Cynober T, *et al*. Ex vivo generation of fully mature human red blood cells from hematopoietic stem cells. *Nat Biotechnol*. 2005;23(1):69–74.
403. Kaufman DS, Hanson ET, Lewis RL, Auerbach R, Thomson JA. Hematopoietic colony-forming cells derived from human embryonic stem cells. *Proc Natl Acad Sci USA*. 2001;98(19):10716–10721.
404. Ma F, Ebihara Y, Umeda K, Sakai H, Hanada S, Zhang H, *et al*. Generation of functional erythrocytes from human embryonic stem cell-derived definitive hematopoiesis. *Proc Natl Acad Sci USA*. 2008;105(35): 13087–13092.
405. Lu SJ, Feng Q, Park JS, Vida L, Lee BS, Strausbauch M, *et al*. Biologic properties and enucleation of red blood cells from human embryonic stem cells. *Blood*. 2008;112(12):4475–4484.
406. Choi KD, Yu J, Smuga-Otto K, Salvagiotto G, Rehrauer W, Vodyanik M, *et al*. Hematopoietic and endothelial differentiation of human induced pluripotent stem cells. *Stem Cells*. 2009;27(3):559–567.
407. Feng Q, Lu SJ, Klimanskaya I, Gomes I, Kim D, Chung Y, *et al*. Hemangioblastic derivatives from human induced pluripotent stem cells exhibit limited expansion and early senescence. *Stem Cells*. 2010;28(4): 704–712.
408. Lapillonne H, Kobari L, Mazurier C, Tropel P, Giarratana MC, Zanella-Cleon I, *et al*. Red blood cell generation from human induced pluripotent stem cells: perspectives for transfusion medicine. *Haematologica*. 2010;95(10):1651–1659.
409. Olsen AL, Stachura DL, Weiss MJ. Designer blood: creating hematopoietic lineages from embryonic stem cells. *Blood*. 2006;107(4):1265–1275.
410. Anstee DJ. Production of erythroid cells from human embryonic stem cells (hESC) and human induced pluripotent stem cells (hiPS). *Transfus Clin Biol*. 2010;17(3):104–109.
411. Lu SJ, Feng Q, Park JS, Lanza R. Directed differentiation of red blood cells from human embryonic stem cells. *Methods Mol Biol*. 2010;636:105–121.
412. Keller G, Kennedy M, Papayannopoulou T, Wiles MV. Hematopoietic commitment during embryonic stem cell differentiation in culture. *Mol Cell Biol*. 1993;13(1):473–486.

413. Keller G. Embryonic stem cell differentiation: emergence of a new era in biology and medicine. *Genes Dev.* 2005;19(10):1129–1155.
414. Douay L, Andreu G. *Ex vivo* production of human red blood cells from hematopoietic stem cells: what is the future in transfusion? *Transfus Med Rev.* 2007;21(2):91–100.
415. Calis JC, Phiri KS, Faragher EB, Brabin BJ, Bates I, Cuevas LE, *et al.* Severe anemia in malawian children. *N Engl J Med.* 2008;358(9):888–899.
416. Boele van Hensbroek M, Calis JC, Phiri KS, Vet R, Munthali F, Kraaijenhagen R, *et al.* Pathophysiological mechanisms of severe anaemia in malawian children. *PLoS One.* 2010;5(9):e12589.
417. Price EA, Schrier SL. Unexplained aspects of anemia of inflammation. *Adv Hematol.* 2010;2010:508739.
418. Price EA, Mehra R, Holmes TH, Schrier SL. Anemia in older persons: Etiology and evaluation. *Blood Cells Mol Dis.* 2011;46(2):159–165.

4
Erythrocyte Senescence

Giel JCGM Bosman,* Frans LA Willekens and Jan M Werre

*Department of Biochemistry,
Radboud University Nijmegen Medical Centre,
P.O. Box 9101, NL-6500 HB Nijmegen, The Netherlands*

4.1 Introduction

Appearance and disappearance of the human erythrocyte from the circulation after approximately 120 days are well-regulated processes. The classical view of the binding of a limited number of naturally occurring autoantibodies, leading to recognition and removal of senescent erythrocytes, has been extended with other mechanisms inducing phagocytosis. Erythrocyte aging starts from the very first appearance of the reticulocyte in the circulation, and consists of changes in volume, surface area, haemoglobin content, membrane structure and composition.

4.2 Erythrocyte Lifespan

When blood of a compatible but discordant blood group is given to a patient, the donor erythrocytes will disappear during the following weeks, as assessed by differential agglutination.[1,2] The measured survival curve is linear and shows the lifespan of transfused erythrocytes as approximately 120 days.[3,4] These observations suggested that erythrocytes have a limited lifespan. This hypothesis was confirmed by cohort labelling of erythrocytes through administration of a labelled precursor of various erythrocyte constituents during a short period of time. Haemoglobin was marked with the stable isotope ^{15}N by oral administration of ^{15}N-glycine

---

* Corresponding author: g.bosman@ncmls.ru.nl

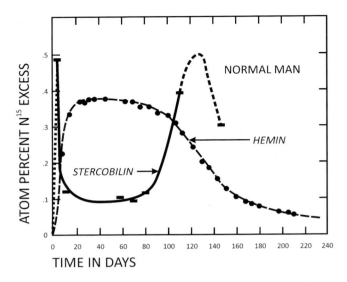

Figure 4.1 N^{15}-concentration in hemin and stercobilin of a normal man after the start of feeding N^{15}-labelled glycine for two days.[5]

for a few days.[5] In these experiments it was established that, after leaving the bone marrow, mature erythrocytes remain in the circulation for 120 ± 15 days (Fig. 4.1).

In a similar fashion, ^{59}Fe bound to transferrin is rapidly taken up by erythroblasts and reticulocytes resulting in a rapid increase in erythrocytic ^{59}Fe.[6,7] Although reutilization of iron prohibits a reliable measurement of the lifespan of the erythrocyte,[8] the ^{59}Fe labelling technique proved to be very useful for the validation of methods used to fractionate old and young erythrocytes.[9–12]

Similarly, the survival of homologous and heterologous erythrocytes can be measured by monitoring the disappearance of erythrocytes that are randomly labelled with $Na_2^{51}CrO_4$, which binds preferentially to haemoglobin without being reutilized.[8,13] Elution of the ^{51}Cr-label from the erythrocyte at a variable rate is a serious drawback of this technique, resulting in curvilinear disappearance curves and the need for correction factors.[8,13] Although in this way clinically significant reductions in erythrocyte survival can be measured, this method does not permit a direct determination of the erythrocyte lifespan.[14]

DF^{32}P (di-isopropyl phosphofluoridate) is another random label, which binds irreversibly to cholinesterase at the outside of the erythrocyte.[15] After reinjection of the marked erythrocytes, 5–10% of the label is lost during the first 24 hours, whereafter the disappearance curve remains linear.[16] As measured with DFP, erythrocyte lifespan ranged from 99 to 130 days.[13]

More recently, random biotinylation of erythrocytes *in vivo* or *ex vivo* has been used to study erythrocyte aging. Fluorochrome-labelled streptavidin allows enumeration of biotinylated erythrocytes with flow cytometry.[17–22] The results thus obtained were virtually identical to the results achieved with ^{51}Cr. However, development of anti-biotin antibodies may interfere with an accurate interpretation of the eventual data.[23] Also, biotin labelling may lead to changes in antigen expression.[24,25] In addition, a steep decrease in the number of biotinylated erythrocytes may occur after the consumption of eggs, which contain high concentrations of avidin.[26] In a recent study with biotin, the lifespan of the erythrocyte was shown to range from 100 to 140 days in healthy subjects.[27]

The results of a rather restricted number of measurements lead to the conclusion that the erythrocyte lifespan must be somewhere between 105 and 135 days in a healthy individual.[13]

4.3 The Fate of Haemoglobin

The use of labelled glycine in the first cohort studies[5] was very appropriate to study formation and breakdown of erythrocytes as well as haemoglobin. Glycine is not only incorporated into the globin moiety of haemoglobin, but is also an essential precursor of protoporphyrin that forms the haem moiety after the addition of iron. This allows measurement of the erythrocyte lifespan by the rate of disappearance of labelled erythrocytes as well as by the rate of appearance of the ^{15}N-label in stercobilin, one of the breakdown products of haem. After administration of ^{15}N-labelled glycine, two peaks in the levels of excreted ^{15}N-labelled bile pigment were found.[28] The early peak probably originates from intramedullarly destructed haemoglobin, indicating that a subset of the erythroblasts is removed early during erythropoiesis. A second peak after 120 days represents the removal of old erythrocytes. During the period between the peaks a low, stable amount of labelled stercobilin was measured (Fig. 4.1). This steady presence of traces of degraded haemoglobin was postulated to be due to continuous erythrocyte destruction.[8] Later measurements on erythrocyte fractions of various cell ages indicated another explanation for this ongoing excretion of haemoglobin degradation products.

4.4 Erythrocytes of Various Ages

In order to study the mechanisms that govern the lifespan of the erythrocyte, methods had to be developed to obtain pure populations of erythrocytes of defined ages. Around 1950 it became apparent that, after centrifugation, the fraction of the most dense cells at the bottom of the centrifuge tube is enriched for old erythrocytes.[9–11,29,30] However, attempts to validate density fractionation by means of cohort labelling of erythrocytes with ^{59}Fe showed the dense fractions to be more heterogeneous with respect to erythrocyte age than the lighter fractions. The latter were strongly enriched for reticulocytes and young erythrocytes.[8,10,31,32] Higher centrifugal forces did not improve the quality of the separation. Slightly better results were achieved with continuous or discontinuous density gradients,[33] but it became clear that age-related purification on the basis of density alone has inherent restrictions.[12,34] However, many changes measured in density-separated erythrocytes are still reported as age-related.[35]

In the meantime it had been recognized that cell shape in combination with cell volume allowed for a physical separation of the individual populations by counterflow centrifugation.[36,37] In this technique a flow of buffer is established from the outer to the inner end of the centrifugation chamber that counteracts the centrifugal force.[38,39] This yields fractions enriched for old and young erythrocytes, as shown with ^{59}Fe labelling.[12] When erythrocytes were separated by counterflow centrifugation and subsequently by density, validation with ^{59}Fe cohort labelling showed clearly that both separation methods are complementary and that this combination yields fractions with larger differences in erythrocyte age.[12]

Simultaneous measurement of erythrocyte volume (MCV) and haemoglobin concentration (MCHC) of individual erythrocytes with haemocytometry[40] showed no overlap between the lighter/larger and the

Erythrocyte Senescence

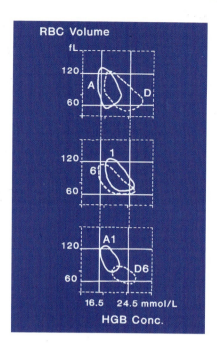

Figure 4.2 Erythrocytograms of erythrocyte fractions. Upper panel: Percoll separation; middle panel: counterflow centrifugation; lower panel: combination. A: lighter fractions in density separations alone; D: densest fraction in density separation alone; 1: larger fraction of the volume separation alone; 6: smaller fraction of the volume separation alone; A1: largest and lightest fraction of combined separation; D6: smallest and densest fraction of combined separation.[34] RBC, red blood cell; HGB conc., hemoglobin concentration.

denser/smaller erythrocyte populations resulting from the combination of separation techniques (Fig. 4.2), in contrast with the fractions obtained by the single separations.

The practical inconveniences and theoretical restrictions of the use of ^{59}Fe necessitate the use of other markers of cell age. Non-enzymatic glycation of haemoglobin (HbA0) leading to the intracellular accumulation of HbA1a, HbA1b and HbA1c has been shown to constitute a solid basis for the use of HbA1c in particular as a marker of cell aging.[41] In a healthy individual with a well-controlled glucose concentration and erythrocyte lifespan, there is a linear and reproducible increase in the percentage of HbA1c. The validity of HbA1c as a cell age marker was corroborated by

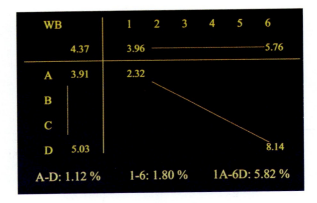

Figure 4.3 HbA1c% of density separation alone (A–D); volume separation alone[1–6]; combined separation: A1–D6[34] WB, whole blood.

measuring the specific ^{59}Fe radioactivity of HbA1c and HbA0 after administration of ^{59}Fe-labelled transferrin. The specific activity of HbA0 rapidly increased, whereas the specific activity of HbA1c showed a slow, linear increase, which flattened after approximately 80 days.[41]

Since the percentage of HbA1c can be accurately determined in very small erythrocyte numbers, it is evident that HbA1c is a very useful marker of cell age.[12,34,42] Measurement of HbA1c showed larger differences in the erythrocyte fractions acquired by combination of the separation methods compared with the erythrocytes separated by the single method (Fig. 4.3).

The experimental data of the cohort labelling with ^{59}Fe, the haemocytometry and the measurement of HbA1c all indicate that a combination of separation techniques yields erythrocyte fractions with a greater difference in mean cell age than separation on the basis of density or volume alone.[12,34]

4.5 Age-Dependent Changes in Erythrocyte Volume, Haemoglobin Content, Haemoglobin Concentration and Surface Area

Measurement of the percentage HbA1c allowed a correlation of the changes of various erythrocyte properties with cell age. A striking correlation is found between MCV and HbA1c percentage (Fig. 4.4). This indicates a strong age-related regulation of volume, although the details remain enigmatic.

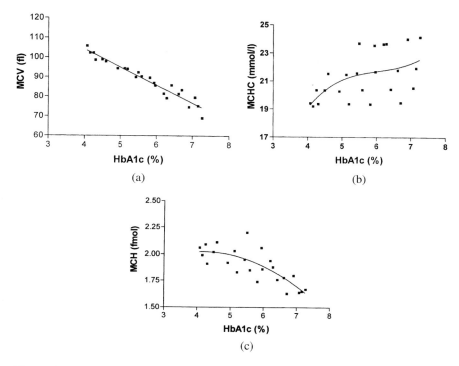

Figure 4.4 The relation between MCV (A), MCHC (B) and MCH (C) on the one hand and percentage HbA1c on the other. The straight line represents the regression line. The curved lines represent a second order curve fit. Each symbol represents the mean of five observations per fraction.[43]

Haemoglobin concentration (MCHC) and content (MCH) correlate to a lesser degree with age (Fig. 4.4). The MCHC mainly increases in the first half of the erythrocyte lifespan, which explains the results of the density separation. The MCH is already decreasing in the first half, but the loss of haemoglobin accelerates in the second half. The decrease in surface area was shown to be approximately 20%. Because the decrease in MCV is larger, the surface to volume ratio increases, which may increase deformability. However, this favourable effect is abolished by an increase of the MCHC and probably by a decreased elasticity of the erythrocyte membrane.[34,43]

Analysis of the various haemoglobin components indicated that the loss of HbA0 in the first half of erythrocyte life is due mainly to posttranslational modification, and in the second half to cellular loss. The amount of glycated haemoglobins increases, but this increase slows down during the second half of the erythrocyte lifespan, also due to cellular loss.[44] The combination of posttranslational modification and loss of cellular material results in a net loss of HbA0 of approximately 450 amol/cell, and a net gain of HbA1-components of approximately 85 amol/cell. The erythrocyte loses 365 haemoglobin amol/cell, which is 20% of the total haemoglobin. This loss explains the tapering increase in ^{59}Fe-labelled HbA1c after 60 days as observed earlier,[41] as well as the loss of ^{15}N-label from erythrocytes and the appearance of ^{15}N-label in stercobilin during the period between the early destruction of erythroblasts and the destruction of erythrocytes at 120 days after labelling (Fig. 4.1; Ref. 28).

In summary, it can be firmly stated that, during its life, the erythrocyte loses 30% of its volume and 20% of its haemoglobin, whereas its haemoglobin concentration increases by 14%. This also implies that, when they become older, erythrocytes lose proportionally more water than haemoglobin.[34] Furthermore, the spleen facilitates this process.[45]

4.6 The Function of the Spleen in the Maturation of Erythrocytes and Its Influence on Erythrocyte Lifespan

Analysis of the events that occur after splenectomy provides compelling evidence for a regulatory function of the spleen in the homeostasis of the erythrocyte.[46–51] After splenectomy, the erythrocytes display a steady, moderate increase in volume together with a larger increase in surface area, resulting in a higher area to volume ratio than observed in nonsplenectomized controls. These processes account for a higher osmotic resistance as well as for the emergence of extremely thin red blood cells, the so-called leptocytes.[46,51,52] Leptocytosis has also been found in patients with iron deficiency, thalassaemia and with diseases of the liver and bile duct.[53] Normally, reticulocytes are significantly larger than mature erythrocytes, and maturation of the reticulocyte is accompanied by a decrease in both cell volume and cell surface area, in association with an increase in haemoglobin concentration.[54–57] During reticulocyte maturation, membrane remodelling is associated with the release of exosomes, which are preformed in multivesicular bodies and contain, among others, the transferrin receptor.[58] In various animal models, splenectomy is followed by a delayed and incomplete reticulocyte maturation with less marked changes in volume and surface area as compared with reticulocytes of animals with a functional spleen.[54–56]

A study on the time course of the changes in reticulocytes and erythrocytes after splenectomy in humans did not show splenectomy-associated differences in the cellular volume of reticulocytes. However, there was a threefold increase in reticulocytes one week after splenectomy. After two months, a stable phase was established with a slightly higher than normal number of reticulocytes. Furthermore, no evidence was found that splenectomy stimulates erythropoiesis. Therefore, these results suggest a delayed maturation of reticulocytes after splenectomy.[49]

As discussed in the previous section, both volume and surface area of mature erythrocytes decline continuously during their lifetime. After splenectomy, old cells display a higher ratio of surface area to volume than younger cells, mainly due to a smaller decrease in surface area. Therefore, older erythrocytes contribute to the higher osmotic resistance of the erythrocyte population of a splenectomized individual.[51] These observations indicate that the effects of splenectomy are not restricted to the reticulocytes and young cells. The delayed reticulocyte maturation and impaired decrease in surface area after splenectomy imply a specialized function of the spleen that is only partially taken over by other organs.[47] Under normal circumstances, this specialized task is presumably exerted by the mononuclear phagocyte system of the spleen.

The proportional contribution of the phagocytes in bone marrow, spleen, liver and lungs to the elimination of old erythrocytes remains a matter of discussion. Since the lifespan of erythrocytes before and after splenectomy is similar,[7,59] it can be concluded that the presence of a spleen is not essential for the removal of old erythrocytes in humans.

4.7 Intra-Erythrocytic Vacuoles and the Spleen

In electron micrographs, vacuoles are found in the cytoplasm of erythrocytes from both normal, healthy subjects and splenectomized patients.[60]

The erythrocytes of splenectomized patients contain more small vacuoles and larger vacuoles than erythrocytes from subjects with a spleen.[61] The larger vacuoles are seen as pits in phase-contrast microscopy.[62] After splenectomy, the percentage of 'pitted' erythrocytes can be as high as 30%, whereas in healthy subjects only a small number of such pitted cells are found.[50,60,63] It should be noted that 4–10% pitted erythrocytes can be found in elderly people, due to an impaired splenic function in 25% of the elderly population.[64] Usually the vacuoles seem empty, but some studies indicate that they may contain amorphous material as well as membrane patches and, more frequently, haemoglobin or haemoglobin degradation products.[60,63,65,66] It has been suggested that the pits in erythrocytes are related to the autophagic vacuoles in erythroblasts and reticulocytes.[65] When erythrocytes from healthy donors with an intact spleen are transfused into an asplenic recipient, pits identical to those seen in the recipient's erythrocytes will emerge in the donor erythrocytes. Conversely, when erythrocytes from a splenectomized individual are transfused to control recipients, the pits disappear rapidly.[63] Hence, it can be concluded that normal erythrocytes may carry pits that are removed or dissolved under the influence of the spleen. It has been argued that haemoglobin or haemoglobin-derived products play a key role in pit formation and elimination.[60,63,65,66] This shows analogy to the pitting of other inclusion bodies such as Heinz bodies, Howell–Jolly bodies and siderotic granules, which constitutes a specialized function of the spleen.[67] The increase in intracellular HbA1 found after splenectomy may be related to the retention of vacuoles within the erythrocytes, suggesting a role of the spleen in the removal of haemoglobin and haemoglobin derivatives.[50] It has been suggested that under normal conditions haemoglobin is eliminated by the release of haemoglobin-containing vesicles.[67–69]

4.8 Removal of Senescent Erythrocytes

Old erythrocytes are removed from the circulation by the mononuclear phagocyte system. The older erythrocytes have indeed been shown to contain IgG, which induces phagocytosis.[70] This IgG has been shown to consist of naturally occurring autoantibodies that are directed to a senescent cell antigen (SCA). SCA is formed by epitopes on band 3 and is thought to be generated by a conformational change of that band, which occurs in the final stages of the lifespan of the erythrocyte.[71–73] This conformational change is probably the result of a breakdown of band 3.[73] High molecular-weight forms of band 3 or band 3-containing complexes that are reactive with autoantibodies have also been described as specific for old erythrocytes.[74–76] Thus, at present there is no consensus on whether the crucial changes leading to senescent cell antigen activity are caused by band 3 degradation or cross-linking.[73,77] Others noticed that binding of deoxyhaemoglobin and denatured haemoglobin to the cytoplasmic part of band 3 induces band 3 aggregation,[78] although this has been disputed.[79] They postulate that this aggregation is the actual trigger for the formation of the senescent cell antigen.

Using erythrocyte fractions obtained by the combined separation method as described above, previous reports[73] on the occurrence of structural and immunological changes in band 3 during erythrocyte aging *in vivo* were basically confirmed.[80] Changes in band 3 were especially apparent with antibodies reactive with epitopes in the N-terminal part of the membrane domain. These antibodies show an age-related increase in immunoreactivity, leading to the conclusion that age-related changes in band 3 occur principally in regions that participate in the generation of senescent cell antigens (Table 4.1). Subtle changes occurring in intact band 3 molecules may generate the activity of senescent cell antigen; breakdown of band 3 may not be necessary for this process.[73]

Table 4.1 Semi-quantitative analysis of the reactivity of anti-band 3 antibodies in red blood cells of various ages and in vesicles.

Band 3 epitope (amino acid number)	Presence in fraction I → V	Presence in vesicles
25–35	++ → ++	0
812–827	++ → +	0
390–550	+ → ++	+
542–555	+ → ++	+
840–911	++ → +++	+
562–565	0 → ++	+++
566–569	0 → 0	++

0: not present; the number of +'s is based on an arbitrary densitometry scale.[80]

A role for complement activation as a cofactor in the phagocytosis of senescent erythrocytes has also been proposed.[81] When erythrocytes of various ages were probed with antibodies against the anti-complement proteins CD55 and CD59, the mean fluorescence intensity decreased with erythrocyte age. However, the mean surface decreased even more, resulting in a small increase with aging in the number of CD55 and CD59 per unit of surface area.[80,82] Thus, full protection against complement activation is still present on old erythrocytes. This indicates a less probable role for activated complement factors in phagocytosis of old erythrocytes.

Recent data have pointed towards CD47 as yet another molecule that may be involved in recognition of senescent erythrocytes by macrophages.[83] CD47, an integrin-associated membrane protein, mediates adhesion-associated processes and modulates phagocytosis as a signaling molecule of self.[84] Erythrocytes from CD47 knockout mice with a normal lifespan rapidly disappeared from the circulation when brought into wild-type recipients. So, regulation of CD47 expression might serve as a mechanism to control elimination of senescent RBC. However, erythrocytes from heterozygous animals survived normally in wild-type mice, which argues against a gradual disappearance of CD47 as the sole mechanism regulating erythrocyte survival. It is more likely that the absence of CD47 triggers erythrocyte removal by other processes not yet identified in mice. In humans, CD47 is part of the Rhesus core complex, and $Rhesus_{null}$ individuals with less than 25% CD47 have haemolytic anaemia and stomatocytosis, but there is no evidence that Rh_{null} erythrocytes are phagocytized more readily than erythrocytes with normal amounts of CD47.[85]

Exposure of phosphatidylserine is widely implicated as a trigger for the removal of senescent erythrocytes,[86] notwithstanding the paucity of data. In fact, flow cytometry data show the virtual absence of erythrocytes exposing phosphatidylserine, as measured by the capacity to bind annexin V, in all age fractions.[80,87] The arguments for this theory are mainly based on pathological data, and on results from manipulation *in vitro*. Indeed, in erythrocyte pathologies such as sickle cell anaemia, thalassaemia and spherocytosis, a decrease in erythrocyte survival is associated with an increase in phosphatidylserine exposure.[86] 'Aging' treatments resulting in an increased calcium influx and oxidation also inhibit flippase and/or activate scramblase, both leading to loss of phospholipid asymmetry.[88] In general, all processes that lead to increased erythrocyte destruction are associated with an increased exposure of phosphatidylserine at the erythrocyte surface. In view of these data, together with the general role that phosphatidylserine plays in recognition and removal of apoptotic cells,[89] a role for phosphatidylserine in physiological erythrocyte aging may get the benefit of the doubt. The possibility that exposure of phosphatidylserine leads to such a fast removal that it is underestimated cannot be excluded.[80,86,87]

In conclusion, the accumulation of autologous, high-affinity and band 3-directed IgG during the final part of the erythrocyte lifespan triggers, above a relatively low threshold, binding and phagocytosis of senescent erythrocytes. Phagocytosis may be promoted by the exposure of phosphatidylserine.

4.9 Erythrocyte-Derived Vesicles and Erythrocyte Senescence

Part of the vesicles that have been found in plasma are derived from erythrocytes.[45,69,90,91]

Freeze-fracture electron microscopy demonstrated that the size of the vesicles shows a large heterogeneity. The plasma-derived vesicle population contains small vesicles (<200 nm) with right-side-out-oriented and inside-out-oriented membranes, as well as vesicles with a wide range in size (200–800 nm) with predominantly right-side-out membranes. This hints at two mechanisms of vesiculation: the larger vesicles may be pinched off and the smaller vesicles may originate by pinching off as well as by fragmentation (Fig. 4.5).

Analysis of the vesicle haemoglobin content shows a striking resemblance to the relative distribution of haemoglobin components in old erythrocytes (Table 4.2).

This supports the conclusion that the continuous loss of haemoglobin in vesicles accelerates during the second half of the erythrocyte lifespan.[45] Vesicles disappear rapidly from the circulation: in a rat model, Kupffer cells in the liver remove naturally occurring, haemoglobin-containing, erythrocyte-derived vesicles from the circulation within minutes, mainly by scavenger receptors and with phosphatidylserine as the principal ligand (Fig. 4.6).

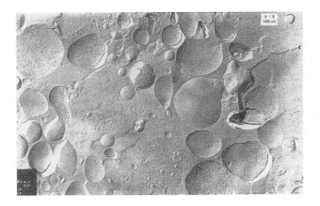

Figure 4.5 Plasma contains small (<200 nm) vesicles with right-side-out and inside-out-oriented membranes, and larger (200–800 nm) vesicles with predominantly right-side-out membranes.[45]

Table 4.2 Haemoglobin components in young and old RBCs and vesicles obtained by plasmapheresis from control individuals.

Hb component	Percentage of total haemoglobin ± SD					
	Young RBCs* (n = 5)		Old RBCs† (n = 5)		Vesicles (n = 4)	
HbA1	9.51	±2.48‡	16.71	±3.93	16.40	±2.69
HbA1a	0.85	±0.34	1.61	±0.75	1.28	±0.37
HbA1b	1.29	±0.62	2.03	±1.33	1.29	±0.17
HbA1c	4.07	±1.22‡	7.27	±0.63	6.83	±0.90
HbA1e1	1.00	±0.78	1.99	±1.82	2.18	±1.38
HbA1e2	2.30	±0.36‡	3.81	±0.49‡	4.90	±0.23
HbA0	82.15	±3.34	74.76	±4.77	78.50	±3.24
HbA2	3.94	±0.73‡	3.44	±0.69‡	2.62	±0.14
HbF	0.40	±0.18	0.43	±0.13	0.27	±0.11
HbX + rest	4.00	±2.17	4.66	±3.63	2.13	±0.59

*, † Fractions with lowest and highest HbA_{1c} values, respectively.
‡ Significantly different ($P < 0.05$) from vesicles (Mann–Whitney U test, one-tailed P value).

Interestingly, both immunoblot and proteomics data fail to demonstrate the presence of the cytoskeletal proteins spectrin and ankyrin in vesicles, but do show the presence of acetylcholinesterase and band 3.[80,92] A similar pattern is observed in vesicles produced during erythrocyte

Erythrocyte Senescence

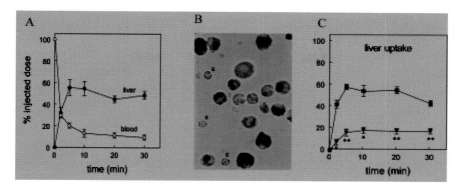

Figure 4.6 A: within 5 minutes, ^{51}Cr-chromate-labelled vesicles are removed from the circulation with a concomitant uptake by the liver; B: nearly all liver activity is found in Kupffer cells, with a concomitant uptake of haemoglobin; C: this uptake is inhibited by preinjection of phosphatidylserine liposomes (t), but not phosphatidylcholine liposomes (■).[92]

Table 4.3 Characterization of (a) glycophorin A-positive* and (b) IgG-positive[†] vesicles.

	Positive for	Percentage Mean ± SD	n
A	IgG	39 ± 11	6
	PS	67 ± 17	4
B	Glycophorin A	85 ± 14	6
	PS	85 ± 6.1	6

* The results are shown as the percentage of glycophorin A-positive vesicles that are positive as well for IgG and PS respectively.
[†] The results are shown as the percentage of IgG positive vesicles that are positive for glycophorin A and PS respectively, as well.[80]

storage in the blood bank.[93] Approximately 70% of the plasma vesicles derived from human erythrocytes expose PS. A considerable portion of the PS-exposing vesicles also contains IgG (Table 4.3, Ref. 80). Clearance of vesicles may not only be facilitated by PS-mediated phagocytosis, but also by the immune-mediated mechanism responsible for the removal of old erythrocytes.[73]

The plasma of healthy individuals contains approximately 170 erythrocyte-derived vesicles per microlitre.[80,94] It can be calculated that every erythrocyte produces approximately two vesicles per day, which implies that in every second 580.10^6 vesicles are produced, and that half of all the vesicles are removed from the circulation in less than one second.

This is at least one order of magnitude shorter than one cycle through the circulation, and much shorter than the half-life found for vesicles in the rat model. As the liver and other organs are equally effective in removing erythrocyte-derived vesicles, it was concluded that most vesicles are directly trapped by the macrophages of the organ in which they originate, i.e. before they can reach the venous circulation.[80] This mechanism restricts the potentially harmful thrombotic effect of vesicles, which arises from the coagulation-promoting activity of the PS-exposing vesicle membrane.[94] It also has been shown that secretory phospholipase A2 uses the PS-exposing membrane to generate lysophosphatidic acid (LPA).[95] LPA has been shown to open Ca^{2+}-channels in erythrocytes, possibly leading to further vesiculation.[96,97]

Almost all vesicles carry the complement-regulating proteins CD55 and CD59. The presence of these complement protection proteins indicates that, under physiological circumstances, complement activation does not play an important role in vesicle formation or vesicle clearance.[80]

Furthermore, vesicle formation is accompanied by the breakdown of band 3. Intact band 3 (95 kDa) is present in vesicles, but proteins with a molecular weight of 70 kDa or less dominated the immunoblot picture. A strong reaction is observed with antibodies directed against membrane areas of band 3 that are involved in senescent cell antigen activity.[80] These data suggest that vesicle formation *in vivo* is associated with changes in the structure and/or breakdown of band 3, a process that probably results in an increased exposure of senescent cell antigen-related epitopes on vesicles.

It can be concluded that vesiculation serves to remove erythrocyte membrane patches associated with damaged cell components, containing phagocytosis-inducing removal signals that facilitate a rapid clearance. In this way, vesiculation protects the erythrocyte against untimely removal.[80] The same strategy may underlie the shedding of vesicles containing the complement membrane attack complex upon influx of Ca^{2+}.[98,99]

The surface area of the shed vesicles considerably exceeds the loss of surface area of the erythrocyte during its lifespan. This extra loss may be compensated for by the incorporation of lipids from plasma lipoproteins into the erythrocyte membrane. Cholesterol and phosphatidylcholine are exchanged readily, whereas phosphatidylethanolamine and phosphatidylserine are exchanged to a lesser degree.[100–103] When the erythrocyte replenishes its deficit of lipids, the lipid composition of the membrane changes due to these differences. This could limit the capacity to vesiculate.

4.10 A New Model: Vesiculation as an Essential Process in Erythrocyte Aging

Taken together, erythrocyte vesiculation should be regarded as a hitherto undiscovered aspect of erythrocyte aging. We integrated vesiculation in a new hypothesis of erythrocyte aging. The first step in this hypothesis is protein modification, especially oxidation and proteolysis of haemoglobin and/or band 3. This results in band 3 breakdown and binding of denatured haemoglobin to band 3 and membrane-bound band 3 breakdown products. This process disturbs the anchorage of the lipid bilayer to the cytoskeleton. The membrane exerts a stretching force on the cytoskeleton network, and a disturbance of the cytoskeleton-membrane interaction diminishes this stretching force. This causes, at the site of membrane damage, relaxation of the cytoskeleton, which drives the plasma membrane to buckling. The buckle pinches off easily, forming a vesicle with a surface area of one cytoskeleton unit. The model of the stretched cytoskeleton predicts the transition of excess surface area from neighbouring cytoskeleton network units to the membrane buckle.[104] This causes the shedding of larger vesicles. In this hypothesis, neither the loss of lipid asymmetry of the membrane nor the binding of autologous IgG is necessary. Alternatively, aggregation of band 3 breakdown products, possibly driven by binding of denatured haemoglobin, may contribute to vesiculation by inducing an increase in membrane curvature. In experimental erythrocyte vesiculation, the shape transition of the erythrocyte from a discoid cell to an echinocyte appears to be essential. The vesicles may pinch off at the top of echinocytic protrusions, where the membrane curvature is extremely large.

The presence in plasma of PS-exposing, erythrocyte-derived vesicles implicates additional mechanisms leading to the loss of lipid asymmetry. An increased intracellular Ca^{2+}-concentration plays a central role in the loss of lipid asymmetry in platelet activation and in various erythrocyte vesiculation models. An increased Ca^{2+}-concentration inhibits the enzyme flippase, which is responsible for the maintenance of lipid asymmetry, and activates scramblase, responsible for the rapid and aspecific exchange of membrane lipids between the inner leaflet and outer leaflet of the lipid bilayer. At the site of band 3-associated, haemoglobin-induced damage, volume-sensitive cation channels could be activated or the activity of Ca^{2+}-ATPase may become impaired. As a result of the locally increased Ca^{2+}-concentration, not only does PS appear at the outer leaflet of the membrane, but Ca^{2+}-dependent proteolysis and breakdown of phosphatidylinositol-4-phosphate and phosphatidylinositol-4,5-biphosphate are also induced. All these effects contribute to a further weakening of the interaction between cytoskeleton and the lipid bilayer, inducing vesicle shedding. Whether the release of K^+ with a concomitant loss of water through the activation of the Gardos channel is as important *in vivo* as it seems to be *in vitro* remains to be established.

The PS-exposing vesicles partially contain IgG, which presumably recognizes senescent cell antigens. Binding of immunoglobulins influences the shape of the plasma membrane by dimerization, or oligomerization of membrane components occurs. In experimental conditions, binding of antibodies to various membrane proteins such as glycophorin A, acetylcholinesterase or glycophorin C causes vesiculation. Thus, binding of natural, autologous IgG to senescent cell antigens may contribute to aging-associated vesiculation.

Bibliography

1. Todd C, White RG. Fate of red blood corpuscles when injected into circulation of animal of same species. *Proc Royal Soc London Series B: Biol Sci.* 1911;84:255–259.
2. Ashby W. The determination of the length of life of transfused blood corpuscles in man. *J Exp Med.* 1919;29:267–281.
3. Callender ST, Powell EO, Witts CJ. Lifespan of red cell in man. *J Pathol Bacteriol.* 1945;57:129–139.

4. Dornhorst AC. Interpretation of red cell survival curves. *Blood.* 1951;6: 1284–1292.
5. Shemin D, Rittenberg D. The lifespan of the human red blood cell. *J Biol Chem.* 1946;166:627–636.
6. Finch CA, Wolff JA, Rath CE, Fluharty RG. Iron metabolism; erythrocyte iron turnover. *J Lab Clin Med.* 1949;34:1480–1490.
7. Berlin NI, Berk PO. The biological life of the red blood cell. In: Surgenor DM, editor. *The Red Blood Cell.* New York: Academic Press; 1975, pp. 957–1019.
8. Berlin NI, Waldmann TA, Weiszman SM. Lifespan of red blood cells. *Physiol Rev.* 1959;39:577–616.
9. Borun ER, Figueroa WG, Perry SM. The distribution of Fe59 tagged human erythrocytes in centrifuged specimens as a function of cell age. *J Clin Invest.* 1957;36:676–679.
10. Hoffman JE. On the relationship of certain erythrocyte characteristics to their physiological age. *J Cell Comp Physiol.* 1958;51:415.
11. Prankerd TA. The ageing of red cells. *J Physiol.* 1958;143:325–331.
12. Van der Vegt SG, Ruben AM, Werre JM, Palsma DM, Verhoef CW, De Gier J *et al.* Counterflow centrifugation of red cell populations: a cell age related separation technique. *Brit J Haematol.* 1985;61:393–403.
13. Mollison PL, Engelfriet CP, Contreras M. The transfusion of red cells. In: Mollison PL, Engelfriet CP, Contreras M. *Blood Transfusion in Clinical Medicine.* Oxford: Blackwell Scientific; 1987, pp. 95–158.
14. Cavill I. Red cell lifespan estimation by ^{51}Cr labelling. *Brit J Haematol.* 2002;117:998.
15. Cohen JA, Warringa MG. The fate of P32 labelled diisopropylfluorophosphonate in the human body and its use as a labelling agent in the study of the turnover of blood plasma and red cells. *J Clin Invest.* 1954;33:459–467.
16. Bratteby LE, Wadman B. Labelling of red blood cells *in vitro* with small amounts of di-iso-propyl-flurophosphonate (DF32P). *Scand J Clin Lab Invest.* 1968;21:197–201.
17. Suzuki T, Dale GL. Biotinylated erythrocytes: *in vivo* survival and *in vitro* recovery. *Blood.* 1987;70:791–795.
18. Hoffmann-Fezer G, Maschke H, Zeitler HJ, Gais P, Heger W, Ellwart J *et al.* Direct *in vivo* biotinylation of erythrocytes as an assay for red cell survival studies. *Ann Hematol.* 1991;63:214–217.

19. Hoffmann-Fezer G, Mysliwietz J, Mortlbauer W, Zeitler HJ, Eberle E, Honle U et al. Biotin labeling as an alternative nonradioactive approach to determination of red cell survival. *Ann Hematol.* 1993;67:81–87.
20. Mock DM, Lankford GL, Widness JA, Burmeister LF, Kahn D, Strauss RG. Measurement of circulating red cell volume using biotin-labeled red cells: validation against ^{51}Cr-labeled red cells. *Transfusion.* 1999;39:149–155.
21. Mock DM, Lankford GL, Widness JA, Burmeister LF, Kahn D, Strauss RG. Measurement of red cell survival using biotin-labeled red cells: validation against ^{51}Cr-labeled red cells. *Transfusion.* 1999;39:156–162.
22. De Jong K, Emerson RK, Butler J, Bastacky J, Mohandas N, Kuypers FA. Short survival of phosphatidylserine-exposing red blood cells in murine sickle cell anemia. *Blood.* 2001;98:1577–1584.
23. Cordle DG, Strauss RG, Lankford G, Mock DM. Antibodies provoked by the transfusion of biotin-labeled red cells. *Transfusion.* 1999;39:1065–1069.
24. King MJ, Hemming N. An assessment of the immunoprecipitation of blood-group antigens using biotinylated red cells. *Transf Med.* 1994;4:195–204.
25. Cowley H, Wojda U, Cipolone KM, Procter JL, Stroncek DF, Miller JL. Biotinylation modifies red cell antigens. *Transfusion.* 1999;39:163–168.
26. Cavill I, Trevett D, Fisher J, Hoy T. The measurement of the total volume of red cells in man: a non-radioactive approach using biotin. *Brit J Haematol.* 1988;70:491–493.
27. Cohen RM, Franco RS, Khera PK, Smith EP, Lindsell CJ, Ciraolo PJ et al. Red cell lifespan heterogeneity in hematologically normal people is sufficient to alter HbA1c. *Blood.* 2008;112:4284–4291.
28. London JM, West R, Shemin D, Rittenberg D. On the origin of bile pigment in normal man. *J Biol Chem.* 1950;184:351–358.
29. Allison AC, Burn GP. Enzyme activity as a function of age in the human erythrocyte. *Brit J Haematol.* 1955;1:291–303.
30. Bernstein RE. Alterations in metabolic energetics and cation transport during aging of red cells. *J Clin Invest.* 1959;38:1572–1586.
31. Garby L, Hjelm M. Ultracentrifugal fractionation of human erythrocytes with respect to cell age. *Blut.* 1963;9:284–291.
32. Van Gastel C. De lipiden en enkele eigenschappen van de menselijke erythrocyt met betrekking tot zijn leeftijd. 1965; thesis, Rijksuniversiteit Utrecht.

33. Corash LM, Piomelli S, Chen HC, Seaman C, Gross E. Separation of erythrocytes according to age on a simplified density gradient. *J Lab Clin Med*. 1974;84:147–151.
34. Bosch FH, Werre JM, Roerdinkholder-Stoelwinder B, Huls TH, Willekens FL, Halie MR. Characteristics of red blood cell populations fractionated with a combination of counterflow centrifugation and Percoll separation. *Blood*. 1992;79:254–260.
35. Glader B. Destruction of erythrocytes. In: Greeg J, Foester J, Lukens J, Rodgers F, Paraskevas F, Glader B, editors. *Wintrobe's Clinical Hematology*. Philadelphia: Lippincott; 2004, pp. 249–265.
36. Lindbergh CA. A method for washing corpuscles in suspension. *Science*. 1932;75:415–416.
37. Lindahl PE. Principle of a counter-streaming centrifuge for the separation of particles of different sizes. *Nature*. 1948;161:648–649.
38. Sanderson RJ, Bird KE, Palmer NF, Brenman J. Design principles for a counterflow centrifugation cell separation chamber. *Anal Biochem*. 1976;71:615–622.
39. Sanderson RJ, Bird KE. Cell separations by counterflow centrifugation. *Meth Cell Biol*. 1977;15:1–14.
40. Mohandas N, Kim YR, Tycko DH, Orlik J, Wyatt J, Groner W. Accurate and independent measurement of volume and hemoglobin concentration of individual red cells by laser light scattering. *Blood*. 1986;68:506–513.
41. Bunn HF, Haney DN, Kamin S, Gabbay KH, Gallop PM. The biosynthesis of human hemoglobin A1c. Slow glycosylation of hemoglobin *in vivo*. *J Clin Invest*. 1976;57:1652–1659.
42. Lasch J, Kullertz G, Opalka JR. Separation of erythrocytes into age-related fractions by density or size? Counterflow centrifugation. *Clin Chem Lab Med*. 2000;38:629–632.
43. Bosch FH, Werre JM, Schipper L, Roerdinkholder-Stoelwinder B, Huls TH, Willekens FLA *et al.* Characteristics of red blood cell populations fractionated with a combination of counterflow centrifugation and Percoll separation. *Eur J Haematol*. 1994;52:35–41.
44. Willekens FL, Bosch FH, Roerdinkholder-Stoelwinder B, Groenen-Dopp YA, Werre JM. Quantification of loss of haemoglobin components from the circulating red blood cell *in vivo*. *Eur J Haematol*. 1997;58:246–250.

45. Willekens FL, Roerdinkholder-Stoelwinder B, Groenen-Dopp YA, Bos HJ, Bosman GJ, van den Bos AG et al. Hemoglobin loss from erythrocytes in vivo results from spleen-facilitated vesiculation. Blood. 2003;101: 747–751.
46. Crosby WH. The pathogenesis of spherocytes and leptocytes (target-cells). Blood. 1952;7:261–274.
47. Crosby WH. Splenic remodeling of red cell surfaces. Blood. 1977;50: 643–645.
48. Clark MR, Shohet SB. Red cell senescence. Clin. Haematol. 1985;14: 223–257.
49. De Haan LD, Werre JM, Ruben AM, Huls AH, De Gier J, Staal GE. Reticulocyte crisis after splenectomy: evidence for delayed red cell maturation? Eur J Haematol. 1988;41:74–79.
50. De Haan LD, Werre JM, Ruben AM, Huls AH, De Gier J, Staal GE. Vacuoles in red cells from splenectomized subjects originate during cell life: association with glycosylated haemoglobin? Eur J Haematol. 1988;41:482–488.
51. De Haan LD, Werre JM, Ruben AM, Huls HA, De Gier J, Staal GE. Alterations in size, shape and osmotic behaviour of red cells after splenectomy: a study of their age dependence. Brit J Haematol. 1988;69:71–80.
52. Cooper RA, Jandl JH. The role of membrane lipids in the survival of red cells in hereditary spherocytosis. J Clin Invest. 1969;48:736–744.
53. Werre JM. Oorzaken van macroplanie van erythrocyten bij lever- en galwegenziekten. 1968; thesis, Rijksuniversiteit Utrecht.
54. Winterbourn CC, Batt RD. Lipid composition of human red cells of different ages. Biochimica et Biophysica Acta. 1970;202:1–8.
55. Shattil SJ, Cooper RA. Maturation of macroreticulocyte membranes in vivo. J Lab Clin Med. 1972;79:215–227.
56. Snyder LM, Fairbanks G, Trainor J, Fortier NL, Jacobs JB, Leb L. Properties and characterization of vesicles released by young and old human red cells. Brit J Haematol. 1985;59:513–522.
57. Houwen B. Reticulocyte maturation. Blood Cells. 1992;18:167–186.
58. Johnstone RM. Revisiting the road to the discovery of exosomes. Blood Cells Mol Dis. 2005;34:214–219.
59. Gevirtz NR, Nathan DG, Berlin NI. Erythrokinetic studies in primary hypersplenism with pancytopenia. Am J Med. 1962;32:148–152.

60. Koyama S, Kihira H, Aoki S, Ohnishi H. Postsplenectomy vacuole, a new erythrocytic inclusion body. *Mie Medical Journal* 1962;11:425–437.
61. Reinhart WH, Chien S. Red cell vacuoles: their size and distribution under normal conditions and after splenectomy. *Am J Hematol*. 1988;27:265–271.
62. Holroyde CP, Oski FA, Gardner FH. The "pocked" erythrocyte. Red-cell surface alterations in reticuloendothelial immaturity of the neonate. *New Engl J Med*. 1969;281:516–520.
63. Holroyde CP, Gardner FH. Acquisition of autophagic vacuoles by human erythrocytes. Physiological role of the spleen. *Blood*. 1970;36:566–575.
64. Markus HS, Toghill PJ. Impaired splenic function in elderly people. *Age Ageing*. 1991;20:287–290.
65. Kent G, Minick OT, Volini FI, Orfei E. Autophagic vacuoles in human red cells. *Am J Pathol*. 1966;48:831–857.
66. Schnitzer B, Rucknagel DL, Spencer HH, Aikawa M. Erythrocytes: pits and vacuoles as seen with transmission and scanning electron microscopy. *Science*. 1971;173:251–252.
67. Crosby WH. Normal functions of the spleen relative to red blood cells: a review. *Blood*. 1959;14:399–408.
68. Lux SE, John KM. Isolation and partial characterization of a high molecular weight red cell membrane protein complex normally removed by the spleen. *Blood*. 1977;50:625–641.
69. Dumaswala UJ, Greenwalt TJ. Human erythrocytes shed exocytic vesicles *in vivo*. *Transfusion*. 1984;24:490–492.
70. Kay MMB. Mechanism of removal of senescent cells by human macrophages in situ. *Proc Natl Acad Sci USA*. 1975;72:3521–3525.
71. Kay MMB. Role of physiologic autoantibody in the removal of senescent human red cells. *J Supramol Struct*. 1978;9:555–567.
72. Kay MMB, Goodman SR, Sorensen K, Whitfield CF, Wong P, Zaki L et al. Senescent cell antigen is immunologically related to band 3. *Proc Natl Acad Sci USA*. 1983;80:1631–1635.
73. Kay MMB. Immunoregulation of cellular lifespan. *Ann New York Acad Sci*. 2005;1057:85–111.
74. Belo L, Rebelo I, Castro EM, Catarino C, Pereira-Leite L, Quintanilha A et al. Band 3 as a marker of erythrocyte changes in pregnancy. *Eur J Haematol*. 2002;69:145–151.

75. Ciccoli L, Rossi V, Leoncini S, Signorini C, Blanco-Garcia J, Aldinucci C et al. Iron release, superoxide production and binding of autologous IgG to band 3 dimers in newborn and adult erythrocytes exposed to hypoxia and hypoxia-reoxygenation. *Biochimica et Biophysica Acta.* 2004;1672:203–213.
76. Rossi V, Leoncini S, Signorini C, Buonocore G, Paffetti P, Tanganelli D et al. Oxidative stress and autologous immunoglobulin G binding to band 3 dimers in newborn erythrocytes. *Free Rad Biol Med.* 2006;40: 907–915.
77. Arese P, Turrini F, Schwarzer E. Band 3/complement-mediated recognition and removal of normally senescent and pathological human erythrocytes. *Cell Physiol Biochem.* 2005;16:133–146.
78. Low PS, Waugh SM, Zinke K, Drenckhahn D. The role of hemoglobin denaturation and band 3 clustering in red blood cell aging. *Science.* 1985;227:531–533.
79. Lelkes G, Fodor I, Lelkes G, Hollan SR, Verkleij AJ. The distribution and aggregatability of intramembrane particles in phenylhydrazine-treated human erythrocytes. *Biochimica et Biophysica Acta.* 1988;945:105–110.
80. Willekens FL, Werre JM, Groenen-Dopp YA, Roerdinkholder-Stoelwinder B, de Pauw B, Bosman GJ. Erythrocyte vesiculation: a self-protective mechanism? *Brit J Haematol.* 2008;141:549–556.
81. Lutz HU. Innate immune and non-immune mediators of erythrocyte clearance. *Cell Mol Biol.* 2004;50:107–116.
82. Willekens FL, Bos HJ, Roerdinkholder-Stoelwinder B, Groenen-Dopp YA, Van der Plas GC, Werre JM. Loss of anticomplement activity from circulating old red cells. *Brit J Haematol.* 1998;102:302.
83. Oldenborg PA, Zheleznyak A, Fang YF, Lagenaur CF, Gresham HD, Lindberg FP. Role of CD47 as a marker of self on red blood cells. *Science.* 2000;288:2051–2054.
84. McDonald JF, Zheleznyak A, Frazier WA. Cholesterol-independent interactions with CD47 enhance alphavbeta3 avidity. *J Biol Chem.* 2004;279: 17301–17311.
85. Arndt PA and Garratty G. Rh(null) red blood cells with reduced CD47 do not show increased interactions with peripheral blood monocytes. *Brit J Haematol.* 2004;125:412–414.
86. Kuypers FA, de Jong K. The role of phosphatidylserine in recognition and removal of erythrocytes. *Cell Mol Biol.* 2004;50:147–158.

87. Willekens FL, Bos HJ, Roerdinkholder-Stoelwinder B, Groenen-Dopp YA, Van der Plas GC, Werre JM. Conservation of lipid asymmetry in circulating old red cells. *Brit J Haematol.* 1998;102:302.
88. Hilarius PM, Ebbing IG, Dekkers DW, Lagerberg JW, de Korte D, Verhoeven AJ. Generation of singlet oxygen induces phospholipid scrambling in human erythrocytes. *Biochemistry.* 2004;43:4012–4019.
89. Krieser RJ, White K. Engulfment mechanism of apoptotic cells. *Curr Opin Cell Biol.* 2002;14:734–738.
90. Berckmans RJ, Nieuwland R, Boing AN, Romijn FP, Hack CE, Sturk A. Cell-derived microparticles circulate in healthy humans and support low grade thrombin generation. *Thromb Haemost.* 2001;85:639–646.
91. Shet AS, Aras O, Gupta K, Hass MJ, Rausch DJ, Saba N *et al.* Sickle blood contains tissue factor-positive microparticles derived from endothelial cells and monocytes. *Blood.* 2003;102:2678–2683.
92. Willekens FL, Werre JM, Kruijt JK, Roerdinkholder-Stoelwinder B, Groenen-Dopp YA, van den Bos AG *et al.* Liver Kupffer cells rapidly remove red blood cell-derived vesicles from the circulation by scavenger receptors. *Blood.* 2005;105:2141–2145.
93. Bosman GJ, Lasonder E, Groenen-Dopp YA, Willekens FL, Werre JM, Novotny VM. Comparative proteomics of erythrocyte aging *in vivo* and *in vitro. J Prot.* 2010;73:396–402.
94. Simak J, Gelderman MP. Cell membrane microparticles in blood and blood products: potentially pathogenic agents and diagnostic markers. *Transf Med Rev.* 2006;20:1–26.
95. Fourcade O, Simon MF, Viode C, Rugani N, Leballe F, Ragab A *et al.* Secretory phospholipase A2 generates the novel lipid mediator lysophosphatidic acid in membrane microvesicles shed from activated cells. *Cell.* 1995;80:919–927.
96. Yang L, Andrews DA, Low PS. Lysophosphatidic acid opens a Ca(++) channel in human erythrocytes. *Blood.* 2000;95:2420–2425.
97. Chung SM, Bae ON, Lim KM, Noh JY, Lee MY, Jung YS *et al.* Lysophosphatidic acid induces thrombogenic activity through phosphatidylserine exposure and procoagulant microvesicle generation in human erythrocytes. *Arterioscl Thromb Vasc Biol.* 2007;27:414–421.
98. Sims PJ, Wiedmer T. Repolarization of the membrane potential of blood platelets after complement damage: evidence for a Ca++-dependent exocytotic elimination of C5b-9 pores. *Blood.* 1986;68:556–561.

99. Iida K, Whitlow MB, Nussenzweig V. Membrane vesiculation protects erythrocytes from destruction by complement. *J Immunol.* 1991;147:2638–2642.
100. Reed CF. Phospholipid exchange between plasma and erythrocytes in man and the dog. *J Clin Invest.* 1968;47:749–760.
101. Van Deenen LL, De Gier J. Lipids of the red cell membrane. In: Surgenor DM, editor. *The Red Blood Cell.* New York: Academic Press; 1974, pp. 147–211.
102. Verkleij AJ, Nauta IL, Werre JM, Mandersloot JG, Reinders B, Ververgaert PH *et al.* The fusion of abnormal plasma lipoprotein (LP-X) and the erythrocyte membrane in patients with cholestasis studied by electron-microscopy. *Biochimica et Biophysica Acta.* 1976;436:366–376.
103. Brossard N, Croset M, Normand S, Pousin J, Lecerf J, Laville M *et al.* Human plasma albumin transports [13C]docosahexaenoic acid in two lipid forms to blood cells. *J Lipid Res.* 1997;38:1571–1582.
104. Sens P, Gov N. Force balance and membrane shedding at the red-blood-cell surface. *Phys Rev Lett.* 2007;98:018102.

5
Eryptosis, the Suicidal Death of Erythrocytes

Florian Lang,[‡] Stephan Huber[†] and Michael Föller[*]

*Departments of *Physiology and †Radiation Oncology, University of Tübingen, Germany*

5.1 Introduction

As amplified in a separate chapter of this book,[1] the lifespan of mature, circulating erythrocytes (some 100–120 days) is usually limited by senescence, which eventually ends with the clearance of the senescent erythrocytes from circulating blood.[2–4] Erythrocyte senescence is paralleled by binding of hemichromes to band 3, clustering of band 3 as well as deposition of complement C3 fragments and anti-band 3 immunoglobulins.[5]

At any time of their life erythrocytes may undergo suicidal death or eryptosis, a phenomenon which resembles apoptosis of nucleated cells and which is independent of senescence.[6–8] Since mature erythrocytes do not have nuclei and mitochondria, they lack several classical features of apoptosis, such as mitochondrial depolarization and condensation of nuclei. Nevertheless, several features of eryptosis are similar to those of apoptosis, such as cell shrinkage, membrane blebbing, and phosphatidylserine exposure.[9–11] Eryptosis has been described in several reviews.[6–8,12–16] The brief synopsis that follows discusses the signaling and significance of eryptosis, diseases leading to enhanced eryptosis, as well as triggers and inhibitors of eryptosis.

[‡] Corresponding author: florian.lang@uni-tuebingen.de

5.2 Clinical Conditions Associated with Eryptosis; Triggers and Inhibitors of Eryptosis

Several clinical disorders have been identified that are associated with excessive eryptosis (Table 5.1). Excessive eryptosis presumably contributes to, or even accounts for the pathophysiology of those disorders, such as anemia and deranged microcirculation. Excessive or decreased eryptosis has further been observed in gene-targeted mice (Table 5.2).

The observations in those mice help to unravel the gene products participating in the regulation of erythrocyte survival. Moreover, a wide variety of xenobiotics and endogenous substances has been identified, which either stimulate (Table 5.3) or inhibit (Table 5.4) eryptosis.

Table 5.1 Disorders associated with enhanced eryptosis. (Mechanisms: Ca^{2+} = stimulation of Ca^{2+} entry; Cer. = stimulation of ceramide formation; other = ATP depletion etc.).

Disorders associated with accelerated eryptosis	Effective through			References
	Ca^{2+}	Cer.	Other	
Iron deficiency	+			17
Phosphate depletion			+	18
Neocytolysis			+	19
Sepsis		+		20
Hyperthermia	+			21
Glucose depletion	+		+	22
Hemolytic anemia	+	+	+	23
Hemolytic uremic syndrome	+	+		24
Renal insufficiency	+			25
Malaria	+			7,26,27
Sickle cell disease			+	14,28–34
Thalassemia			+	29,32,33,35,36
Glucose-6-phosphate dehydrogenase (G6PD) deficiency			+	33,37
Paroxysmal nocturnal hemoglobinuria			+	38
Myelodysplastic syndrome			+	38
Wilson's disease		+		16
AE1 mutation	+			39
GLUT1 mutation	+			40

Eryptosis, the Suicidal Death of Erythrocytes

Table 5.2 Enhanced or decreased eryptosis in gene-targeted mice. (Mechanisms: Ca^{2+} = stimulation of Ca^{2+} entry; Cer. = stimulation of ceramide formation; other = ATP depletion etc.).

Targeted gene	Effective through			References
	Ca^{2+}	Cer.	Other	
Enhanced eryptosis				
Defective hemoglobin (sickle cell, thalassemia)	+			31,32
cGMP-dependent protein kinase type I (cGKI) deficiency	+			41
AMP-activated protein kinase deficiency	+			42
Klotho deficiency	+			43
EPO excess	+			44
AE1 deficiency	+			45
Annexin A7 deficiency	+			12,28
Endothelin B receptor deficiency	+			47
Reduced eryptosis				
PDK1 deficiency	+			48
PAF receptor deficiency		+		46
TRPC6 deficiency	+			49
TAUT deficiency			+	50,51

Table 5.3 Stimulators of eryptosis. (Mechanisms: Ca^{2+} = stimulation of Ca^{2+} entry; Cer. = stimulation of ceramide formation; other = ATP depletion etc.).

Stimulators	Concentration	Effective through			References
		Ca^{2+}	Cer.	Other	
Aluminum	10–30 μM	+			52
Amantadine	0.2 μg/ml	+			53
Amiodarone	0.1 μM	+			54,55
Amphotericin B	0.5 μg/ml	+		+	56,57
Amyloid	0.5–1 μM		+		58
Anandamide	2.5 μM	+			59,60
Anti-A IgG	0.5 μg/ml	+			61
Arsenic	7–10 μM	+	+	+	62,63
Aurothiomalate	1 μg/ml	+			64
Azathioprine	2 μg/ml	+			65,66

(*Continued*)

Table 5.3 (*Continued*)

Stimulators	Concentration	Effective through Ca^{2+}	Cer.	Other	References
Bay-Y5884	20 μM	+			67
Bismuth chloride	0.5 μg/ml	+	+		68
Cadmium	5.5 μM	+		+	69
Ceramide (acylsphingosine)	50 μM			+	70
Chlorpromazine	10 μM	+		+	71
Ciglitazone	5–10 μM	+			72
Cisplatin	1 μM	+			73
Copper	3 μM		+		16
Cordycepin	60 μM	+			74
Curcumin	1 μM	+	+		75
Cyclosporine	10 μM		+	+	60,76
Dimethylfumarate	50 μM			+	77
Glycation	not applicable	+			78
Ligation of glycophorin-C	12 μg/ml			+	79
Gold chloride	0.75 μg/ml	+			80
Hemin	1–10 μM	+	+		81
Hemolysin	1 U/ml	+			82
IgA	not applicable			+	83
Lead	0.1 μM	+			84
Leukotriene C(4)	10 nM	+			85
Lipopeptides	1 μM			+	86
Listeriolysin	10 ng/ml	+		+	87
Lithium	1 mM	+			88
Menadione (vitamin K$_3$)	1 μM		+		89
Mercury	1 μM		+	+	90
Methyldopa	6 μg/ml		+	+	91
Methylglyoxal	0.3 μM			+	92
Monensin	1 μM	+			93
Paclitaxel	10 μM	+	+		94,95
PAF	0.5–1 μM		+		46
Peptidoglycan	50 μg/ml	+	+		96
Phytic acid	1 mM			+	97
Prostaglandin E2	100 pM	+			98
Radiocontrast agents	5 mM	+		+	99
Retinoic acid	3 μM	+			100

(*Continued*)

Table 5.3 (*Continued*)

Stimulators	Concentration	Ca^{2+}	Cer.	Other	References
Selenium (sodium selenite)	200 ng/ml	+	+		101
Silver ions	100 nM			+	102
Thrombospondin-1-receptor CD47 ligation	20 µg/ml			+	103
Thymoquinone	3 µM			+	104
Tin	30 µM	+	+	+	105
Valinomycin	1 nM			+	106
Sodium vanadate	10 µg/ml	+			15
Zinc	25 µM	+	+		107

Table 5.4 Inhibitors of eryptosis. (Mechanisms: Ca^{2+} = inhibition of Ca^{2+} entry; Cer. = inhibition of ceramide formation; other = inhibition of ATP depletion etc.).

Inhibitors	Concentration	Ca^{2+}	Cer.	Other	References
Adenosine	10–30 µM			+	108
Amitriptyline	50 µM		+	+	109
Caffeine	50–500 µM	+		+	110
Catecholamines (isoproterenol)	IC50: 1 µM	+			111
Chloride		+			112
EIPA	IC50: 0.2 µM	+			113
Endothelin	500 nM	+			47
Epo	1 U/ml	+			25
Flufenamic acid	10 µM	+			114
NBQX/CNQX	10–50 µM	+			115
Niflumic acid	100 µM			+	116
NO (nitroprusside)	1 µM			+	117
NPPB	100 µM			+	116
Resveratrol	5 µM	+		+	118
Sarafotoxin 6c	10 nM	+			47
Staurosporine	500 nM			+	119
Thymol	2.5 µg/ml	+		+	120
Urea	650 µM		+	+	121
Xanthohumol	1 µM	+		+	122
Zidovudine	2 µg/ml	+			123

Stimulation of eryptosis may contribute to the side-effects of drugs or add to the toxicity of poisons. The effect of several of those xenobiotics is mild. However, the possibility must be considered that patients suffering from one of the eryptosis-stimulating disorders may be particularly sensitive to the effects of other eryptosis-stimulating substances.

5.3 Signaling of Eryptosis

Eryptosis is stimulated by an increase in cytosolic Ca^{2+} activity,[9–11] which triggers cell membrane vesiculation and cell membrane scrambling.[8,124] Enhanced intracellular Ca^{2+} activity is further followed by activation of the cysteine endopeptidase calpain, an enzyme degrading the cytoskeleton and thus causing cell membrane blebbing.[8,125] Ca^{2+} stimulates Ca^{2+}-sensitive K^+ channels[14,126,127] with subsequent efflux of K^+, hyperpolarization of the cell membrane and Cl^- exit.[128] The cellular loss of KCl with osmotically obliged water results in cell shrinkage.[128,129]

The cytosolic Ca^{2+} activity may increase due to Ca^{2+} entry through non-selective cation channels.[98,130–133] The molecular identity of the cation channels is still ill-defined but apparently involves TRPC6.[49] Stimulators of the cation channels include osmotic shock,[112,134] oxidative stress[134,135] and Cl^- removal (i.e. the iso-osmotic replacement of chloride by organic anions such as gluconate in the extracellular fluid).[98,112,135]

Eryptosis is further triggered by ceramide[70] which increases the Ca^{2+} sensitivity of erythrocyte cell membrane scrambling.[70] The enzyme accounting for the formation of ceramide has not been defined yet. In several disorders including sepsis[20] and hemolytic uremic syndrome[24] the stimulation of ceramide formation is due to a component in the serum from the respective patients, which could be sphingomyelinase. Sphingomyelinase activity has been observed in the serum of patients suffering from Wilson's disease.[16] Stimulators of ceramide formation include platelet-activating factor (PAF).[46] Osmotic cell shrinkage stimulates PAF release from erythrocytes.[46] In erythrocytes PAF stimulates PAF receptors leading to breakdown of sphingomyelin and subsequent ceramide formation.[46] PAF-stimulated ceramide formation and eryptosis were expectedly blunted in gene-targeted mice lacking the PAF receptor.[46]

Additional mechanisms stimulating eryptosis include energy depletion,[119] oxidative stress[33,136,137] and impaired antioxidative defense.[138–140] In erythrocytes, oxidative stress activates Ca^{2+}-permeable nonselective cation channels,[135] K^+-[129] and Cl^- channels,[141,142] the latter two contributing to eryptotic cell shrinkage.[116] Oxidative stress further triggers eryptosis by activation of caspases.[10,143,144] The death receptor CD95 (Fas) is expressed in erythrocytes but its role in the triggering of suicidal erythrocyte death has been debated.[145] According to an earlier study,[143] oxidative stress-induced phosphatidylserine exposure of human erythrocytes does involve Fas/caspase 8/caspase 3-dependent signaling.

5.4 Significance of Eryptosis

Erythrocytes, which expose phosphatidylserine at their surface, are bound to receptors of phagocytosing cells leading to engulfment and subsequent degradation.[146,147] Accordingly, phosphatidylserine-exposing erythrocytes are rapidly cleared from circulating blood.[17] Stimulation of eryptosis may thus cause anemia, as long as the accelerated loss of erythrocytes is not compensated by enhanced formation of new erythrocytes. Phosphatidylserine-exposing erythrocytes further adhere to the vascular wall.[34,148–153] Accordingly, excessive eryptosis may interfere with microcirculation.[154]

Eryptosis may, on the other hand, protect against hemolysis of defective erythrocytes. Cell injury, such as energy depletion, defective Na^+/K^+ATPase, or leakiness of the cell membrane result in cellular gain of Na^+ and Cl^- and osmotically obliged water with eventual cell swelling.[155] In the early stages of cell injury, the entry of Na^+ is compensated by cellular loss of K^+. However, the cellular K^+ loss dissipates the chemical K^+ gradient eventually resulting in a decrease of the K^+ equilibrium potential and gradual depolarization. The depolarization decreases the electrical gradient across the cell membrane and thus favors Cl^- entry, which is followed by osmotically obliged water. The subsequent cell swelling jeopardizes the integrity of the cell membrane and may eventually lead to cell membrane rupture with release of cellular hemoglobin. The hemoglobin thus released may be filtered in the renal glomerula and subsequently may occlude renal tubules. Phosphatidylserine at the cell surface of eryptotic cells is recognized by macrophages, which clear phosphatidylserine-exposing defective erythrocytes from circulating blood prior to hemolysis. The activation of Ca^{2+}-sensitive K^+ channels counteracts cell swelling and delays the disruption of defective erythrocytes, thus expanding the time allowed for macrophages to clear the injured erythrocytes from circulating blood. The removal of injured erythrocytes prior to hemolysis is thus an efficient and important physiological mechanism.

Eryptosis may further play a significant role in malaria. The malaria pathogen, such as *Plasmodium falciparum*, induces oxidative stress resulting in activation of ion channels in the host cell membrane,[135,141,142] which are involved in the uptake of nutrients and the disposal of waste products.[156] Activation of the oxidant-sensitive Ca^{2+}-permeable cation channels leads to Ca^{2+} and Na^+ entry.[141,142,157] Both Na^+ and Ca^{2+}, taken up by the parasitized erythrocyte, are essential for the intraerythrocytic survival of the parasite.[158] The Ca^{2+} uptake does not lead to execution of eryptosis in most of the parasitized erythrocytes because of rapid Ca^{2+} siphoning by the parasite combined with powerful Ca^{2+} extrusion by the erythrocyte Ca^{2+} pump.[159,160]

Acceleration of eryptosis, however, may limit the lifespan of the intraerythrocytic parasite and thus counteracts the amplification of the parasites.[28] Diseases associated with accelerated eryptosis, such as sickle-cell trait, beta-thalassemia-trait, homozygous hemoglobin (Hb) C, and glucose-6-phosphate deficiency (Table 5.1) lead to premature senescence and/or eryptosis upon infection with *Plasmodium*, thus leading to accelerated clearance of ring stage-infected erythrocytes.[29,31–33,36,37] Moreover, iron deficiency[161] and lead[162] delay the increase in parasitemia and thus enhance the survival of *Plasmodium berghei*-infected mice, presumably by accelerating erythrocyte death.

Eryptosis is further instrumental in the replacement of the neonatal HbF-containing erythrocytes[163] after birth. During intrauterine life, the high affinity of HbF ascertains adequate oxygen uptake in the placenta. Following birth, however, the high oxygen affinity of HbF is not required for full oxygenation of hemoglobin in the inflated lung, but impairs the oxygen delivery to tissues. Accordingly, HbF is not appropriate for efficient gas exchange after birth. Fetal erythrocytes are more resistant to Cl^- removal, osmotic shock, PGE_2, and PAF, but more sensitive to oxidative stress.[164]

The exquisite oxygen sensitivity of fetal erythrocytes fosters their removal, as soon as the blood is exposed to inspired oxygen in the lung after birth.

5.5 Neocytolysis

The term neocytolysis has been coined to describe the particular sensitivity of newly formed erythrocytes to suicidal death[19] leading to accelerated death of young erythrocytes following a limited exposure to high altitude or after a space flight. At this stage, we do not know whether or not neocytolysis involves the same mechanisms as eryptosis. Moreover, the causes for the enhanced sensitivity of newly formed erythrocytes to suicidal cell death are not known. It is tempting to speculate that neocytolysis occurs following decline of erythropoietin. Erythropoietin inhibits eryptosis.[25] However, erythrocytes drawn from erythropoietin-overexpressing transgenic mice were significantly more resistant to osmotic lysis than wild type erythrocytes, but more sensitive to the eryptotic effects of Cl^- removal and exposure to the Ca^{2+} ionophore ionomycin.[44] Possibly, erythropoietin inhibits eryptosis and apoptosis of progenitor cells but by the same token stimulates the expression of genes in progenitor cells which render the erythrocytes more sensitive to eryptosis. Accordingly, erythrocyte death occurs as soon as the erythropoietin concentration declines. The increase in the sensitivity to eryptosis under the influence of a high erythropoietin concentration would lead to the rapid removal of excessive erythrocytes, as soon as erythropoietin decreases, i.e. when a high erythrocyte concentration is no longer needed. Accordingly, a feedback regulation is shortened, which otherwise would take 120 days. Neocytolysis may thus reflect the death of those erythrocytes which have been generated under a high erythropoietin concentration, and are thus more vulnerable to eryptosis. Additional investigation is required to test that hypothesis.

Bibliography

1. Willekens FLA, Bosman GJCGM, Werre JM. Erythrocyte senescence. In: Lang F, Föller M, editors. *Erythrocytes: Physiology and Pathophysiology.* London: Imperial College Press; 2012, pp. 301–326.

2. Arese P, Turrini F, Schwarzer E. Band 3/complement-mediated recognition and removal of normally senescent and pathological human erythrocytes. *Cell Physiol Biochem*. 2005;16:133–146.
3. Bosman GJ, Willekens FL, Werre JM. Erythrocyte aging: a more than superficial resemblance to apoptosis? *Cell Physiol Biochem*. 2005;16:1–8.
4. Kiefer CR, Snyder LM. Oxidation and erythrocyte senescence. *Curr Opin Hematol*. 2000;7:113–116.
5. Lutz HU. Innate immune and non-immune mediators of erythrocyte clearance. *Cell Mol Biol* (Noisy-le-grand). 2004;50:107–116.
6. Cimen MY. Free radical metabolism in human erythrocytes. *Clin Chim Acta*. 2008;390:1–11.
7. Foller M, Bobbala D, Koka S, Huber SM, Gulbins E, Lang F. Suicide for survival — death of infected erythrocytes as a host mechanism to survive malaria. *Cell Physiol Biochem*. 2009;24:133–140.
8. Lang F, Gulbins E, Lerche H, Huber SM, Kempe DS, Foller M. Eryptosis, a window to systemic disease. *Cell Physiol Biochem*. 2008;22:373–380.
9. Berg CP, Engels IH, Rothbart A, Lauber K, Renz A, Schlosser SF, *et al*. Human mature red blood cells express caspase-3 and caspase-8, but are devoid of mitochondrial regulators of apoptosis. *Cell Death Differ*. 2001;8:1197–1206.
10. Bratosin D, Estaquier J, Petit F, Arnoult D, Quatannens B, Tissier JP, *et al*. Programmed cell death in mature erythrocytes: a model for investigating death effector pathways operating in the absence of mitochondria. *Cell Death Differ*. 2001;8:1143–1156.
11. Daugas E, Cande C, Kroemer G. Erythrocytes: death of a mummy. *Cell Death Differ*. 2001;8:1131–1133.
12. Lang E, Lang PA, Shumilina E, Qadri SM, Kucherenko Y, Kempe DS, *et al*. Enhanced eryptosis of erythrocytes from gene-targeted mice lacking annexin A7. *Pflügers Arch Eur J Physiol*. 2010;460:667–676.
13. Bao GQ, Ju AZ. [Signal pathways of eryptosis-review]. *Zhongguo Shi Yan Xue Ye Xue Za Zhi*. 2009;17:1097–1100.
14. Browning JA, Robinson HC, Ellory JC, Gibson JS. Deoxygenation-induced non-electrolyte pathway in red cells from sickle cell patients. *Cell Physiol Biochem*. 2007;19:165–174.
15. Foller M, Sopjani M, Mahmud H, Lang F. Vanadate induced suicidal erythrocyte death. *Kidney Blood Pres Res*. 2008;21:87–93.

16. Lang PA, Schenck M, Nicolay JP, Becker JU, Kempe DS, Lupescu A, *et al.* Liver cell death and anemia in Wilson disease involve acid sphingomyelinase and ceramide. *Nat Med.* 2007;13:164–170.
17. Kempe DS, Lang PA, Duranton C, Akel A, Lang KS, Huber SM, *et al.* Enhanced programmed cell death of iron-deficient erythrocytes. *FASEB J.* 2006;20:368–370.
18. Birka C, Lang PA, Kempe DS, Hoefling L, Tanneur V, Duranton C, *et al.* Enhanced susceptibility to erythrocyte "apoptosis" following phosphate depletion. *Pflügers Arch Eur J Physiol.* 2004;448:471–477.
19. Rice L, Alfrey CP. The negative regulation of red cell mass by neocytolysis: physiologic and pathophysiologic manifestations. *Cell Physiol Biochem.* 2005;15:245–250.
20. Kempe DS, Akel A, Lang PA, Hermle T, Biswas R, Muresanu J, *et al.* Suicidal erythrocyte death in sepsis. *J Mol Med.* 2007;85:273–281.
21. Foller M, Braun M, Qadri SM, Lang E, Mahmud H, Lang F. Temperature sensitivity of suicidal erythrocyte death. *Eur J Clin Invest.* 2010; 40:534–540.
22. Vasudevan S, Chen GC, Andika M, Agarwal S, Chen P, Olivo M. Dynamic quantitative photothermal monitoring of cell death of individual human red blood cells upon glucose depletion. *J Biomed Opt.* 2010;15:057001.
23. Banerjee D, Saha S, Basu S, Chakrabarti A. Porous red cell ultrastructure and loss of membrane asymmetry in a novel case of hemolytic anemia. *Eur J Haematol.* 2008;81:399–402.
24. Lang PA, Beringer O, Nicolay JP, Amon O, Kempe DS, Hermle T, *et al.* Suicidal death of erythrocytes in recurrent hemolytic uremic syndrome. *J Mol Med.* 2006;84:378–388.
25. Myssina S, Huber SM, Birka C, Lang PA, Lang KS, Friedrich B, *et al.* Inhibition of erythrocyte cation channels by erythropoietin. *J Am Soc Nephrol.* 2003;14:2750–2757.
26. Koka S, Lang C, Boini KM, Bobbala D, Huber S, Lang F. Influence of chlorpromazine on eryptosis, parasitemia and survival of Plasmodium berghei infected mice. *Cell Physiol Biochem.* 2008;22:261–268.
27. Koka S, Lang C, Niemoeller OM, Boini KM, Nicolay JP, Huber SM, *et al.* Influence of NO synthase inhibitor L-NAME on parasitemia and survival of Plasmodium berghei infected mice. *Cell Physiol Biochem.* 2008; 21:481–488.

28. Lang PA, Kasinathan RS, Brand VB, Duranton C, Lang C, Koka S, *et al.* Accelerated clearance of Plasmodium-infected erythrocytes in sickle cell trait and annexin-A7 deficiency. *Cell Physiol Biochem.* 2009;24:415–428.
29. Ayi K, Turrini F, Piga A, Arese P. Enhanced phagocytosis of ring-parasitized mutant erythrocytes: a common mechanism that may explain protection against falciparum malaria in sickle trait and beta-thalassemia trait. *Blood.* 2004;104:3364–3371.
30. Chadebech P, Habibi A, Nzouakou R, Bachir D, Meunier-Costes N, Bonin P, *et al.* Delayed hemolytic transfusion reaction in sickle cell disease patients: evidence of an emerging syndrome with suicidal red blood cell death. *Transfusion.* 2009;49:1785–1792.
31. de Jong K, Emerson RK, Butler J, Bastacky J, Mohandas N, Kuypers FA. Short survival of phosphatidylserine-exposing red blood cells in murine sickle cell anemia. *Blood.* 2001;98:1577–1584.
32. Kean LS, Brown LE, Nichols JW, Mohandas N, Archer DR, Hsu LL. Comparison of mechanisms of anemia in mice with sickle cell disease and beta-thalassemia: peripheral destruction, ineffective erythropoiesis, and phospholipid scramblase-mediated phosphatidylserine exposure. *Exp Hematol.* 2002;30:394–402.
33. Lang KS, Roll B, Myssina S, Schittenhelm M, Scheel-Walter HG, Kanz L, *et al.* Enhanced erythrocyte apoptosis in sickle cell anemia, thalassemia and glucose-6-phosphate dehydrogenase deficiency. *Cell Physiol Biochem.* 2002;12:365–372.
34. Wood BL, Gibson DF, Tait JF. Increased erythrocyte phosphatidylserine exposure in sickle cell disease: flow-cytometric measurement and clinical associations. *Blood.* 1996;88:1873–1880.
35. Basu S, Banerjee D, Chandra S, Chakrabarti A. Eryptosis in hereditary spherocytosis and thalassemia: role of glycoconjugates. *Glycoconj J.* 2010;27(7–9):717–722.
36. Kuypers FA, Yuan J, Lewis RA, Snyder LM, Kiefer CR, Bunyaratvej A, *et al.* Membrane phospholipid asymmetry in human thalassemia. *Blood.* 1998;91:3044–3051.
37. Cappadoro M, Giribaldi G, O'Brien E, Turrini F, Mannu F, Ulliers D, *et al.* Early phagocytosis of glucose-6-phosphate dehydrogenase (G6PD)-deficient erythrocytes parasitized by Plasmodium falciparum may explain malaria protection in G6PD deficiency. *Blood.* 1998;92:2527–2534.

38. Basu S, Banerjee D, Ghosh M, Chakrabarti A. Erythrocyte membrane defects and asymmetry in paroxysmal nocturnal hemoglobinuria and myelodysplastic syndrome. *Hematology.* 2010;15:236–239.
39. Bruce LJ, Robinson HC, Guizouarn H, Borgese F, Harrison P, King MJ, *et al.* Monovalent cation leaks in human red cells caused by single amino-acid substitutions in the transport domain of the band 3 chloride-bicarbonate exchanger, AE1. *Nat Genet.* 2005;37:1258–1263.
40. Weber YG, Storch A, Wuttke TV, Brockmann K, Kempfle J, Maljevic S, *et al.* GLUT1 mutations are a cause of paroxysmal exertion-induced dyskinesias and induce hemolytic anemia by a cation leak. *J Clin Invest.* 2008;118:2157–2168.
41. Foller M, Feil S, Hofmann F, Koka S, Kasinathan R, Nicolay J, *et al.* Anemia of gene targeted mice lacking functional cGMP-dependent protein kinase type I. *Proc Natl Acad Sci USA.* 2008;105:6771–6778.
42. Foller M, Sopjani M, Koka S, Gu S, Mahmud H, Wang K, *et al.* Regulation of erythrocyte survival by AMP-activated protein kinase. *FASEB J.* 2009;23:1072–1080.
43. Kempe DS, Ackermann TF, Fischer SS, Koka S, Boini KM, Mahmud H, *et al.* Accelerated suicidal erythrocyte death in Klotho-deficient mice. *Pflügers Arch Eur J Physiol.* 2009;458:503–512.
44. Foller M, Kasinathan RS, Koka S, Huber SM, Schuler B, Vogel J, *et al.* Enhanced susceptibility to suicidal death of erythrocytes from transgenic mice overexpressing erythropoietin. *Am J Physiol Regul Integr Comp Physiol.* 2007;293:R1127–R1134.
45. Akel A, Wagner CA, Kovacikova J, Kasinathan RS, Kiedaisch V, Koka S, *et al.* Enhanced suicidal death of erythrocytes from gene-targeted mice lacking the Cl–/HCO(3)(–) exchanger AE1. *Am J Physiol Cell Physiol.* 2007;292:C1759–C1767.
46. Lang PA, Kempe DS, Tanneur V, Eisele K, Klarl BA, Myssina S, *et al.* Stimulation of erythrocyte ceramide formation by platelet-activating factor. *J Cell Sci.* 2005;118:1233–1243.
47. Foller M, Mahmud H, Qadri SM, Gu S, Braun M, Bobbala D, *et al.* Endothelin B receptor stimulation inhibits suicidal erythrocyte death. *FASEB J.* 2010;24:3351–3359.
48. Foller M, Mahmud H, Koka S, Lang F. Reduced Ca^{2+} entry and suicidal death of erythrocytes in PDK1 hypomorphic mice. *Pflügers Arch Eur J Physiol.* 2008;455:939–949.

49. Foller M, Kasinathan RS, Koka S, Lang C, Shumilina E, Birnbaumer L, *et al*. TRPC6 contributes to the Ca(2+) leak of human erythrocytes. *Cell Physiol Biochem*. 2008;21:183–192.
50. Delic D, Warskulat U, Borsch E, Al Qahtani S, Al Quraishi S, Haussinger D, *et al*. Loss of ability to self-heal malaria upon taurine transporter deletion. *Infect Immun*. 2010;78:1642–1649.
51. Lang PA, Warskulat U, Heller-Stilb B, Huang DY, Grenz A, Myssina S, *et al*. Blunted apoptosis of erythrocytes from taurine transporter deficient mice. *Cell Physiol Biochem*. 2003;13:337–346.
52. Niemoeller OM, Kiedaisch V, Dreischer P, Wieder T, Lang F. Stimulation of eryptosis by aluminium ions. *Toxicol Appl Pharmacol*. 2006; 217:168–175.
53. Foller M, Geiger C, Mahmud H, Nicolay J, Lang F. Stimulation of suicidal erythrocyte death by amantadine. *Eur J Pharmacol*. 2008;581:13–18.
54. Bobbala D, Alesutan I, Foller M, Tschan S, Huber SM, Lang F. Protective effect of amiodarone in malaria. *Acta Trop*. 2010;116:39–44.
55. Nicolay JP, Bentzen PJ, Ghashghaeinia M, Wieder T, Lang F. Stimulation of erythrocyte cell membrane scrambling by amiodarone. *Cell Physiol Biochem*. 2007;20:1043–1050.
56. Siraskar B, Ballal A, Bobbala D, Foller M, Lang F. Effect of amphotericin B on parasitemia and survival of Plasmodium berghei-infected mice. *Cell Physiol Biochem*. 2010;26:347–354.
57. Mahmud H, Mauro D, Qadri SM, Foller M, Lang F. Triggering of suicidal erythrocyte death by amphotericin B. *Cell Physiol Biochem*. 2009; 24:263–270.
58. Nicolay JP, Gatz S, Liebig G, Gulbins E, Lang F. Amyloid induced suicidal erythrocyte death. *Cell Physiol Biochem*. 2007;19:175–184.
59. Bobbala D, Alesutan I, Foller M, Huber SM, Lang F. Effect of anandamide in Plasmodium berghei-infected mice. *Cell Physiol Biochem*. 2010; 26:355–362.
60. Bentzen PJ, Lang F. Effect of anandamide on erythrocyte survival. *Cell Physiol Biochem*. 2007;20:1033–1042.
61. Attanasio P, Shumilina E, Hermle T, Kiedaisch V, Lang PA, Huber SM, *et al*. Stimulation of eryptosis by anti-A IgG antibodies. *Cell Physiol Biochem*. 2007;20:591–600.

62. Biswas D, Banerjee M, Sen G, Das JK, Banerjee A, Sau TJ, et al. Mechanism of erythrocyte death in human population exposed to arsenic through drinking water. *Toxicol Appl Pharmacol.* 2008; 230:57–66.
63. Mahmud H, Foller M, Lang F. Arsenic-induced suicidal erythrocyte death. *Arch Toxicol.* 2009;83:107–113.
64. Alesutan I, Bobbala D, Qadri SM, Estremera A, Foller M, Lang F. Beneficial effect of aurothiomalate on murine malaria. *Malar J.* 2010;9:118.
65. Bobbala D, Koka S, Geiger C, Foller M, Huber SM, Lang F. Azathioprine favourably influences the course of malaria. *Malar J.* 2009;8:102.
66. Geiger C, Foller M, Herrlinger KR, Lang F. Azathioprine-induced suicidal erythrocyte death. *Inflamm Bowel Dis.* 2008;14:1027–1032.
67. Shumilina E, Kiedaisch V, Akkel A, Lang P, Hermle T, Kempe DS, et al. Stimulation of suicidal erythrocyte death by lipoxygenase inhibitor Bay-Y5884. *Cell Physiol Biochem.* 2006;18:233–242.
68. Braun M, Foller M, Gulbins E, Lang F. Eryptosis triggered by bismuth. *Biometals.* 2009;22:453–460.
69. Sopjani M, Foller M, Dreischer P, Lang F. Stimulation of eryptosis by cadmium ions. *Cell Physiol Biochem.* 2008;22:245–252.
70. Lang KS, Myssina S, Brand V, Sandu C, Lang PA, Berchtold S, et al. Involvement of ceramide in hyperosmotic shock-induced death of erythrocytes. *Cell Death Differ.* 2004;11:231–243.
71. Akel A, Hermle T, Niemoeller OM, Kempe DS, Lang PA, Attanasio P, et al. Stimulation of erythrocyte phosphatidylserine exposure by chlorpromazine. *Eur J Pharmacol.* 2006;532:11–17.
72. Niemoeller O, Mahmud H, Foller M, Wieder T, Lang F. Ciglitazone and 15d-PGJ2 induced suicidal erythrocyte death. *Cell Physiol Biochem.* 2008;22:237–244.
73. Mahmud H, Foller M, Lang F. Suicidal erythrocyte death triggered by cisplatin. *Toxicology.* 2008;249:40–44.
74. Lui JC, Wong JW, Suen YK, Kwok TT, Fung KP, Kong SK. Cordycepin induced eryptosis in mouse erythrocytes through a Ca2+-dependent pathway without caspase-3 activation. *Arch Toxicol.* 2007;81:859–865.
75. Bentzen PJ, Lang E, Lang F. Curcumin induced suicidal erythrocyte death. *Cell Physiol Biochem.* 2007;19:153–164.

76. Niemoeller OM, Akel A, Lang PA, Attanasio P, Kempe DS, Hermle T, *et al*. Induction of eryptosis by cyclosporine. *Naunyn-Schmiebergs Arch Pharmacol*. 2006;374:41–49.
77. Ghashghaeinia M, Bobbala D, Wieder T, Koka S, Bruck J, Fehrenbacher B, *et al*. Targeting glutathione by dimethylfumarate protects against experimental malaria by enhancing erythrocyte cell membrane scrambling. *Am J Physiol Cell Physiol*. 2010;299:C791–C804.
78. Kucherenko YV, Bhavsar SK, Grischenko VI, Fischer UR, Huber SM, Lang F. Increased cation conductance in human erythrocytes artificially aged by glycation. *J Membr Biol*. 2010;235:177–189.
79. Head DJ, Lee ZE, Poole J, Avent ND. Expression of phosphatidylserine (PS) on wild-type and Gerbich variant erythrocytes following glycophorin-C (GPC) ligation. *Br J Haematol*. 2005;129:130–137.
80. Sopjani M, Foller M, Lang F. Gold stimulates Ca2+ entry into and subsequent suicidal death of erythrocytes. *Toxicology*. 2008;244:271–279.
81. Gatidis S, Foller M, Lang F. Hemin-induced suicidal erythrocyte death. *Ann Hematol*. 2009;88:721–726.
82. Lang PA, Kaiser S, Myssina S, Birka C, Weinstock C, Northoff H, *et al*. Effect of Vibrio parahaemolyticus haemolysin on human erythrocytes. *Cell Microbiol*. 2004;6:391–400.
83. Chadebech P, Michel M, Janvier D, Yamada K, Copie-Bergman C, Bodivit G, *et al*. IgA-mediated human autoimmune hemolytic anemia as a result of hemagglutination in the spleen, but independent of complement activation and Fc{alpha}RI. *Blood*. 2010;116:4141–4147.
84. Kempe DS, Lang PA, Eisele K, Klarl BA, Wieder T, Huber SM, *et al*. Stimulation of erythrocyte phosphatidylserine exposure by lead ions. *Am J Physiol Cell Physiol*. 2005;288:C396–C402.
85. Foller M, Mahmud H, Gu S, Wang K, Floride E, Kucherenko Y, *et al*. Participation of leukotriene C(4) in the regulation of suicidal erythrocyte death. *J Physiol Pharmacol*. 2009;60:135–143.
86. Wang K, Mahmud H, Foller M, Biswas R, Lang KS, Bohn E, *et al*. Lipopeptides in the triggering of erythrocyte cell membrane scrambling. *Cell Physiol Biochem*. 2008;22:381–386.
87. Foller M, Shumilina E, Lam R, Mohamed W, Kasinathan R, Huber S, *et al*. Induction of suicidal erythrocyte death by listeriolysin from Listeria monocytogenes. *Cell Physiol Biochem*. 2007;20:1051–1060.

88. Nicolay J, Gatz S, Lang F, Lang U. Lithium-induced suicidal erythrocyte death. *J Psychopharmacol.* 2009;24:1533–1539.
89. Qadri SM, Eberhard M, Mahmud H, Foller M, Lang F. Stimulation of ceramide formation and suicidal erythrocyte death by vitamin K(3) (menadione). *Eur J Pharmacol.* 2009;623:10–13.
90. Eisele K, Lang PA, Kempe DS, Klarl BA, Niemoller O, Wieder T, et al. Stimulation of erythrocyte phosphatidylserine exposure by mercury ions. *Toxicol Appl Pharmacol.* 2006;210:116–122.
91. Mahmud H, Foller M, Lang F. Stimulation of erythrocyte cell membrane scrambling by methyldopa. *Kidney Blood Press Res.* 2008;31:299–306.
92. Nicolay JP, Schneider J, Niemoeller OM, Artunc F, Portero-Otin M, Haik G, Jr, et al. Stimulation of suicidal erythrocyte death by methylglyoxal. *Cell Physiol Biochem.* 2006;18:223–232.
93. Bhavsar SK, Eberhard M, Bobbala D, Lang F. Monensin induced suicidal erythrocyte death. *Cell Physiol Biochem.* 2010;25:745–752.
94. Koka S, Bobbala D, Lang C, Boini KM, Huber SM, Lang F. Influence of paclitaxel on parasitemia and survival of Plasmodium berghei infected mice. *Cell Physiol Biochem.* 2009;23:191–198.
95. Lang PA, Huober J, Bachmann C, Kempe DS, Sobiesiak M, Akel A, et al. Stimulation of erythrocyte phosphatidylserine exposure by paclitaxel. *Cell Physiol Biochem.* 2006;18:151–164.
96. Foller M, Biswas R, Mahmud H, Akel A, Shumilina E, Wieder T, et al. Effect of peptidoglycans on erythrocyte survival. *Int J Med Microbiol.* 2009;299:75–85.
97. Eberhard M, Foller M, Lang F. Effect of phytic acid on suicidal erythrocyte death. *J Agric Food Chem.* 2010;58:2028–2033.
98. Lang PA, Kempe DS, Myssina S, Tanneur V, Birka C, Laufer S, et al. PGE(2) in the regulation of programmed erythrocyte death. *Cell Death Differ.* 2005;12:415–428.
99. Foller M, Sopjani M, Schlemmer HP, Claussen CD, Lang F. Triggering of suicidal erythrocyte death by radiocontrast agents. *Eur J Clin Invest.* 2009;39:576–583.
100. Niemoeller OM, Foller M, Lang C, Huber SM, Lang F. Retinoic acid induced suicidal erythrocyte death. *Cell Physiol Biochem.* 2008;21:193–202.
101. Sopjani M, Foller M, Gulbins E, Lang F. Suicidal death of erythrocytes due to selenium-compounds. *Cell Physiol Biochem.* 2008;22:387–394.

102. Sopjani M, Foller M, Haendeler J, Gotz F, Lang F. Silver ion-induced suicidal erythrocyte death. *J Appl Toxicol.* 2009;29:531–536.
103. Head DJ, Lee ZE, Swallah MM, Avent ND. Ligation of CD47 mediates phosphatidylserine expression on erythrocytes and a concomitant loss of viability *in vitro. Br J Haematol.* 2005;130:788–790.
104. Qadri SM, Mahmud H, Foller M, Lang F. Thymoquinone-induced suicidal erythrocyte death. *Food Chem Toxicol.* 2009;47:1545–1549.
105. Nguyen TT, Foller M, Lang F. Tin triggers suicidal death of erythrocytes. *J Appl Toxicol.* 2009;29:79–83.
106. Schneider J, Nicolay JP, Foller M, Wieder T, Lang F. Suicidal erythrocyte death following cellular K+ loss. *Cell Physiol Biochem.* 2007;20:35–44.
107. Kiedaisch V, Akel A, Niemoeller OM, Wieder T, Lang F. Zinc-induced suicidal erythrocyte death. *Am J Clin Nutr.* 2008;87:1530–1534.
108. Niemoeller OM, Bentzen PJ, Lang E, Lang F. Adenosine protects against suicidal erythrocyte death. *Pflügers Arch Eur J Physiol.* 2007;454:427–439.
109. Brand V, Koka S, Lang C, Jendrossek V, Huber SM, Gulbins E, *et al.* Influence of amitriptyline on eryptosis, parasitemia and survival of Plasmodium berghei-infected mice. *Cell Physiol Biochem.* 2008;22:405–412.
110. Floride E, Foller M, Ritter M, Lang F. Caffeine inhibits suicidal erythrocyte death. *Cell Physiol Biochem.* 2008;22:253–260.
111. Lang PA, Kempe DS, Akel A, Klarl BA, Eisele K, Podolski M, *et al.* Inhibition of erythrocyte "apoptosis" by catecholamines. *Naunyn-Schmiebergs Arch Pharmacol.* 2005;372:228–235.
112. Huber SM, Gamper N, Lang F. Chloride conductance and volume-regulatory nonselective cation conductance in human red blood cell ghosts. *Pflügers Arch Eur J Physiol.* 2001;441:551–558.
113. Lang KS, Myssina S, Tanneur V, Wieder T, Huber SM, Lang F, *et al.* Inhibition of erythrocyte cation channels and apoptosis by ethylisopropylamiloride. *Naunyn-Schmiebergs Arch Pharmacol.* 2003;367:391–396.
114. Kasinathan RS, Foller M, Koka S, Huber SM, Lang F. Inhibition of eryptosis and intraerythrocytic growth of Plasmodium falciparum by flufenamic acid. *Naunyn-Schmiebergs Arch Pharmacol.* 2007;374:255–264.
115. Foller M, Mahmud H, Gu S, Kucherenko Y, Gehring EM, Shumilina E, *et al.* Modulation of suicidal erythrocyte cation channels by an AMPA antagonist. *J Cell Mol Med.* 2009;13:3680–3686.

116. Myssina S, Lang PA, Kempe DS, Kaiser S, Huber SM, Wieder T, et al. Cl-channel blockers NPPB and niflumic acid blunt Ca(2+)-induced erythrocyte 'apoptosis'. *Cell Physiol Biochem.* 2004;14:241–248.

117. Nicolay JP, Liebig G, Niemoeller OM, Koka S, Ghashghaeinia M, Wieder T, et al. Inhibition of suicidal erythrocyte death by nitric oxide. *Pflügers Arch Eur J Physiol.* 2008;456:293–305.

118. Qadri SM, Foller M, Lang F. Inhibition of suicidal erythrocyte death by resveratrol. *Life Sci.* 2009;85:33–38.

119. Klarl BA, Lang PA, Kempe DS, Niemoeller OM, Akel A, Sobiesiak M, et al. Protein kinase C mediates erythrocyte "programmed cell death" following glucose depletion. *Am J Physiol Cell Physiol.* 2006;290: C244–C253.

120. Mahmud H, Mauro D, Foller M, Lang F. Inhibitory effect of thymol on suicidal erythrocyte death. *Cell Physiol Biochem.* 2009;24:407–414.

121. Lang KS, Myssina S, Lang PA, Tanneur V, Kempe DS, Mack AF, et al. Inhibition of erythrocyte phosphatidylserine exposure by urea and Cl-. *Am J Physiol Renal Physiol.* 2004;286:F1046-F1053.

122. Qadri SM, Mahmud H, Foller M, Lang F. Inhibition of suicidal erythrocyte death by xanthohumol. *J Agric Food Chem.* 2009;57:7591–7595.

123. Kucherenko Y, Geiger C, Shumilina E, Foller M, Lang F. Inhibition of cation channels and suicidal death of human erythrocytes by zidovudine. *Toxicology.* 2008;253:62–69.

124. Allan D, Michell RH. Calcium ion-dependent diacylglycerol accumulation in erythrocytes is associated with microvesiculation but not with efflux of potassium ions. *Biochem J.* 1977;166:495–459.

125. Pant HC, Virmani M, Gallant PE. Calcium-induced proteolysis of spectrin and band 3 protein in rat erythrocyte membranes. *Biochem Biophys Res Commun.* 1983;117:372–377.

126. Bookchin RM, Ortiz OE, Lew VL. Activation of calcium-dependent potassium channels in deoxygenated sickled red cells. *Prog Clin Biol Res.* 1987;240:193–200.

127. Franco RS, Palascak M, Thompson H, Rucknagel DL, Joiner CH. Dehydration of transferrin receptor-positive sickle reticulocytes during continuous or cyclic deoxygenation: role of KCl cotransport and extracellular calcium. *Blood.* 1996;88:4359–4365.

128. Lang PA, Kaiser S, Myssina S, Wieder T, Lang F, Huber SM. Role of Ca2+-activated K+ channels in human erythrocyte apoptosis. *Am J Physiol Cell Physiol*. 2003;285:C1553–C1560.
129. Foller M, Bobbala D, Koka S, Boini KM, Mahmud H, Kasinathan RS, *et al*. Functional significance of the intermediate conductance Ca2+-activated K+ channel for the short-term survival of injured erythrocytes. *Pflügers Arch Eur J Physiol*. 2010;460:1029–1044.
130. Bernhardt I, Weiss E, Robinson HC, Wilkins R, Bennekou P. Differential effect of HOE642 on two separate monovalent cation transporters in the human red cell membrane. *Cell Physiol Biochem*. 2007; 20:601–606.
131. Ivanova L, Bernhardt R, Bernhardt I. Nongenomic effect of aldosterone on ion transport pathways of red blood cells. *Cell Physiol Biochem*. 2008;22:269–278.
132. Kaestner L, Bernhardt I. Ion channels in the human red blood cell membrane: their further investigation and physiological relevance. *Bioelectrochemistry*. 2002;55:71–74.
133. Kaestner L, Tabellion W, Lipp P, Bernhardt I. Prostaglandin E2 activates channel-mediated calcium entry in human erythrocytes: an indication for a blood clot formation supporting process. *Thromb Haemost*. 2004;92:1269–1272.
134. Lang KS, Duranton C, Poehlmann H, Myssina S, Bauer C, Lang F, *et al*. Cation channels trigger apoptotic death of erythrocytes. *Cell Death Differ*. 2003;10:249–256.
135. Duranton C, Huber SM, Lang F. Oxidation induces a Cl(−)-dependent cation conductance in human red blood cells. *J Physiol*. 2002;539:847–855.
136. Barvitenko NN, Adragna NC, Weber RE. Erythrocyte signal transduction pathways, their oxygenation dependence and functional significance. *Cell Physiol Biochem*. 2005;15:1–18.
137. Bracci R, Perrone S, Buonocore G. Oxidant injury in neonatal erythrocytes during the perinatal period. *Acta Paediatr Suppl*. 2002;91:130–134.
138. Bilmen S, Aksu TA, Gumuslu S, Korgun DK, Canatan D. Antioxidant capacity of G-6-PD-deficient erythrocytes. *Clin Chim Acta*. 2001;303:83–86.
139. Damonte G, Guida L, Sdraffa A, Benatti U, Melloni E, Forteleoni G, *et al*. Mechanisms of perturbation of erythrocyte calcium homeostasis in favism. *Cell Calcium*. 1992;13:649–658.

140. Mavelli I, Ciriolo M, Rossi L, Meloni T, Forteleoni G, De Flora A, et al. Favism: a hemolytic disease associated with increased superoxide dismutase and decreased glutathione peroxidase activities in red blood cells. *Eur J Biochem*. 1984;139:8–13.

141. Huber SM, Uhlemann AC, Gamper NL, Duranton C, Kremsner PG, Lang F. Plasmodium falciparum activates endogenous Cl(-) channels of human erythrocytes by membrane oxidation. *EMBO J*. 2002;21:22–30.

142. Tanneur V, Duranton C, Brand VB, Sandu CD, Akkaya C, Kasinathan RS, et al. Purinoceptors are involved in the induction of an osmolyte permeability in malaria-infected and oxidized human erythrocytes. *FASEB J*. 2006;20:133–135.

143. Mandal D, Mazumder A, Das P, Kundu M, Basu J. Fas-, caspase 8-, and caspase 3-dependent signaling regulates the activity of the aminophospholipid translocase and phosphatidylserine externalization in human erythrocytes. *J Biol Chem*. 2005;280:39460–39467.

144. Matarrese P, Straface E, Pietraforte D, Gambardella L, Vona R, Maccaglia A, et al. Peroxynitrite induces senescence and apoptosis of red blood cells through the activation of aspartyl and cysteinyl proteases. *FASEB J*. 2005;19:416–418.

145. Sagan D, Jermnim N, Tangvarasittichai O. CD95 is not functional in human erythrocytes. *Int J Lab Hematol*. 2010;32:e244–e247.

146. Boas FE, Forman L, Beutler E. Phosphatidylserine exposure and red cell viability in red cell aging and in hemolytic anemia. *Proc Natl Acad Sci USA*. 1998;95:3077–3081.

147. Fadok VA, Bratton DL, Rose DM, Pearson A, Ezekewitz RA, Henson PM. A receptor for phosphatidylserine-specific clearance of apoptotic cells. *Nature*. 2000;405:85–90.

148. Andrews DA, Low PS. Role of red blood cells in thrombosis. Curr Opin *Hematol*. 1999;6:76–82.

149. Chung SM, Bae ON, Lim KM, Noh JY, Lee MY, Jung YS, et al. Lysophosphatidic acid induces thrombogenic activity through phosphatidylserine exposure and procoagulant microvesicle generation in human erythrocytes. *Arterioscler Thromb Vasc Biol*. 2007;27:414–421.

150. Closse C, Dachary-Prigent J, Boisseau MR. Phosphatidylserine-related adhesion of human erythrocytes to vascular endothelium. *Br J Haematol*. 1999;107:300–302.

151. Gallagher PG, Chang SH, Rettig MP, Neely JE, Hillery CA, Smith BD, et al. Altered erythrocyte endothelial adherence and membrane phospholipid asymmetry in hereditary hydrocytosis. *Blood.* 2003;101:4625–4627.
152. Pandolfi A, Di Pietro N, Sirolli V, Giardinelli A, Di Silvestre S, Amoroso L, et al. Mechanisms of uremic erythrocyte-induced adhesion of human monocytes to cultured endothelial cells. *J Cell Physiol.* 2007;213:699–709.
153. Zwaal RF, Comfurius P, Bevers EM. Surface exposure of phosphatidylserine in pathological cells. *Cell Mol Life Sci.* 2005;62:971–988.
154. Zappulla D. Environmental stress, erythrocyte dysfunctions, inflammation, and the metabolic syndrome: adaptations to CO2 increases? *J Cardiometab Syndr.* 2008;3:30–34.
155. Lang F, Busch GL, Ritter M, Volkl H, Waldegger S, Gulbins E, et al. Functional significance of cell volume regulatory mechanisms. *Physiol Rev.* 1998;78:247–306.
156. Kirk K. Membrane transport in the malaria-infected erythrocyte. *Physiol Rev.* 2001;81:495–537.
157. Duranton C, Huber S, Tanneur V, Lang K, Brand V, Sandu C, et al. Electrophysiological properties of the Plasmodium falciparum-induced cation conductance of human erythrocytes. *Cell Physiol Biochem.* 2003;13:189–198.
158. Brand VB, Sandu CD, Duranton C, Tanneur V, Lang KS, Huber SM, et al. Dependence of Plasmodium falciparum *in vitro* growth on the cation permeability of the human host erythrocyte. *Cell Physiol Biochem.* 2003; 13:347–356.
159. Staines HM, Chang W, Ellory JC, Tiffert T, Kirk K, Lew VL. Passive Ca(2+) transport and Ca(2+)-dependent K(+) transport in Plasmodium falciparum-infected red cells. *J Membr Biol.* 1999;172:13–24.
160. Tiffert T, Staines HM, Ellory JC, Lew VL. Functional state of the plasma membrane Ca2+ pump in Plasmodium falciparum-infected human red blood cells. *J Physiol.* 2000;525 Pt 1:125–134.
161. Koka S, Foller M, Lamprecht G, Boini KM, Lang C, Huber SM, et al. Iron deficiency influences the course of malaria in Plasmodium berghei infected mice. *Biochem Biophys Res Commun.* 2007;357:608–614.
162. Koka S, Huber SM, Boini KM, Lang C, Foller M, Lang F. Lead decreases parasitemia and enhances survival of Plasmodium berghei-infected mice. *Biochem Biophys Res Commun.* 2007;363:484–489.

163. Egberts J, Van Pelt J. Evaluation of the blood analyzer ABL 735 radiometer for determination of the percentage of fetal hemoglobin in fetal and neonatal blood. *Scand J Clin Lab Invest*. 2004;64:128–131.
164. Hermle T, Shumilina E, Attanasio P, Akel A, Kempe DS, Lang PA, *et al*. Decreased cation channel activity and blunted channel-dependent eryptosis in neonatal erythrocytes. *Am J Physiol Cell Physiol*. 2006;291:C710–C717.

6

Regulation of Red Cell Mass by Erythropoietin

Johannes Vogel and Max Gassmann*

Institute of Veterinary Physiology, Vetsuisse Faculty University of Zürich and Zürich Center for Integrative Human Physiology (ZIHP) Winterthurerstrasse 260, CH-8057 Zürich, Switzerland

6.1 Origin of Systemic Erythropoietin

Prior to the discovery of erythropoietin (Epo) it was believed that the bone marrow might sense hypoxemia directly. Then, in the 1950s, it was realized that hypoxemia, either hypoxic or anemic, results in stimulation of the bone marrow to produce more red blood cells by the hormone Epo that is mainly released from the kidneys.[1,2] These early papers represent the basis of the feedback loop model of the regulation of red cell mass (summarized in Fig. 6.1).

In recent decades, however, this picture became more and more detailed and complex: human Epo consists of 165 amino acids in its circulating form, is heavily glycosylated, and has a molecular mass of 30 kDa, 40% of which is derived from its carbohydrate portion. About 90% of plasma Epo is produced in the kidney, and in this organ the Epo-producing cells are now believed to be fibroblast-like type-1 interstitial cells located in the renal cortex and outer medulla.[3,4] Epo up-regulation in the kidney during hypoxemia is achieved by recruitment of cells producing a fixed amount of Epo mRNA.[5] In addition to the kidney, a

---

*Corresponding author: maxg@access.uzh.ch

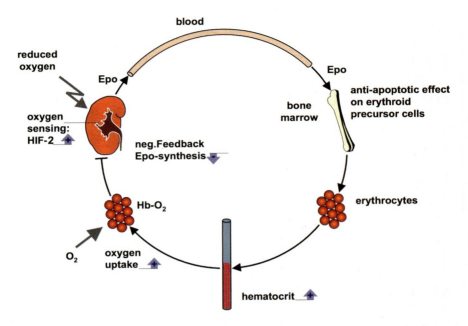

Figure 6.1 Classical model of red cell mass regulation in mammalian organisms. Hypoxia is sensed by renal cells resulting in increased Epo secretion into the blood. Epo exerts anti-apoptotic effects on erythroid precursor cells in the bone marrow, leading to an increased number of erythrocytes and elevated hematocrit. This in turn increases oxygen delivery to the tissues including the kidney. The increased oxygen availability to the kidney results in a decrease of Epo production.

considerable amount of Epo is also produced in the liver, the main source of Epo during fetal development.[6] In humans, the amount of plasma Epo originating from the liver is around 10%, but has been estimated to be as high as 40% in rats.[7] The shift from hepatic towards renal Epo production occurs in late gestation, with considerable species differences in onset,[6,8] suggesting that this shift is not due to physiological changes in circulation or oxygenation during birth. In the liver, both hepatocytes and Ito cells produce Epo, but in contrast to the kidney these hepatic cells are able to increase Epo expression levels.[9] Interestingly, the hepatic Epo expression appears to be suppressed in a paracrin manner by vascular endothelial growth factor (VEGF) although the VHL/HIF pathway (see below) might be more dominant.[10,11]

When considering the source of Epo in the adult organism, the question that arises is what makes the kidney the favorite organ for sensing systemic oxygen shortage? The answer might lie in the interplay between blood flow and tissue oxygenation as speculated several years ago.[12] Increased hematocrit results in increased blood viscosity; this impairs blood flow and thus should increase tissue hypoxia. In most organs the oxygen flow to tissues is a function of hematocrit with an optimum of 40–45%. Above that the oxygen delivery drops despite increase in oxygen transport capacity. The kidney however appears to be unique among the organs, since after basal requirements have been met its oxygen consumption parallels blood flow, and consequently tissue oxygen tension is constant as long as the oxygen content of the blood is not altered. This is due to the fact that both sodium filtration and sodium re-absorption rates are linear functions of blood flow.[13]

In a recent study it has been suggested that oxygen sensing by the skin influences Epo production.[14] This effect appears to rely on angiogenesis and NO-triggered alterations in skin blood flow that induces hepatic Epo production. Moreover, the skin also seems to be capable of influencing Epo production *via* modulation of NO signaling during acute hypoxia in mice with unaltered skin vasculature.[14]

6.2 Regulatory Principles

The main stimulus for Epo production is a decreased tissue oxygen partial pressure (pO_2). As synthesized Epo is not stored by renal cells, increased production correlates with increased mRNA due to increased stability and transcription. Apart from hypoxia and anemia a leftwards shift of the oxygen dissociation curve due to hemoglobin mutations or low levels of 2,3-di-phosphogycerate (2,3-DPG) can also be causal. In addition, several studies show altered Epo plasma levels during normoxia as a result of alterations of the central venous pressure. Specifically, a reduction of central blood volume results in increased Epo plasma levels and *vice versa*. Thus, Epo can also be considered as a plasma volume-regulation hormone, thereby indirectly regulating the hematocrit (for review see Kirsch *et al.*[15]). How Epo expression is induced under conditions of altered central volume remains elusive. Therefore we will focus on Epo expression as a result of altered tissue pO_2.

6.3 Cellular Oxygen Sensing

Up-regulation of Epo depends critically on the hypoxia-inducible transcription factor (HIF). HIF is a heterodimeric DNA-binding complex composed of two basic helix-loop-helix proteins of the PER-ARNT-SIM (PAS) family. The prominent family member is the aryl hydrocarbon receptor nuclear translocator (ARNT) — also termed HIF-1β — that is constitutively expressed in cells and is not oxygen-responsive. The other partner in this complex is one of the hypoxia-inducible α-subunits HIF-1α, HIF-2α, or HIF-3α.[16,17] Of note, a HIF-3α splice variant exists that gives rise to an inhibitory PAS protein (iPAS). This protein lacks the trans-activation domain and thus cannot bind the DNA, thus serving as a natural HIF-α antagonist. iPAS is strongly expressed in the continuously hypoxic corneal epithelium of the eye in order to suppress HIF-α-dependent angiogenesis.[18]

HIF-1α and HIF-2α subunits are continuously produced by the cell but rapidly degraded during normoxia. As this degradation process requires molecular oxygen, it is inhibited during hypoxia, resulting in accumulation of the HIF-α subunits that then translocate into the nucleus and dimerize with HIF-1β.[16] Prolyl hydroxylase domain (PHD) enzymes, of which three isoforms exist, initiate HIF-α degradation.[19] These enzymes are non-heme Fe(II) and 2-oxoglutarate-dependent dioxygenases modifying prolyl residues at positions 402 and 564 of the HIF-α molecules in the presence of molecular oxygen. In addition, HIF-α is also hydroxylated at position 803 (asparagin) by another non-heme Fe(II), 2-oxoglutarate-dependent, and oxygen-dependent enzyme called factor inhibiting HIF (FIH). The latter hydroxylation alone can reduce HIF-α activity even when not degraded, since this modification prevents binding of CBP/p300, an important co-activator of the transcription complex, thus reducing transcriptional activity (cf. next paragraph).[20] Subsequently, the hydroxylated prolins in position 402 and 564 are recognized by the von Hippel–Lindau tumor suppressor protein (VHL) that recruits a number of other proteins including the E3 ubiquitin-ligase that ubiquitinilates HIF-α for proteolysis by the ubiquitin-proteasome pathway.[21]

During hypoxia HIF-α is not degraded, but accumulates and translocates to the nucleus where it is phosphorylated and sulfonylated. Then HIF-α dimerizes with ARNT/HIF-1β and recruits several other factors that finally form the functional transcription complex. One important factor in this complex is the CBP/p300 coactivator that is important for initiation of transcription and can bind the HIF-α/HIF-1β heterodimer only when the asparagine in position 803 of HIF-α is not hydroxylated. This pathway is summarized in Fig. 6.2.

Regarding the continuous production and degradation of HIF-α, one might wonder why nature has chosen such a costly way to sense cellular oxygen. A sufficient oxygen-dependent energy supply is of utmost importance for all cells of the mammalian organism, and an immediate response to hypoxia appears to be mandatory.

Indeed, it could be shown that HIF-α accumulates and translocates to the nucleus in less than two minutes after onset of hypoxia, and also has a similarly fast switching-off kinetic upon re-oxygenation.[22]

Thus, the energy burden to synthesize HIF-α continuously and instantly degrade it is outweighed by increased survival, due to a faster emergency response to hypoxia-induced energy shortage.

6.4 The Epo-Promoter

The Epo promoter has been studied extensively, but although the mechanism of Epo expression is not completely elucidated, the hypoxia-inducible up-regulation of Epo is understood in considerable detail (Fig. 6.3).

Interestingly, despite HIF-1 having been initially identified as binding to the hypoxia-responsive elements of the Epo promoter[24] there is now convincing evidence that HIF-2 controls Epo expression.[25] However, the molecular basis for Epo's regulation by HIF-2 is not well understood. It is likely that HIF-2 binding to Epo's HRE requires additional nuclear factors that associate with the EPO gene, most likely members of the nuclear steroid hormone receptor family. Hepatocyte nuclear factor-4 (HNF-4) binds to a DR-2 element near the HRE and is a potential candidate factor that may specifically cooperate with HIF-2α.[26] Similar to HIF-2α, HNF-4 expression coincides with sites of EPO production in the liver and renal cortex, and is required for the hypoxic induction of EPO in Hep3B cells.[26–28]

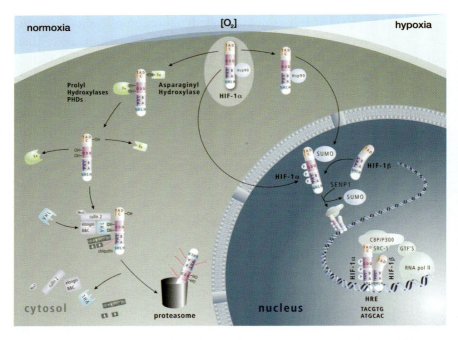

Figure 6.2 Schematic illustration of hypoxia-inducible gene transcription. HIF-1α or HIF-2α is continuously produced and immediately targeted for proteasomal degradation in normoxic conditions by hydroxylation of two prolyl and one asparagyl residues. The enzymes responsible for this (PHD and FIH) are non-heme Fe(II) hydroxylases that need molecular oxygen and 2-oxoglutarate as substrates. Hydroxylated HIF-α is recognized by the von Hippel–Lindau tumor suppressor (VHL) promoting HIF-α ubiquitinilation that targets this protein for proteasomal degradation. During hypoxia, HIF-α degradation is less efficient. Consequently HIF-α accumulates and translocates into the nucleus, where it is phosphorylated and sulfonylated to be fully activated for binding the constitutively expressed ARNT/HIF-1β transcription factor as well as other nuclear factors, to form a functional transcription complex. In turn, this complex binds to specific DNA motifs present in the promoter of hypoxia inducible genes — the so-called hypoxia responsive elements (HRE).[23] Of note, HIF-α can also be phosphorylated and translocated into the nucleus during normoxia.

In addition, HIF-α can undergo posttranslational modification by SUMOylation. SUMO (small ubiquitin-like modifier) proteins are structurally related to ubiquitin and reversibly modify function and cellular localization of targeted proteins. One of the enzymes that removes SUMO is

Regulation of Red Cell Mass by Erythropoietin

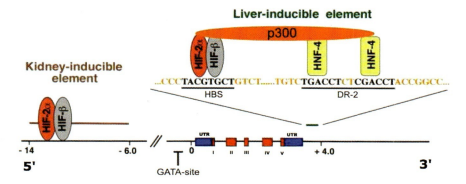

Figure 6.3 Schematic overview of the Epo gene. The kidney-inducible element lies far upstream at the 5′ end of the gene between −9.5 and −14 kb. Not shown is a negative regulatory element between −0.4 and −6 kb that prevents ectopic Epo expression. The functional GATA site lies within the minimal promoter. The liver-inducible element is located at the 3′ end of the gene and the hypoxia response element with the HIF binding site (HBS) is shown in more detail. Close to the HBS the nuclear receptor half site (DR-2) is located, to which the hepatic nuclear factor 4 (HNF-4) can bind. Most likely the recruitment of p300 to these proteins determines the HIF-2α specificity of this motif. The untranslated regions (UTR) of the Epo gene are shown in blue and its five exons in red.

the sentrin/SUMO-specific protease (SENP). SENP1 knockout mice developed severe anemia and died during midgestation.[29] In these mice, de-SUMOylation under hypoxic conditions did not occur and prevented the activation of HIF signaling in the nucleus. Instead, SUMOylated HIF-1α was targeted for proteasomal degradation in a pVHL- and ubiquitin-dependent, but PHD-independent manner, resulting in a strong reduction of hepatic Epo mRNA levels.[29] Although SUMOylation of HIF-α was specifically investigated with regard to hypoxic HIF-1α signaling, the presence of anemia strongly suggests that SENP de-SUMOylates HIF-2α as well.

Epo expression in renal cells other than fibroblast-like interstitial cells, such as tubular epithelial cells, appears to be suppressed by GATA transcription factors, in particular GATA-2 and GATA-3. The GATA site might also play a role in the transition from liver as the primary site of Epo production during fetal development to the kidney. Although this switch remains poorly understood it may involve transcriptional repression and/or reduced expression of GATA-4.[30] It is also remarkable that the kidney, as mentioned above, does not respond to hypoxia by dynamically increasing Epo message in individual cells, but rather recruits in a pO_2-dependent manner additional cells that generate a fixed amount of Epo. As such, renal Epo production differs from Epo synthesis in hepatic, hepatoma, neuroblastoma, or melanoma cell lines. Of note, HIF-mediated induction of Epo in the liver and kidney is controlled by distinct regulatory elements that are located on the opposite ends of the Epo gene, namely the kidney-inducibility element in the 5′-region, and the liver-inducibility element in the 3′-region.[31–33]

6.5 Epo-Receptor Signaling

Binding of Epo to its receptor (EpoR) results in receptor homodimerization and autophosphorylation of the receptor-associated Janus kinase 2 (JAK2), a cytoplasmic protein tyrosine kinase. In turn, activated JAK2 phosphorylates key tyrosine residues on the distal cytoplasmic region of EpoR[34] that then serve as docking sites for the signal transducer and activator of transcription protein 5 (STAT5) and phosphatidylinositol 3 kinase (PI3K). Once activated, STAT5 homodimerizes and translocates to the nucleus where it affects gene transcription. In addition, PI3K activates AKT serine/threonine kinase and the GATA-1 transcription factor. As a result, basic survival and proliferation of late colony-forming units-erythroid (CFUe)-like pro-erythroblasts is increased. This occurs predominately by activation of the anti-apoptotic protein Bcl-xl[35] together with mitogen-activated protein kinase (MAPK), p85-a/PI3K effects on AKT, and downstream effectors[36,37] (Fig. 6.4).

However, this simple model is complicated by some additional facts. In particular, studies of erythropoiesis in Bcl-xl$^{-/-}$ mice indicate survival roles selectively during late erythroblast development,[38] and effects of Bcl-xl on late-stage erythroblasts also have been suggested to be independent of Epo.[39] In addition, recent studies of EpoR response genes in EPO-dependent primary bone marrow progenitors[40–42] tend to discount Bcl-xl as a target in CFU-like pro-erythroblasts (but point to several alternate new Epo/EpoR-modulated regulators). For PI3K a possible non-essential role is based on congenital polycytemic patients exhibiting

Regulation of Red Cell Mass by Erythropoietin

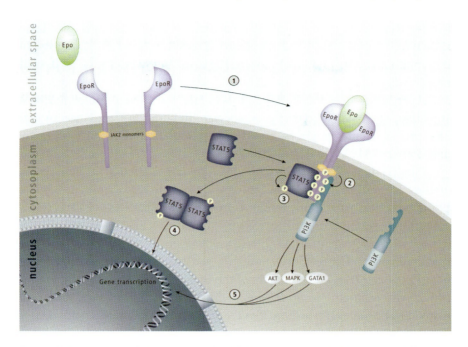

Figure 6.4 Erythropoietin receptor (EpoR) signaling. Epo binding results in homodimerization (1) of the EpoR and the receptor-associated Janus kinase 2 (JAK2) resulting in autophosphorylation of JAK2 (red dot). In turn, the activated JAK2 phosphorylates tyrosine residues on the distal cytoplasmic region of the EpoR (2, red dots) serving as docking sites for signal transducer and activator of transcription protein 5 (STAT5) and phosphatidylinositol 3 kinase (PI3K).[3] Activated STAT5 homodimerizes and translocates to the nucleus[4] where it affects gene transcription. In addition, PI3K activates AKT, MAPK, and the GATA-1 transcription factor and thereby also influences gene transcription.[5]

EpoR alleles that lack a distal docking site for p85-alpha[43] (p85-alpha knockout mice however display anemia during embryogenesis[44]). In addition, Epo may affect cell cycle progression by stimulation of type-D cyclin expression[41] (and D1 as well as D2 cyclin, also known STAT5 targets).[45,46] Epo-dependent regulation of cyclin G2 has also been discovered in primary bone marrow progenitors,[41] specifically through EpoR/JAK2/STAT5 signaling, but interestingly involves repression of this inhibitory cyclin. Repressive actions of STAT5 are atypical but have also been shown to be important, for example, for B-cell lymphoma 6 regulation in B-lymphoma cells,[47] and interleukin (IL)-4-mediated inhibition of IL-2 in helper T cells.[48]

Apart from these mechanisms, others might also be involved in EpoR signaling, such as reactive oxygen species. In addition, regulation of cell adhesion molecules might be involved in modulation of erythropoiesis by defining erythroblastic island stability.

Finally, the EpoR is tightly regulated by modulation of its activity, as well as by trafficking. At least three different protein families regulate EpoR activity: specific protein tyrosine phosphatases (PTP), SH2-containing suppressors of cytokine signaling (SOCS), and protein inhibitors of activated STAT (PIAS). Members of the PTP family are recruited to the EpoR shortly after its activation and de-phosphorylate both EpoR and JAK2.[49] Members of the SOCS family can directly inhibit JAK2 and promote its proteasomal degradation.[50] However there are many uncertainties regarding these additional mechanisms and consequently we need to await future research to understand them in detail.

6.6 Down-Regulation of Red Cell Mass

Under physiological conditions inadequately high hematocrit levels result in a decrease of the red cell mass. This occurs either during descent from high altitude and also in astronauts when subjected to microgravity.[51] The latter is due to pooling of central blood, resulting in a loss of plasma volume that secondarily increases hematocrit. Normally erythrocyte clearance is well controlled and the probability for red cells to be removed from the circulation increases with their age. The recognition process of senescent red cells relies on phosphatidylserine exposure, presumably the predominant mechanism because it is greatly increased in sickle cell anemia.[52,53] In addition, oxidative damage, accumulation of surface immunoglobulins,[54] loss of surface sialic acids,[55] and CD47[56,57] mediate recognition of senescent red cells by macrophages. Interestingly, this mechanism is enhanced during excessive erythrocytosis in mice overexpressing Epo.[58,59] Whether erythrophagocytosis is also enhanced during physiological down-regulation of the red cell mass, e.g. after descent from high altitude, remains to be determined. Nonetheless, at present the mechanism shown in Fig. 6.1 appears to be sufficient to explain red cell mass down-regulation by normal elimination rate, but reduced red cell production.

However there are ideas that during down-regulation additional Epo actions might play a role in a mechanism called "neocytolysis".[60] According to this mechanism Epo might prevent the interaction of red cells with phagocytes. In conditions where Epo expression is suppressed, such as descent from high altitude, the youngest cells are specifically removed from the circulation. A prerequisite for such a mechanism would be the existence of Epo receptors on erythrocytes. Epo binding has been shown to occur, and moreover to prevent the breakdown of phosphatidylserine asymmetry at the surface of mature red cells[61] that is an important step for initiation of eryptosis (see Chapter 5), at the surface of mature red cells. For neocytolysis to be specific for the youngest erythrocytes, the Epo receptor should be most abundant on the youngest red cells, as has been demonstrated recently.[62] However the question remains regarding how older red cells, most likely lacking the Epo receptor, escape phagocytosis for quite long periods of time.

Bibliography

1. Erslev A. Humoral regulation of red cell production. *Blood*. 1953;8:349–357.
2. Jacobson LO, Goldwasser E, Fried W, Plzak L. Role of the kidney in erythropoiesis. *Nature*. 1957;179:633–634.
3. Bachmann S, Le Hir M, Eckardt KU. Co-localization of erythropoietin mRNA and ecto-5'-nucleotidase immunoreactivity in peritubular cells of rat renal cortex indicates that fibroblasts produce erythropoietin. *J Histochem Cytochem*. 1993;41:335–341.
4. Maxwell PH, Osmond MK, Pugh CW, Heryet A, Nicholls LG, Tan CC, et al. Identification of the renal erythropoietin-producing cells using transgenic mice. *Kidney Int*. 1993;44:1149–1162.
5. Koury ST, Bondurant MC, Koury MJ. Localization of erythropoietin synthesizing cells in murine kidneys by in situ hybridization. *Blood*. 1988;71:524–527.
6. Zanjani ED, Ascensao JL, McGlave PB, Banisadre M, Ash RC. Studies on the liver to kidney switch of erythropoietin production. *J Clin Invest*. 1981;67:1183–1188.

7. Tan CC, Eckardt KU, Ratcliffe PJ. Organ distribution of erythropoietin messenger RNA in normal and uremic rats. *Kidney Int.* 1991;40:69–76.
8. Eckardt KU, Ratcliffe PJ, Tan CC, Bauer C, Kurtz A. Age-dependent expression of the erythropoietin gene in rat liver and kidneys. *J Clin Invest.* 1992;89:753–760.
9. Koury ST, Bondurant MC, Koury MJ, Semenza GL. Localization of cells producing erythropoietin in murine liver by *in situ* hybridization. *Blood.* 1991;77:2497–2503.
10. Rankin EB, Higgins DF, Walisser JA, Johnson RS, Bradfield CA, Haase VH. Inactivation of the arylhydrocarbon receptor nuclear translocator (Arnt) suppresses von Hippel-Lindau disease-associated vascular tumors in mice. *Mol Cell Biol.* 2005;25:3163–3172.
11. Tam BY, Wei K, Rudge JS, Hoffman J, Holash J, Park SK, *et al.* VEGF modulates erythropoiesis through regulation of adult hepatic erythropoietin synthesis. *Nat Med.* 2006;12:793–800.
12. Boutin AT, Weidemann A, Fu Z, Mesropian L, Gradin K, Jamora C, *et al.* Epidermal sensing of oxygen is essential for systemic hypoxic response. *Cell.* 2008;133:223–234.
13. Erslev AJ, Caro J, Besarab A. Why the kidney? *Nephron.* 1985;41: 213–216.
14. Deetjen P, Kramer K. [The relation of O2 consumption by the kidney to Na re-resorption.]. *Pflügers Archiv für die gesamte Physiologie des Menschen Und der Tiere.* 1961;273: 636–650.
15. Kirsch KA, Schlemmer M, De Santo NG, Cirillo M, Perna A, Gunga HC. Erythropoietin as a volume-regulating hormone: an integrated view. *Semin Nephrol.* 2005;25:388–391.
16. Wenger RH. Cellular adaptation to hypoxia: O2-sensing protein hydroxylases, hypoxia-inducible transcription factors, and O2-regulated gene expression. *FASEB J.* 2002;16:1151–1162.
17. Fandrey J, Gorr TA, Gassman M. Regulating cellular oxygen sensing by hydroxylation. *Cardiovasc Res.* 2006;71:642–651.
18. Makino Y, Cao R, Svensson K, Bertilsson G, Asman M, Tanaka H, *et al.* Inhibitory PAS domain protein is a negative regulator of hypoxia-inducible gene expression. *Nature.* 2001;414:550–554.
19. Schofield CJ, Ratcliffe PJ. Oxygen sensing by HIF hydroxylases. *Nat Rev Mol Cell Biol.* 2004;5:343–354.

20. Lando D, Peet DJ, Gorman JJ, Whelan DA, Whitelaw ML, Bruick RK. FIH-1 is an asparaginyl hydroxylase enzyme that regulates the transcriptional activity of hypoxia-inducible factor. *Genes Dev.* 2002;16:1466–1471.
21. Maxwell PH, Wiesener MS, Chang GW, Clifford SC, Vaux EC, Cockman ME, et al. The tumour suppressor protein VHL targets hypoxia-inducible factors for oxygen-dependent proteolysis. *Nature.* 1999;399:271–275.
22. Jewell UR, Kvietikova I, Scheid A, Bauer C, Wenger RH, Gassmann M. Induction of HIF-1alpha in response to hypoxia is instantaneous. *FASEB J.* 2001;15:1312–1314.
23. Wenger RH, Stiehl DP, Camenisch G. Integration of oxygen signaling at the consensus HRE. *Sci STKE.* 2005;2005:re12.
24. Wang GL, Semenza GL. Characterization of hypoxia-inducible factor 1 and regulation of DNA binding activity by hypoxia. *J Biol Chem.* 1993;268: 21513–2158.
25. Rosenberger C, Mandriota S, Jurgensen JS, Wiesener MS, Horstrup JH, Frei U, et al. Expression of hypoxia-inducible factor-1alpha and -2alpha in hypoxic and ischemic rat kidneys. *J Am Soc Nephrol.* 2002;13:1721–1732.
26. Warnecke C, Zaborowska Z, Kurreck J, Erdmann VA, Frei U, Wiesener M, et al. Differentiating the functional role of hypoxia-inducible factor (HIF)-1alpha and HIF-2alpha (EPAS-1) by the use of RNA interference: erythropoietin is a HIF-2alpha target gene in Hep3B and Kelly cells. *FASEB J.* 2004;18:1462–1464.
27. Blanchard KL, Acquaviva AM, Galson DL, Bunn HF. Hypoxic induction of the human erythropoietin gene: cooperation between the promoter and enhancer, each of which contains steroid receptor response elements. *Mol Cell Biol.* 1992;12:5373–5385.
28. Galson DL, Tsuchiya T, Tendler DS, Huang LE, Ren Y, Ogura T, et al. The orphan receptor hepatic nuclear factor 4 functions as a transcriptional activator for tissue-specific and hypoxia-specific erythropoietin gene expression and is antagonized by EAR3/COUP-TF1. *Mol Cell Biol.* 1995;15:2135–2144.
29. Cheng J, Kang X, Zhang S, Yeh ET. SUMO-specific protease 1 is essential for stabilization of HIF-1alpha during hypoxia. *Cell.* 2007;131:584–595.
30. Dame C, Sola MC, Lim KC, Leach KM, Fandrey J, Ma Y, et al. Hepatic erythropoietin gene regulation by GATA-4. *J Biol Chem.* 2004;279: 2955–2961.

31. Semenza GL, Traystman MD, Gearhart JD, Antonarakis SE. Polycythemia in transgenic mice expressing the human erythropoietin gene. *Proc Natl Acad Sci USA*. 1989;86:2301–2305.
32. Semenza GL, Koury ST, Nejfelt MK, Gearhart JD, Antonarakis SE. Cell-type-specific and hypoxia-inducible expression of the human erythropoietin gene in transgenic mice. *Proc Natl Acad Sci USA*. 1991;88:8725–8729.
33. Semenza GL, Dureza RC, Traystman MD, Gearhart JD, Antonarakis SE. Human erythropoietin gene expression in transgenic mice: multiple transcription initiation sites and cis-acting regulatory elements. *Mol Cell Biol*. 1990;10:930–938.
34. Constantinescu SN, Ghaffari S, Lodish HF. The erythropoietin receptor: structure, activation and intracellular signal transduction. *Trends Endocrinol Metab*. 1999;10:18–23.
35. Socolovsky M, Fallon AE, Wang S, Brugnara C, Lodish HF. Fetal anemia and apoptosis of red cell progenitors in Stat5a-/-5b-/- mice: a direct role for Stat5 in Bcl-X(L) induction. *Cell*. 1999;98:181–191.
36. Klingmuller U, Wu H, Hsiao JG, Toker A, Duckworth BC, Cantley LC, et al. Identification of a novel pathway important for proliferation and differentiation of primary erythroid progenitors. *Proc Natl Acad Sci USA*. 1997;94:3016–3021.
37. Haseyama Y, Sawada K, Oda A, Koizumi K, Takano H, Tarumi T, et al. Phosphatidylinositol 3-kinase is involved in the protection of primary cultured human erythroid precursor cells from apoptosis. *Blood*. 1999;94:1568–1577.
38. Wagner KU, Claudio E, Rucker EB, 3rd, Riedlinger G, Broussard C, Schwartzberg PL, et al. Conditional deletion of the Bcl-x gene from erythroid cells results in hemolytic anemia and profound splenomegaly. *Development*. 2000;127:4949–4958.
39. Rhodes MM, Kopsombut P, Bondurant MC, Price JO, Koury MJ. Bcl-x(L) prevents apoptosis of late-stage erythroblasts but does not mediate the antiapoptotic effect of erythropoietin. *Blood*. 2005;106:1857–1863.
40. Sathyanarayana P, Dev A, Fang J, Houde E, Bogacheva O, Bogachev O, et al. EPO receptor circuits for primary erythroblast survival. *Blood*. 2008;111:5390–5399.
41. Fang J, Menon M, Kapelle W, Bogacheva O, Bogachev O, Houde E, et al. EPO modulation of cell-cycle regulatory genes, and cell division, in primary bone marrow erythroblasts. *Blood*. 2007;110:2361–2370.

42. Sathyanarayana P, Menon MP, Bogacheva O, Bogachev O, Niss K, Kapelle WS, et al. Erythropoietin modulation of podocalyxin and a proposed erythroblast niche. *Blood.* 2007;110:509–518.
43. Arcasoy MO, Karayal AF. Erythropoietin hypersensitivity in primary familial and congenital polycythemia: role of tyrosines Y285 and Y344 in erythropoietin receptor cytoplasmic domain. *Biochimica et Biophysica Acta.* 2005;1740:17–28.
44. Huddleston H, Tan B, Yang FC, White H, Wenning MJ, Orazi A, et al. Functional p85alpha gene is required for normal murine fetal erythropoiesis. *Blood.* 2003;102:142–145.
45. Mziaut H, Kersting S, Knoch KP, Fan WH, Trajkovski M, Erdmann K, et al. ICA512 signaling enhances pancreatic beta-cell proliferation by regulating cyclins D through STATs. *Proc Natl Acad Sci USA.* 2008;105:674–679.
46. Friedrichsen BN, Richter HE, Hansen JA, Rhodes CJ, Nielsen JH, Billestrup N, et al. Signal transducer and activator of transcription 5 activation is sufficient to drive transcriptional induction of cyclin D2 gene and proliferation of rat pancreatic beta-cells. *Mol Endocrinol.* 2003;17:945–958.
47. Walker SR, Nelson EA, Frank DA. STAT5 represses BCL6 expression by binding to a regulatory region frequently mutated in lymphomas. *Oncogene.* 2007;26:224–233.
48. Villarino AV, Tato CM, Stumhofer JS, Yao Z, Cui YK, Hennighausen L, et al. Helper T cell IL-2 production is limited by negative feedback and STAT-dependent cytokine signals. *J Exp Med.* 2007;204:65–71.
49. Wormald S, Hilton DJ. Inhibitors of cytokine signal transduction. *J Biol Chem.* 2004;279:821–824.
50. Tong W, Zhang J, Lodish HF. Lnk inhibits erythropoiesis and Epo-dependent JAK2 activation and downstream signaling pathways. *Blood.* 2005;105:4604–4612.
51. De SNG, Cirillo M, Kirsch KA, Correale G, Drummer C, Frassl W, et al. Anemia and erythropoietin in space flights. *Semin Nephrol.* 2005;25:379–387.
52. Dasgupta SK, Thiagarajan P. The role of lactadherin in the phagocytosis of phosphatidylserine-expressing sickle red blood cells by macrophages. *Haematologica.* 2005;90:1267–1268.
53. Wandersee NJ, Tait JF, Barker JE. Erythroid phosphatidyl serine exposure is not predictive of thrombotic risk in mice with hemolytic anemia. *Blood Cells Mol Dis.* 2000;26:75–83.

54. Dale GL, Daniels RB. Quantitation of immunoglobulin associated with senescent erythrocytes from the rabbit. *Blood.* 1991;77:1096–1099.
55. Kiehne K, Schauer R. The influence of alpha- and beta-galactose residues and sialic acid O-acetyl groups of rat erythrocytes on the interaction with peritoneal macrophages. *Biological Chemistry Hoppe-Seyler.* 1992;373:1117–1123.
56. Oldenborg PA, Gresham HD, Lindberg FP. CD47-signal regulatory protein alpha (SIRPalpha) regulates Fcgamma and complement receptor-mediated phagocytosis. *J Exp Med.* 2001;193:855–862.
57. Oldenborg PA, Zheleznyak A, Fang YF, Lagenaur CF, Gresham HD, Lindberg FP. Role of CD47 as a marker of self on red blood cells. *Science.* 2000;288:2051–2054.
58. Vogel J, Gassman M. Erythropoietic and non-erythropoietic functions of erythropoietin (Epo) in mouse models. *J Physiol.* 2011:1259–1264.
59. Bogdanova A, Mihov D, Lutz H, Saam B, Gassmann M, Vogel J. Enhanced erythro-phagocytosis in polycythemic mice overexpressing erythropoietin. *Blood.* 2007;110:762–769.
60. Alfrey CP, Rice L, Udden MM, Driscoll TB. Neocytolysis: physiological down-regulator of red-cell mass. *Lancet.* 1997;349:1389–1390.
61. Myssina S, Huber SM, Birka C, Lang PA, Lang KS, Friedrich B, *et al.* Inhibition of erythrocyte cation channels by erythropoietin. *J Am Soc Nephrol.* 2003;14:2750–2757.
62. Mihov D, Vogel J, Gassmann M, Bogdanova A. Erythropoietin activates nitric oxide synthase in murine erythrocytes. *Am J Physiol Cell Physiol.* 2009;297:C378–388.

7
Anaemia

Gordon W Stewart* and Michael Watts[†]

Divisions of Medicine and [†]Haematology, University College London, Rayne Building, University Street, London WC1E 6JF, UK

7.1 Introduction

This chapter is designed to be a brief introduction to the idea of 'anaemia' for basic scientists. For more detailed descriptions of this haematology, see[1] or (even more detailed).[2] For a useful primer in general clinical medicine, covering all systems and suitable for basic scientists, you could look at this one.[3]

Let's start with the briefest of descriptions of the structure and function of the human red cell, with a particular view to clinical relevance. The red cell is a circulating cell with two main functions:

1. the carriage of oxygen bound to its principal intracellular protein, haemoglobin (Hb: itself composed of two copies each of two chains, alpha and beta, with a haem group adduct); and
2. the facilitation by intra-erythrocytic carbonic anhydrase of the interconversion of CO_2 and bicarbonate, so that the CO_2 can be carried from tissue to lung in aqueous solution as the soluble anion bicarbonate.

The red cell is very flexible and robust. Its structure is highly edited. There are neither mitochondria, nor endoplasmic reticulum, nor nucleus. There is neither protein synthesis nor capacity for cell division. Energy is derived from anaerobic glycolysis only. There is neither fatty acid nor amino-acid metabolism. There is some ability to replenish membrane lipids from the plasma. It controls its volume by the manipulation of intracellular sodium and potassium, all powered by an NaK pump. In all

---
* Corresponding author: g.stewart@ucl.ac.uk

of these respects, the red cell resembles the fibre cells of the lens of the eye, which are likewise packed with a single major protein (crystallin). Containing iron, the red cell is vulnerable to oxidation, and has energy-consuming defensive anti-oxidant systems.

Many red cell proteins are directly shared with other tissues (actin, band 3, stomatin, ankyrin), although most are unique to the red cell (Hb, spectrin). It is probably true that all red cell proteins have homologues outside the red cell (e.g. the spectrin homologue fodrin in the brain). Proteomic and gene expression studies show the presence of proteins for which there is no obvious role, and suggest that there is a lot more happening in the red cell than we currently understand.[4] Examples might include the 'raft' proteins stomatin and flotillin.

In normal adults, the red cell develops in the bone marrow, where the stem cells reside. These give rise to 'lymphoid' and 'myeloid' lines. The myeloid progenitor divides into erythroid, granulocyte (neutrophil), megakaryocyte (platelet) and monocyte (macrophage) lines. The erythroid cell passes through a series of stages (pronormoblast, basophilic normoblast, polychromatic normoblast, orthochromic normoblast, reticulocyte) before it finally loses all of its intracellular organisation, cuts its moorings and becomes a loose floating bag of haemoglobin and carbonic anhydrase.

Erythropoiesis is stimulated by the circulating hormone erythropoietin, made in the kidney and liver, and is under the control of the oxygen level.

7.2 Anaemia: Definition, Symptoms and Signs

'Anaemia' means a shortage of red cells, conventionally measured as the concentration of haemoglobin (Hb) in the blood. It is diagnosed when levels are below about 13.0 g/dl in a male and 12 g/dl in a female. As anaemia worsens, symptoms of fatigue, dizziness and shortness of breath begin to develop at Hb levels of less than about 8 g/dl. The patient is very tired at a level of about 6 g/dl, and cardiac arrest can occur as the level drops below 5 g/dl.

7.3 Causes: General Summary

Anaemia has many possible causes. There may be insufficient materials (amino acids, vitamins, iron) for red cell manufacture. The growth factors required for erythropoiesis may be lacking. The architecture of the bone marrow, the 'factory', may be faulty in one way or another; or the red cells that are made may have a shortened life, either because they carry a genetic fault, or because they are mistakenly attacked by a faulty immune system.

7.4 Investigations: How it is Sorted Out

For doctors, the red cell is highly accessible. Both its quantity and many of its properties are easily measurable. It is the cellular victim of many clinical problems and is a very useful 'biomarker' for many diseases outside the bone marrow.

The red cell can 'react' in different ways. If there is insufficient material to fill the cytoplasm (basically, insufficient Hb due to lack of iron for the haem group, or a mutation in a globin gene as in thalassaemia), the body responds by making smaller-sized cells than normal, down to a rough minimum of 66 fl. If there is a defect in the developmental process of cell division, then the body makes fewer, but larger cells, up to a rough maximum of 120 fl. The shape of the cell may be abnormal, and a keen-eyed haematologist can become very adept at the visual assessment of red cell morphology. These different causes can often quickly be distinguished by some simple tests. Table 7.1 shows an idealised matrix that illustrates the diagnostic patterns of abnormality that can be found using relatively simple tests available in a haematology lab. The 'Hb' (i.e. the concentration of Hb in the blood) tells us the simple severity of the anaemia. The cell counter gives us a very accurate and precise measure of average red cell size (the 'mean cell volume', or 'MCV'). This should be between 80 and about 98 fl. The cells are large ('macrocytic') when the vitamins B_{12} and/or folate are deficient. Other causes of macrocytosis are heavy ethanol intake and the presence of a large proportion of immature cells. The cells are small ('microcytic', less than 80 fl) when the marrow cannot fill the dividing cells with Hb, either because of lack of iron to make the complete Hb protein including the haem group, or because there is a genetic flaw in the globin genes which reduces mRNA availability for globin synthesis (this is 'thalassaemia').

Table 7.1 A diagnostic matrix for 'anaemia'. This is an idealised picture. In very many instances the clinical picture is complex and the tests do not have such clarity. But nevertheless these are the basics of diagnosis in anaemia. Results are expected to be normal unless otherwise stated. The different tests are listed down the two left columns. The extreme left column shows the classification types of the tests. 'General' tests are to be done in every case. 'Haemolysis' comprises a triad that is commonly abnormal in shortened-lifespan conditions. 'Iron' and 'vitamins' are the two groups that make up 'haematinics'. The 'specific' tests are aimed at single diseases.

Class of test	Test	Anaemia	Iron deficiency	PA	Myelofibrosis	Chronic disease	AHA	PNH	Sickle	Thalassaemia	G6PDH	Hered. sphero.
General	Hb	↓	↓	↓	↓	↓	↓	↓	↓	↓	↓	↓
	MCV		↓	↑		N or ↓	N or ↑			↓		
	WBC				↓							
	Platelets				↓							
	Film		1,2	3	4,5	N	6 ± 7	6	8,6	1,2,6	N	9,6
Haemolysis	Retics						↑		↑	↑	↑	↑
	Haptos						↓		↓	↓	↓	↓
	Bilirubin						↑		↑	↑	↑	↑
Iron	Iron		↓			↓						
	TIBC		↑			↓						
	Ferritin		↓			↑						
Vitamins	B_{12}			↓								
	Folate											
Specific	DAG			+								
	Ham's							+				
	Osmotic fragility											↑
	Hb eph								Abn	Abn		
	Context				Older	Infl					Male	
	G6PDH assay										↓	

Blood film abnormalities: 1, Microcytic; 2, Hypochromic; 3, Macrocytes; 4, Poikilocytes; 5, Tear drop forms; 6, Reticulocytes; 7, Nucleated red cells; 8, Sickle cells; 9, Spherocytes; N, Normal appearance.

The appearance of the red cells under the microscope (the 'blood film') can be telling. Classic abnormalities include spherocytes, elliptocytes, stomatocytes and sickle cells.[5] Some of these are illustrated in Fig. 7.1.

The maturity of the red cells in the peripheral circulation (i.e. those which you sample with a needle in a vein, as opposed to those which you sample by aspirating the bone marrow itself) can be assessed either by a protein stain for the amount of intracellular structure still present, or by staining with a fluorescent RNA dye such as acridine orange. The common term for an immature red cell is a 'reticulocyte', so named because the degrading intracellular organelles take on the appearance of a 'network' or 'reticulum'. If the lifespan of the cells is unduly short (that is, 'haemolysis' is present), then the number of immature red cells in the peripheral circulation can be increased ('reticulocytosis' occurs). In addition, circulating plasma proteins known as 'haptoglobins', whose job it is to mop up free Hb in the plasma,

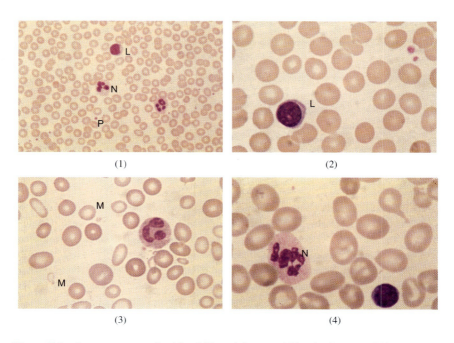

Figure 7.1 Some representative blood films. 1,2, normal blood at lower and higher power. The cells are roughly circular and have a less dense central area, known as 'central pallor'; 3, blood film from a patient recently started on treatment for iron deficiency. The older cells,

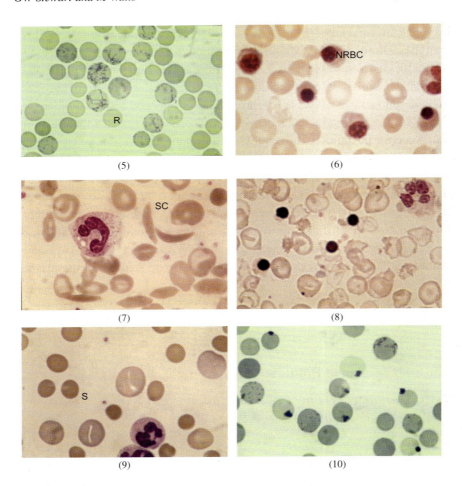

Figure 7.1 (*Continued*) still circulating, are small and pale, while the newer ones are larger and contain more Hb; 4, macrocytes (large cells), which might be seen in B$_{12}$ deficiency; 5, reticulocytes (R): relatively large young red cells expelled from the marrow when the lifespan of circulating red cells is reduced. The 'network'-like pattern of material is the degrading of intracellular organelles; 6, nucleated red blood cells (NRBC), even less mature than reticulocytes; 7, sickle cell anaemia, showing some classical 'sickle cells' (SC), and many other malformed cells; 8, thalassaemia major, a severe thalassaemia, showing small, pale misshapen cells; 9, hereditary spherocytosis, showing small cells lacking central pallor (S), the result of repeated removal of membrane area by the spleen; 10, Heinz bodies, dense granules of oxidised protein normally removed by the spleen but which persist if that organ is absent. L, lymphocyte; M, microcytic; N, neutrophil; NRBC, nucleated red blood cells; P, platelet (the pin-head type cells); R, reticulocyte; S, spherocytes; SC, sickle cells.

are reduced in concentration. The level of bilirubin, the insoluble yellow degradation product of haem, tends to rise, making the patient jaundiced.

Other investigations can follow. The bone marrow can be aspirated *via* one of the iliac crests in the back. This will give information on many aspects of erythropoiesis: the architecture of the marrow; abnormalities in the morphology of cells in the developmental stages of erythropoiesis; and the presence or absence of abnormal cells such as malignant cells in the marrow. Specialised tests aimed at specific diseases, e.g. Ham's acid haemolysis test (for paroxysmal nocturnal haemoglobinuria, see below), direct anti-globulin test (for antibodies directed against red cells), haemoglobin electrophoresis (for genetically-encoded variations in the haemoglobin protein) can all be conducted, if indicated.

The clinical context can be very telling. As has been said, anaemia can occur in a number of clinical conditions and it is important to diagnose these. Renal failure, chronic infective or inflammatory conditions, gut diseases, endocrine and hepatic conditions can all be associated with some kind of anaemia. Drug treatment can have a profound effect on red cell production, especially chemotherapy for malignant disease.

7.5 Some General Points About Red Cells and Anaemia

7.5.1 *Iron and ferritin: key associations with red cell disease and anaemia*

All cells need some iron but the red cells need the most. Iron deficiency leads to microcytic anaemia (see below), but the links do not stop there. The commonest cause of iron deficiency is loss of red cells themselves from the circulation, usually by bleeding from the gut or the female genital tract. The amount of iron in the body is controlled by variation in the rate of iron absorption: primates have no means of excretion of surplus iron. The rate of absorption of iron is determined by the perceived needs of the bone marrow. In many haemolytic conditions, especially those characterised by 'ineffective erythropoiesis' (where red cell manufacture gets started but cannot effectively finish), signals are sent such that iron absorption through the gut is caused to increase, beyond the body's actual needs.[6] Since the body has no effective way of excreting iron, the metal accumulates, causing damage to the liver, the heart and other organs. In addition, patients with severe haemolysis are very often transfused, and this treatment itself leads to yet more iron overload.

There is yet another link between iron metabolism and red cells. Patients with a number of long-standing (i.e. 'chronic') conditions suffer from a specific anaemia, the 'anaemia of chronic disease', also known as the 'anaemia of inflammation'. The medical conditions with which this anaemia is associated can be infective, inflammatory or malignant. The blood tests usually show an 'acute phase reaction', a series of abnormalities which are thought to be a physiological defence against infection. The c-reactive protein, or CRP, is high; the ferritin level is high; the plasma albumin concentration is usually low. The blood count shows anaemia with red cells of normal size and microscopic appearance. It seems likely that in this very common condition, which occurs in a number of general medical patients in many wards of hospitals, the key molecular link is the secretion into the blood of hepcidin, which has the effect of inhibiting an iron-exporting protein, ferroportin, limiting iron export from cells into the plasma.[7] This is almost certainly part of an evolutionary 'iron withholding' response to infection.

The principal role of the protein ferritin is intracellular iron storage, but it is also a secreted, circulating plasma protein whose plasma concentration is very useful to measure. Low ferritin levels in the plasma occur in iron deficiency, while high plasma ferritin levels are found both in states of iron overload and in states of inflammation. In both of these clinical states, it may be that the plasma level of ferritin is increased in order to soak up excess iron.

7.5.2 *Vitamins and the red cell*

Deficiency in four major nutritional elements has a marked effect on red cell production. As we have said, iron is a necessary building block for haem and therefore haemoglobin. Iron deficiency anaemia is characterised by small, microcytic, red cells. The lack of iron may be due to an inadequate, or poor absorption by a diseased gut, or more likely because blood itself is being lost through the gut (seeping from a quietly bleeding ulcer or tumour somewhere).

Both of the vitamins B_{12} and folate are concerned with one-carbon metabolic steps in the synthesis of DNA, a metabolic process that is required in any rapidly dividing tissue, of which the haematopoietic system is a prime example (epithelia and then skin come second). The developing red cells cannot effectively divide. The body reacts by making larger macrocytic red cells. The bone marrow shows 'megaloblastic' change (i.e. characterised by large precursor cells).

Although its deficiency is very rare, pyridoxine is also vital for haem synthesis during red cell production. Pyridoxine is important in many other reactions (e.g. amino-acid metabolism) but anaemia is one result of its deficiency.[8,9]

Let's not forget the simple state of nutrition of the patient. Red cells, like all other cells, fail to be produced when the global supply of energy and amino acids is reduced.

7.5.3 *The red cell and thrombosis*

A number of haematological conditions are associated with thrombosis. The broken red cell is pro-thrombotic. The interior leaflet of the membrane contains negatively charged phospholipids which trigger the intrinsic pathway in the clotting cascades. This is probably the mechanism behind thrombosis in paroxysmal nocturnal haemoglobinuria[10] and in continuing haemolysis after splenectomy.[11] Sickle cells have a tendency to dehydrate and shrink, causing the Hb to come out of solution inside the cell, with adverse effects on red cell flexibility and the viscosity of the blood. Red cells may be a source of thrombogenic 'microparticles'.[12,13]

7.5.4 *Nitric oxide and red cells*

It is a relatively recent observation that the red cell has the ability to generate the vasodilator nitric oxide in response to hypoxia. The corollary is that in disease states this ability may be lacking, leading to unwanted vasoconstriction. This may be important in the pathogenesis of pulmonary hypertension in sickle cell anaemia.[14]

7.5.5 Inherited red cell disorders and malaria

Malaria can be considered as 'an infection of the red cell'. Indeed, it is the *only* infection of the mature red cell. (Parvovirus infection temporarily stops red cell production.) Inherited haematological disease is much, much commoner in tropical peoples, where malaria is endemic. It seems that many minor changes in the red cell structure or function, that just makes the red cell slightly less efficient, can hamper the parasite. Examples include the enzyme condition G6PDH deficiency, haemoglobin conditions such as sickle and thalassaemias and the membrane abnormality Southeast Asian ovalocytosis.[15,16] Lew has shown that the act of parasitisation places a very severe osmotic stress on the red cell.[17]

By and large the malaria-protective condition is the heterozygous state while the homozygous conditions are either lethal or very deleterious. The classic example is sickle. The heterozygous state, which is presumed to be the malaria-protective state, is virtually asymptomatic. The homozygous state is the unpleasant 'sickle cell anaemia', which is described below.

In the tropics, invasion and premature destruction of red cells by malarial parasites are a major cause of anaemia in its own right.

7.6 The Ten Most Instructive Anaemias

We will now consider ten clinical conditions that illustrate different forms of anaemia.

7.6.1 *Iron deficiency*

Iron deficiency is perhaps the commonest form of anaemia. As we have said, a deficiency of iron causes a microcytic anaemia. The iron may be lacking because of dietary insufficiency, or more commonly because of loss of iron in the form of blood itself, usually from the GI tract or the female genital tract. The condition is diagnosed by the combination of low plasma iron, high iron binding capacity and low ferritin. Bone marrow examination should not be required. If the cause is not immediately obvious then it must be sought, typically by endoscopic examination of the oesophagus, stomach and duodenum, and the colon. The jejunum and ileum cannot easily be reached by an endoscope but are only rarely the source of occult gastrointestinal bleeding. Iron deficiency can be treated with oral iron: sometimes blood transfusion and/or intravenous iron infusion are necessary.

7.6.2 *Pernicious anaemia*

Pernicious anaemia is caused by the deficiency of vitamin B_{12}. Autoimmune attack on the gastric parietal cells in the stomach destroys the cells which secrete 'intrinsic factor' into the lumen of the gut. Intrinsic factor is a cofactor required for the absorption on B_{12} in the terminal ileum. Very gradually, over years, the patient becomes anaemic, sometimes also showing neurological dysfunction. The blood picture shows macrocytic anaemia, the B_{12} level is found to be low and antibodies to gastric parietal cells can be found in the blood. The condition is easily treated by injections of B_{12}, which circumvent the gut absorption problem, although theoretically the condition could be treated by pills containing both B_{12} (which was the original 'extrinsic factor') and intrinsic factor.

A similar anaemia is caused by deficiency of folate. Both B_{12} and folate can be deficient because of gut disorders which impair their absorption, e.g. coeliac disease or Crohn's disease.

7.6.3 *Myelodysplasia/myelofibrosis*

Myelodysplasia is a bone marrow disorder of the more elderly patient. The marrow gradually becomes replaced by fibrotic tissue, displacing the haematopoietic tissue to its reserve locations, the liver and spleen (which in normal physiology are used for erythropoiesis only in foetal life). The lack of marrow access reduces the ability of the haematopoietic tissue to make the cells of the blood, and 'pancytopenia' occurs, a shortage of the three main cellular elements of the blood: red and white cells and platelets. The liver and particularly the spleen can enlarge to huge sizes.

Myelodysplasia is a disorder of haematopoietic stem cells. Many cases of myelodysplasia are associated with somatic mutations in the Janus kinase, JAK2.[18,19] The condition can progress to leukaemia.

Other conditions can cause replacement of the space in the marrow. Malignant conditions of the white cells (leukaemias) and plasma cells (multiple myeloma) can do this, and occasionally metastatic malignant cells from other primaries such as prostate or lung or breast can take over the marrow.

7.6.4 Anaemia of chronic disease

Also known as the 'anaemia of inflammation', the anaemia of chronic disease is a common diagnosis in a hospital ward. A patient suffering from a chronic infective or inflammatory illness (with a prominent 'acute phase response': high CRP, low albumin, high ferritin) shows a moderate anaemia (Hb, 8–9 g/dl, say) with normal-sized red cells and no signs of haemolysis. B_{12} and folate are normal; the iron level is normal or low and the iron binding capacity is likewise normal or low. (This combination also occurs in renal failure.) It is thought that the anaemia of chronic disease represents a physiological response to infection, designed to withhold iron from the attacking infection. As stated above, the key signalling molecule seems to be hepcidin.

These above conditions are all disorders of red cell production. We will now consider some haemolytic conditions.

7.6.5 Auto-immune haemolytic anaemia

Auto-immune pathology can be directed directly against the red cell. Antibodies may bind to the red cell membrane and recruit complement, which lyses the cell. There is a haemolytic blood picture (high reticulocytes, high bilirubin, low haptoglobins) but crucially the so-called direct anti-globulin test (Coombs' test) is abnormal. In this test, washed red cells are incubated with a reagent containing anti-human immunoglobulin antibody. If the red cells under test are themselves coated with a human IgG, then the anti-human immunoglobulin antibody will cause the red cells to aggregate together, which can be seen with the naked eye. Treatment is with immunosuppression. This is a classic example of an acquired (as opposed to congenital) haemolytic anaemia.

7.6.6 Paroxysmal nocturnal haemoglobinuria

By comparison with the other nine conditions in this section, paroxysmal nocturnal haemoglobinuria (PNH) is very, very rare. It is important to consider, because it is another acquired haemolytic anaemia but in this case the cause is quite different from the abnormal immunology of autoimmune haemolytic anaemia. The disease is caused by the development of a clone of erythropoietic stem cells carrying a somatic mutation in the gene *PIGA* coding for a subunit of a protein assembly aimed at glucosylaminyltransferase activity, which catalyses the attachment of GPA-linked proteins to their glycolipid anchor.[10] The cell develops without its GPA-linked proteins and it becomes exceptionally sensitive to attack by complement: this property is embodied in the disease-specific 'Ham's acid haemolysis test' which, when positive, diagnoses PNH. PNH is an example of an *acquired* clonal membrane protein disorder.

7.6.7 *Sickle cell disease*

Sickle cell anaemia is the archetypal haemoglobinopathy.[20] First described 100 years ago, sickle cell anaemia was shown to be a disease of haemoglobin in 1949, by Pauling and others. In 1958, long before cloning, the amino acid abnormality (glutamate 6 in beta globin changed to valine) was found by chemical methods. Most haemoglobinopathies lead to a mild, moderate or severe chronic haemolytic anaemia, subject to some variations which are typically infrequent. The clinical disease associated with sickle is greatly amplified by episodic derangements in cellular hydration, which have huge consequences for the rheological quality of the blood. The dehydrated cells block capillaries and cause local ischaemia, commonly in the bone marrow. These 'sickle crises' are driven by hypoxia and infection. Infarction of the marrow leads to almost unbearable bone pain. The patients can suffer strokes, bone problems, skin ulcers, eye problems, renal failure, and cardiac and chest problems.

7.6.8 *Thalassaemia*

While sickle cell anaemia is associated with a unique molecular defect, the thalassaemias are a very heterogeneous group of diseases, ranging from very mild through moderate to severe, requiring constant transfusion. However, the common feature behind all cases is a defect in transcription of one or other globin genes (alpha or beta), affecting the availability of mRNA for the globin chain in question. Thalassaemic patients do not have the ischaemic problems seen in sickle, but severe cases of thalassaemia show major problems with iron overloading, even in the absence of transfusion. (Thalassaemia is a prime example of 'ineffective erythropoiesis'.) Cardiac failure is a major consequence of this iron overloading.

7.6.9 *G6PDH deficiency*

Glucose-6-phosphate-dehydrogenase is the first step in the pentose phosphate shunt, which makes reduced NADPH to fuel the reduction of oxidised glutathione (GSSG) to reduced glutathione (GSH). The gene is on the X chromosome. When the red cells of patients carrying mutations are exposed to oxidant stress, there is rapid and severe haemolysis, which is episodic, as in sickle, but triggered by different stimuli. The classic stress in medical history is the fava bean, which contains the oxidants vicine, divicine, convicine and isouramil.

G6PDH is a classic example of an enzyme deficiency. It is unusual among enzymopathies in giving anaemia of such variable severity, and is also unusual in being X-linked in its inheritance. There are many other enzymopathies, especially in the glycolytic pathway, e.g. pyruvate kinase deficiency.

7.6.10 *Hereditary spherocytosis*

Hereditary spherocytosis is the classic inherited disorder of the red cell membrane, causing lifelong haemolytic anaemia. Caused by mutations in any one of a number of membrane proteins (band 3, ankyrin, spectrin, protein 4.2), the basic pathophysiological mechanism lies in a mechanical weakness in the protein-protein interactions that link the protein-based cytoskeleton with the overlying lipid bilayer facing the plasma.[21–23] It is almost always inherited as an autosomal dominant. The cells are 'spherocytic' because the mechanical weakness of the membrane allows blebs of lipid bilayer to peel up from the cytoskeleton. These blebs are amputated in the spleen, gradually reducing the available surface area of the red cell, changing its shape from disc to sphere. The condition can be ameliorated by splenectomy.

There are many other inherited membrane disorders. Our lab has been interested in the 'hereditary stomatocytoses', in which the cellular defect lies not in the mechanical instability but in membrane transport 'leaks' to sodium and potassium, which disrupt volume regulation. These can be caused by mutations in at least two genes, *SLC4A1* and *RHAG*, coding for the anion exchanger band 3^{24} and the gas transporter Rhesus-associated glycoprotein (RHAG).[25] The mutations cause both of these proteins to change function into cation leaks.

7.7 Concluding Remarks

This has a been a very swift tour around red cell haematology. We have seen how the red cell develops in the marrow in company with other circulating cells; how it requires iron, vitamins and erythropoietin to grow; how anaemia can be caused by lack of space in the marrow, by lack of building blocks and vitamins necessary for its growth, by acquired and congenital genetic defects which shorten its lifespan in the circulation. Anaemia often occurs because of systemic inflammatory or renal disease. Both iron deficiency and overload can be important in red cell haematology. Thrombosis can be a problem in some red cell diseases. The evolutionary pressure of the red cell infection malaria has made inherited haematological disease common in tropical peoples.

Glossary

Term	Meaning, significance
Acute	Of sudden onset. Note: does not necessarily mean 'severe'.
Anaemia	Deficiency of red blood cells in the circulation.
Autoimmune	Pathology in which the immune system attacks 'self'.
B_{12}	Cobalt-containing vitamin essential for erythropoiesis.

Bilirubin	Yellow pigment, poorly soluble in water, a degradation product of haem and therefore of red cells. Is excreted from the plasma by the liver but at a limited rate. If red cell breakdown increases it accumulates in the blood, and the patients turns yellow: 'jaundice'.
Blood count	Collection of measurements on the cells of blood. Includes: Hb, red cell size (MCV), reticulocyte count, total white cell count, plus breakdown of different cell types ('differential' white cell count), platelet count.
Chronic	Long-lasting: opposite of 'acute'.
Clinical context	The general state of the patient; the assembly of other illnesses that are present.
Cobalamin	Essentially the same as vitamin B_{12}.
Coombs' test	See 'Direct anti-globulin test'.
CRP	C-reactive protein, very useful plasma marker for inflammation in all its forms; a 25kD protein.
Direct anti-globulin test (DAT)	A test on peripheral blood which detects antibodies in the plasma directed against patient's own red cells.
Erythropoietin	Hormone that stimulates red cell production in the marrow. Largely made in the kidney; a 34kd protein.
Ferritin	Iron storage protein. Useful to measure in plasma. Level is low in iron deficiency, high in iron overload and in the acute phase reaction.
Folic acid ('folate')	441 dalton vitamin necessary for one-carbon transfers in DNA synthesis. Essential for erythropoiesis and many other functions. Unlike B_{12}, bodily stores are not high.
Haemoglobin electrophoresis	In which Hb is electrophoresed as a diagnostic investigation. The protein is red, so no stain is required

Haemoglobinuria	Passing of Hb in urine. Occurs with severe haemolysis
Haemolysis	State where the lifespan of the red cell in the circulation is reduced.
Ham's test	Test aimed at diagnosis of PNH. Red cells under test are exposed to complement under acid conditions. PNH cells will lyse before normal cells.
Haptoglobins	Plasma proteins whose task it is to mop up free Hb if it should get loose. Levels fall when red cells break down quickly.
Hepcidin	Polypeptide regulator of iron metabolism. 25 amino-acid form binds to and inhibits ferroportin, a gut iron transporter, limiting supply of iron.
Ineffective erythropoiesis	Situation in which erythropoiesis starts but cannot finish. Associated with increased ironabsorption from the gut.
Intrinsic factor	As-yet-unidentified enzyme-like substance secreted by 'parietal cells' of the gastric lining, that is a co-factor for B_{12} absorption.
Macrocytosis	When red cells are abnormally large (MCV>99 fl).
Malaria	Tropical infection caused by parasite of the *Plasmodium* class, transmitted to humans by mosquito bite. During its life cycle, invades and multiplies within red cells, destroying them in the process.
MCV	Mean (red) cell volume.
Microcytosis	When red cells are abnormally small (MCV<80 fl).
Osmotic fragility	Lab test on red cells in which aliquots of a red cell sample are exposed to solutions containing reducing concentrations of NaCl. Cells that have a low surface area:volume ratio will lyse at higher NaCl concentrations than normal cells.

Ovalocytosis	Oval-shaped red cells.
Pancytopenia	Shortage of all cell types in peripheral blood: red cells, white cells, platelets.
Pernicious anaemia	A B_{12} deficiency conditions caused by auto-immune attack on the parietal cells of the gastric mucosa, which make intrinsic factor: B_{12} cannot be absorbed without intrinsic factor.
PIGA	Phosphatidylinositol N-acetylglucosaminyl transferase subunit A. Catalyses first step in formation of GPI anchor.
Poikilocyte	Deformed, abnormally-shaped red cell.
Pyridoxine	Vitamin B_6. Pyridine-based vitamin, metabolite of which is a cofactor for aromatic amino acid decarboxylase.
Reticulocyte	A red cell seen in peripheral blood showing a 'network' of intracellular material, representing degrading intracellular organelles. An immature red cell.
Sideroblastic	A 'blast' (very immature) cell in the bone marrow laden with excess iron.
Sickle cell	A crescent-shaped cell seen on a blood film.
Sickle cell anaemia	Homozygous state for mutation giving glutamate-6 to valine change in beta globin.
Sickle cell trait	Heterozygous state for sickle.
Splenectomy	Removal of the spleen.
Thalassaemia	Inherited condition in which control over globin synthesis (alpha or beta) is faulty.
Total iron binding capacity (TIBC)	A measurement of the capacity of the plasma to bind iron; proportional to the concentration of transferrin in the blood.
Transferrin	Iron binding protein of the plasma and extracellular space, the main carrier of iron around the body.
Vitamin B_{12}	See 'B_{12}'.

White blood count (WBC)	A cell count, usually expressed as $N \times 10^9/l$, representing the sum of all 'white cells' (neutrophils, lymphocytes, monocytes, eosinophils, basophils), which are the cells in the peripheral blood that are not red cells or platelets.

Bibliography

1. Green AR, Hoffbrand AV, Catovsky D, Tuddenham EG, editors. *Postgraduate Haematology*. London: Wiley-Blackwell; 2010.
2. Hoffman R, Heslop H, Furie B, Benz EJ, McGlave P, Silberstein LE et al., editors. *Hematology: Basic Principles and Practice*. New York; Edinburgh: Churchill Livingstone; 2008.
3. Stewart GW. *Core Clinical Medicine*. London: Imperial College Press; 2010.
4. Pasini EM, Kirkegaard M, Mortensen P, Lutz HU, Thomas AW, Mann M. In-depth analysis of the membrane and cytosolic proteome of red blood cells. *Blood*. 2006;108(3):791–801.
5. Bain BJ. Diagnosis from the blood smear. *N Engl J Med*. 2005; 353(5):498–507.
6. Porter JB. Pathophysiology of transfusional iron overload: contrasting patterns in thalassemia major and sickle cell disease. *Hemoglobin*. 2009;33 (Suppl 1): S37–S45.
7. Ganz T, Nemeth E. Hepcidin and disorders of iron metabolism. *Annu Rev Med*. 2010;62:347–360.
8. Camaschella C. Recent advances in the understanding of inherited sideroblastic anaemia. *Br J Haematol*. 2008;143(1):27–38.
9. Clayton PT. B6-responsive disorders: a model of vitamin dependency. *J Inherit Metab Dis*. 2006:29(23):317–326.
10. Bessler M, Hiken J. The pathophysiology of disease in patients with paroxysmal nocturnal hemoglobinuria. *Hematology Am Soc Hematol Educ Program*. 2008;104–110.
11. Hirsh J, Dacie JV. Persistent post-splenectomy thrombocytosis and thromboembolism: a consequence of continuing anaemia. *Br J Haematol*. 1966;12: 44–53.

12. Verduzco LA, Nathan DG. Sickle cell disease and stroke. *Blood.* 2009; 114(25):5117–5125.
13. Rubin O, Crettaz D, Tissot JD, Lion N. Microparticles in stored red blood cells: submicron clotting bombs? *Blood Transfus.* 2010;8(Suppl 3): S31–S38.
14. Allen BW, Piantadosi CA. How do red blood cells cause hypoxic vasodilation? The SNO-hemoglobin paradigm. *Am J Physiol Heart Circ Physiol.* 2006;291(4):H1507–H1512.
15. Durand PM, Coetzer TL. Hereditary red cell disorders and malaria resistance. *Haematologica.* 2008;93(7):961–963.
16. Lopez C, Saravia C, Gomez A, Hoebeke J, Patarroyo MA. Mechanisms of genetically-based resistance to malaria. *Gene.* 2010;467(1–2):1–12.
17. Lew VL, Macdonald L, Ginsburg H, Krugliak M, Tiffert T. Excess haemoglobin digestion by malaria parasites: a strategy to prevent premature host cell lysis. *Blood Cells Mol Dis.* 2004;32(3):353–359.
18. Reuther GW. JAK2 activation in myeloproliferative neoplasms: a potential role for heterodimeric receptors. *Cell Cycle.* 2008;7(6):714–719.
19. Klco JM, Vij R, Kreisel FH, Hassan A, Frater JL. Molecular pathology of myeloproliferative neoplasms. *Am J Clin Pathol.* 2010;133(4):602–615.
20. Serjeant GR. One hundred years of sickle cell disease. *Br J Haematol.* 2010;425–429.
21. An X, Mohandas N. Disorders of red cell membrane. *Br J Haematol.* 2008;141(3):367–375.
22. Tse WT, Lux SE. Red blood cell membrane disorders. *Br J Haematol.* 1999;104(1):2–13.
23. Eber S, Lux SE. Hereditary spherocytosis — defects in proteins that connect the membrane skeleton to the lipid bilayer. *Sem Hematol.* 2004;41(2): 118–141.
24. Bruce LJ, Robinson H, Guizouarn H, Harrison P, King M-J, Goede JS *et al.* Monovalent cation leaks in human red cells caused by single amino-acid substitutions in the transport domain of the band 3 chloride-bicarbonate exchanger, AE1. *Nature Genetics.* 2005;37:1258–1263.
25. Bruce LJ, Guizouarn H, Burton NM, Gabillat N, Poole J, Flatt JF *et al.* The monovalent cation leak in overhydrated stomatocytic red blood cells results from amino acid substitutions in the Rh-associated glycoprotein. *Blood.* 2009;113(6):1350–1357.

8
Erythrocytes and Malaria

Henry M Staines,[*,†] Elvira T Derbyshire,[†] Farrah A Fatih,[†] Amy K Bei[‡] and Manoj T Duraisingh[‡]

[†]*Centre for Infection and Immunity, Division of Clinical Sciences, St George's, University of London, Cranmer Terrace, London SW17 0RE, UK;* [‡]*Harvard School of Public Health, 665 Huntington Avenue, Building 1, Room 715, Boston, MA 02115, USA*

8.1 Introduction

Erythrocytes play host to a range of intracellular organisms, including some of those from the phylum *Apicomplexa*. The *Apicomplexa* are a large and diverse group of unicellular, eukaryotic, obligate, intracellular parasites. They can cause serious illness in humans and other animals, resulting in diseases such as malaria and babesiosis. Malaria, the most studied of these diseases, is caused by parasites belonging to the genus *Plasmodium*. Focusing mainly on the human malarial parasite *P. falciparum*, this chapter will:

i introduce the disease, the parasite and its life cycle (Section 8.1);
ii consider how malarial parasites interact with and alter their host erythrocytes during invasion, development and egress (Section 8.2); and
iii describe the erythrocyte polymorphisms that may have been selected for by malaria and highlight the mechanisms by which they might thwart infection (Section 8.3).

---

[*] Corresponding author: hstaines@sgul.ac.uk

8.2 Malaria and Plasmodial Parasites

Each year there are approximately 300 million clinical episodes of malaria, leading to between one and two million deaths.[1] The majority of these occur in sub-Saharan Africa, where mortality rates are highest in children under five years of age. There are five plasmodial species known to infect humans: *P. falciparum* (causing the most morbidity and mortality), *P. vivax* (the most widely distributed plasmodial parasite),[2] *P. ovale*, *P. malariae* and *P. knowlesi* (a simian parasite that has recently been identified to cause clinically relevant infections in humans).[3] *P. vivax*, *P. malariae* and *P. ovale* infections normally result in uncomplicated malaria, which is characterised by fatigue, nausea, headaches and fever that is often cyclical. Complicated or severe malaria is associated generally with *P. falciparum* infection but can also arise during *P. vivax* and *P. knowlesi* infections.[4–6] This is characterised by severe anaemia, respiratory distress, hypoglycaemia, organ failure and cerebral complications (cerebral malaria, including brain damage and coma.[7]

The complex life cycles of *Plasmodium* parasites involve mosquito vectors and vertebrate hosts. In the vertebrate host, it is the asexual blood-stage of the life cycle (Fig. 8.1) that is responsible for the clinical manifestations of the disease.

Human infection by *Plasmodium* parasites is initiated following the bite from an infected female *Anopheles* mosquito. After invasion of and replication within the host liver, parasites in the form of merozoites are released into the bloodstream, where they invade erythrocytes. Inside the erythrocyte, the parasite resides within a parasitophorous vacuole membrane (PVM), which originates from both host and parasite components.[8] The intraerythrocytic parasite matures through the ring-, trophozoite-, and final schizont-stage. In the case of *P. falciparum*, this process takes place over a period of 48 hours, with the parasite replicating to produce up to 32 daughter merozoites. These are released from the infected cell and the freed merozoites continue the asexual blood-stage cycle by invasion of new erythrocytes. A small proportion of parasites undergo gametocytogenesis, developing into male or female gametocytes. These sexual forms of the parasite circulate in the bloodstream until they are

Erythrocytes and Malaria

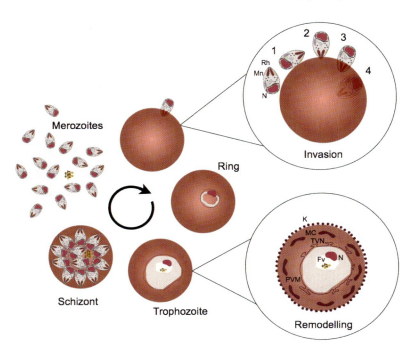

Figure 8.1 A schematic representation of the asexual blood-stage of the *Plasmodium falciparum* life cycle. Following release from infected hepatocytes, *Plasmodium* merozoites enter the blood stream and invade erythrocytes. The merozoite establishes itself within a vacuole in the infected erythrocyte and then develops from the relatively inert ring-stage into the metabolically active trophozoite-stage. The trophozoite then undergoes division, developing into a schizont. This process produces up to 32 daughter merozoites. Merozoites egress from the infected erythrocyte and go on to invade more erythrocytes, resulting in an exponential increase in parasitaemia.

Invasion expanded panel: the process of merozoite invasion of the human erythrocyte is a multi-step process that consists of (1) initial attachment, (2) apical re-orientation, (3) tight junction formation, an irreversible step in which the merozoite is committed to invade the erythrocyte, and (4) establishment of the intraerythrocytic parasite within its vacuole. Merozoite organelles are denoted as N — nucleus, Mn — micronemes, and Rh — rhoptries.

Remodelling expanded panel: As the parasite develops, it remodels the erythrocyte. This includes the development of protrusions from the parasitophorous vacuole membrane (PVM), termed the tubulovesicular network (TVN), the appearance of Mauer's clefts (MC) and the development of knobs (K) on the erythrocyte surface. N — nucleus, Fv — food vacuole.

taken up by a female mosquito during a blood meal and allow the life cycle to continue.

The intraerythrocytic malarial parasite has evolved complex secretory mechanisms to export numerous proteins into the host cytosol and plasma membrane, which allow the developing parasite to alter radically the host cell. These changes are essential for parasite survival within the host cell and include structural, permeability and antigenic modification of the host plasma membrane.[9,10] There are also changes in the phosphorylation status of the host cell, as well as the continual uptake, by endocytosis, and digestion of the host cell cytosol (predominantly haemoglobin) to provide nutrients and possibly space for the growing parasite.[11-13] These processes are discussed in greater detail in Section 8.2.

8.3 Parasite Invasion, Erythrocyte Remodelling and Egress

8.3.1 *Parasite invasion*

8.3.1.1 *Cell biology of erythrocyte invasion*

Following egress from the erythrocyte, invasive blood-stage merozoite forms are free in the bloodstream until they encounter a new erythrocyte. Parasites of the phylum *Apicomplexa* contain secretory organelles in the unique 'apical complex' within which reside many of the molecules that mediate the different steps of parasite invasion.[14] These include the pear-shaped rhoptry organelles and tubular-shaped micronemes. Invasion of erythrocytes by all *Plasmodium* spp. is a rapid process involving (i) a reversible step, with a weak interaction with the erythrocyte; (ii) an irreversible step with a high affinity interaction with the erythrocyte; (iii) active movement into the erythrocyte; and (iv) establishment of the parasite in a nascent parasitophorous vacuole.[15] The sequential steps of merozoite invasion (Fig. 8.1) are tightly regulated, with signal transduction through secondary messengers, including Ca^{2+}, cAMP and IP_3 playing a central role in activating the different stages of invasion.[16] Specialised machinery, including a motor complex, allows for active and rapid movement into host cells through a process known as gliding motility.[17] The merozoite is attached to the erythrocyte through multiple ligand-receptor interactions. Invasion is achieved through an active process of ratcheting, where the tight junction is moved from the anterior to the posterior end of the parasite. During this process the parasite is enveloped by the erythrocyte, resulting in the formation of a nascent parasitophorous vacuole.[18]

8.3.1.2 *Parasite ligands*

Merozoites utilise a number of parasite molecules for the recognition of the erythrocyte at different stages of invasion. Several proteins involved in the low affinity attachment stage include proteins spread across the surface of the merozoite, many of which are glycophosphatidylinositol (GPI)-anchored. These include the vaccine antigens merozoite surface protein 1 and 2, MSP1 and MSP2.[19–21] These molecules contain potential cysteine-rich adhesive domains, such as epidermal growth factor (EGF) and 6-cysteine domains. Peripheral proteins include expanded families of MSP3 and MSP7, which are polymorphic and associated with the GPI-anchored proteins.[22]

Following the initial attachment to the erythrocyte, an apical reorientation occurs, whereby the apical end of the merozoite is juxtaposed to the surface of the erythrocyte. Proteins stored within the secretory organelles interact with erythrocyte surface proteins, forming an irreversible host–parasite junction and committing the parasite to cellular invasion. These include the apical membrane antigen 1 (AMA-1), which is uniquely conserved across apicomplexans and which appears to be a link between apical reorientation and release of the contents of the secretory organelles.[23,24] Following apical orientation, specific high affinity interactions are formed that signal for the subsequent steps of the invasion process to occur. These involve two superfamilies of *P. falciparum* proteins — the Duffy binding-like (DBL) proteins and the reticulocyte binding-like (RBL) proteins. These are large type-1 transmembrane proteins that bind to different cognate erythrocyte receptors with high affinity.[25] However, it does not appear that the DBL and RBL proteins are involved in linking the erythrocyte to the actinomyosin motor complex.

Members of thrombospondin-related adhesive protein (TRAP) family expressed in merozoites may play this role.[18]

8.3.1.3 *Erythrocyte receptors for* Plasmodium *spp.*

Much biochemical and genetic evidence has implicated several proteins, which encode human blood groups, as receptors for parasite invasion. The interaction between the *P. falciparum* DBL-containing invasion protein, erythrocyte binding antigen-175 (EBA-175), and its cognate receptor glycophorin A is well characterised.[26] Recent biochemical evidence suggests that the EBA-140 ligand binds to glycophorin C (27) and the EBL-1 ligand binds to glycophorin B.[28] Aside from the *Plasmodium falciparum* reticulocyte homology protein 4 (PfRh4) that binds to complement receptor 1 (CR1),[29] the erythrocyte binding partners for the other *P. falciparum* RBL, or PfRh, proteins remain unknown. Binding of specific PfRh ectodomain regions to different receptors, characterised by their sensitivity to enzymatic digestion, has been demonstrated.[30–34] However, the functional importance of the utilisation of different erythrocyte receptors for invasion remains poorly understood. Some evidence suggests that MSP1 binds to Band 3[20] and AMA-1 binds to Kx.[35]

For the second most important parasite *P. vivax,* it is clear that a major receptor is the Duffy receptor that is largely absent in sub-Saharan Africans, where *P. vivax* infection is not present.[36] However, the identities of other critical receptors, which account for Duffy-independent invasion, have not been elucidated, but presumably are those that are bound by *P. vivax* RBL proteins, PvRB1 and PvRBP2.[37] The receptors for the other human malarial parasites, *P. ovale* and *P. malariae*, are unknown.

8.3.1.4 *Alternative invasion pathways in* Plasmodium *spp. for host tropism*

P. falciparum invades human erythrocytes using multiple receptor-ligand interactions that have been defined as invasion pathways.[38–40] The current model of alternative invasion pathway utilisation postulates that members of the DBL or RBL families bind different erythrocyte surface receptors *via* their diversified ectodomains. Multiple studies with both laboratory strains and primary isolates of *P. falciparum* have shown that different parasite strains utilise different ligand-receptor interactions for erythrocyte invasion, known as invasion pathways, due to variant expression of members of the DBL and RBL families.[32,33,41–49] This variant expression is in part a result of regulation by epigenetic mechanisms. This variation in parasite ligands can direct the merozoites to polymorphic receptors, including to erythrocytes of different ages, to genetic mutations between individuals and to receptors that are variant between species. Variant expression is also a potential mechanism for immune evasion.[50] Following ligand-receptor binding, parasite proteins trigger the downstream steps of invasion. In a few regions, Duffy-independent invasion has been described,[51–53] providing evidence that *P. vivax* parasites have the potential to utilise alternative invasion pathways.

8.3.2 *Protein trafficking to the host*

Directly after invasion the intraerythrocytic malarial parasite begins the process of remodelling the host cell (Fig. 8.1). As part of this, two novel structures develop in the host cytosol as the parasite matures: (i) apparent extensions of the PVM are observed,[54] termed the tubulovesicular network (TVN); and (ii) thin, disc-shaped organelles (Maurer's clefts) appear, distinct from the PVM-TVN.[55,56] In addition, several types of vesicle have been characterised (e.g. Hanssen *et al.* 2008[57] and Kulzer *et al.* 2010[58]). All these structures are thought to be involved in the trafficking of parasite-derived proteins to the host, which occurs *via* novel mechanisms. The majority of parasite proteins exported to the host contain a pentameric amino acid sequence (R/KxLxE/Q/D) called the protein export element/host targeting (PEXEL/HT) motif,[59,60] which is cleaved in the parasite endoplasmic reticulum to initiate export.[61,62] It should be noted that some exported parasite proteins do not contain a PEXEL/HT motif but have a PEXEL/HT-like motif or, in some cases, no discernable export motif, although whether this suggests that multiple export mechanisms exist is a matter of debate. Analysis of the *P. falciparum* genome using the PEXEL/HT sequence has identified in excess of 300 proteins in the 'exportome'.[63] However, the exportomes of other plasmodial species are far smaller, due primarily to the absence of several gene families involved in virulence (see sections below) and the expansion of other gene families in *P. falciparum*. Another recent trafficking discovery is the identification of a protein complex for the translocation of, at least, soluble parasite proteins across the PVM to the host cytosol,[64] termed the *Plasmodium* translocon of exported proteins (PTEX). Our understanding of parasite protein trafficking to the host has increased dramatically over the last decade but is far from complete.[65,66]

8.3.3 Remodelling the erythrocyte membrane and cytoskeleton

There are marked changes in the surface topology and deformability characteristics of *Plasmodium*-infected erythrocytes. A good deal of our knowledge of host membrane and cytoskeletal alterations by intra-erythrocytic plasmodial parasites has come from the study of *P. falciparum* virulence associated with cytoadhesion by infected erythrocytes (see Section 8.2.4). A major determinant of *P. falciparum* virulence is the expression of adhesive variant surface antigens on infected erythrocytes that are encoded by a family of *var* genes.[67,68] *Var* genes encode for an integral membrane protein known as the *P. falciparum* erythrocyte membrane protein 1 (PfEMP1), and nearly 60 are present within the genome but only one is expressed per infected cell. Switching *var* genes allows parasites to evade protective host antibody responses and bind to different host receptors.[69] Other integral *P. falciparum* proteins that are expressed on the host cell surface and may be involved in antigenic variation at a late stage of development are from the repetitive interspersed family (*rif*) of genes, and the subtelomeric variant open reading frame (*stevor*) genes.[70,71]

Knob-like protrusions on the surface of *P. falciparum*-infected erythrocytes appear approximately 16 hours post-invasion and act as platforms to promote the presentation of PfEMP1.[72] The protrusions are due to the presence at the cytoplasmic surface of the infected erythrocyte of knob associated histidine-rich protein (KAHRP), which is essential for knob formation and which reduces erythrocyte deformability.[73,74] In addition to containing binding sites for the cytoplasmic domain of PfEMP1, KAHRP has been reported to interact with host spectrin, actin and ankyrin. Of these, the interaction with spectrin has been studied in greatest detail, with KAHRP binding to repeat 4 of α-spectrin.[75] Interestingly, host casein kinase II has been suggested to increase KAHRP binding to PfEMP1 by phosphorylation of the latter.[76]

The ring-infected erythrocyte surface antigen (RESA) is a 127 kDa soluble parasite phosphoprotein, which is secreted into the parasitophorous vacuole of infected erythrocytes during or shortly after invasion. From here, RESA is trafficked to the host cytoskeleton and specifically interacts with spectrin, reducing erythrocyte deformability. An approximately 100 amino acid region of RESA (residues 663–770) binds to β-spectrin at repeat 16, close to the dimer-dimer self-association site. This was shown to stabilise spectrin tetramers and strengthen cells against both mechanical and thermal degradation.[77,78] It was proposed that RESA may (i) protect infected cells from further infection; and (ii) protect infected cells during periods of fever, and thus the higher than normal temperatures that are associated with malaria infection.[77–79] Another exported soluble parasite phosphoprotein is the 168 kDa mature parasite-infected surface antigen (MESA, also known as PfEMP2), which, unlike RESA, is expressed during the late developmental stages of parasitised erythrocytes.[80] While the exact function of this protein is unknown, its interaction with the host cytoskeleton has been characterised. An N-terminal 19 amino acid region of MESA binds to a 51 residue region in the N-terminal of host protein 4.1R, which forms part of a ternary complex with p55 and glycophorin C, and specifically interferes with 4.1R binding to p55.[81–83]

P. falciparum erythrocyte membrane protein 3 (PfEMP3) is a soluble, highly charged 274 kDa exported protein, which is expressed in mature infected erythrocytes and which localises to the cytoplasmic surface of the host plasma membrane.[84] PfEMP3 binds to spectrin, specifically a 60 amino acid domain (residues 38–97) of the N-terminus of PfEMP3 binds to a region near the C-terminus of α-spectrin at the site of a ternary complex between spectrin, actin and protein 4.1R.[85,86] This junctional complex is disrupted by PfEMP3 and leads to decreased membrane mechanical stability, which is postulated to aid subsequent parasite egress (see Section 8.2.6). Several other exported parasite proteins that alter erythrocyte rigidity have been identified (many in a large-scale project that generated mutant *P. falciparum* parasites lacking exported proteins)[87] and are reviewed elsewhere.[88]

Integral plasma membrane host proteins are also affected by the developing intra-erythrocytic plasmodial parasite. The mobility of both Band 3 and glycophorin is reduced significantly by maturing *P. falciparum* parasites.[89] Furthermore, Band 3 is known to aggregate in infected erythrocytes, a process which is caused by oxidative stress.[90] Modification of Band 3 is thought to promote adhesion (see Section 8.2.4).

There is evidence for alteration of the lipid composition of the host plasma membrane in infected erythrocytes (*P. falciparum* in particular), which might explain increased membrane fluidity.[91] However, this is not supported by other reports (reviewed in).[92] Previously debated,[92–94] there are also data that suggest the phospholipid asymmetry of the erythrocyte plasma membrane is modified by infection.[95–98] In particular, phosphatidylserine, which is found with phosphatidylethanolamine predominantly in the inner leaflet, is exposed increasingly on the outer leaflet of infected erythrocytes as they mature, an effect which is enhanced by exposing infected erythrocytes to febrile temperatures.[99] It has been proposed that phosphatidylserine exposure aids the adherence of infected erythrocytes (see Section 8.2.4) and/or is an effect of the process of eryptosis, a host defence mechanism (see Chapter 5).

Intracellular *Plasmodium* parasites alter the phosphorylation status of the host erythrocyte. Hyperphosphorylation of the host plasma membrane and cytoskeleton has been reported,[12,100] starting during the ring stage and increasing greatly as the parasite matures. Both serine/threonine and tyrosine phosphorylation sites are observed and phosphorylated host proteins include Band 3 and 4.1, spectrin and ankyrin.[12,100,101] *P. falciparum* has a large number of kinases within its genome, of which some are exported to the host cytosol and phosphorylate the erythrocyte membrane/cytoskeleton (e.g. FIK kinases).[102,103] It should be noted that parasite phosphatases are also exported into the host, playing possible roles in processes such as parasite egress[104] and nutrient uptake.[105] Furthermore, there is evidence that host kinases are co-opted by the parasite, including casein kinase II (see above) and tyrosine kinases (representatives of which are not found in the plasmodial kinomes).[103] While parasite-induced host phosphorylation may account for some of the physical properties of infected erythrocytes, the regulation and functional consequences of hyperphosphorylation await detailed characterisation.

8.3.4 Adhesion

P. falciparum-infected human erythrocytes, in particular, become adherent as they mature, due to the expression of parasite-derived adhesive proteins on the host cell surface.[106–110] There are three types of reported adhesion: cytoadhesion (the binding of infected erythrocytes to endothelial cells),[111] rosetting (the binding of infected erythrocytes to uninfected erythrocytes)[112] and platelet interaction.[113] These mechanisms can lead to parasitised erythrocytes obstructing the microvasculature by forming rosettes or platelet-mediated clumps, or by adhering to endothelial cells, and have long been associated with the more severe complications that occur during malaria infection.[108,114–116] It is hypothesised that the parasite evolved the mechanism of sequestration as an immune-avoidance strategy, allowing the removal of mature infected erythrocytes from the circulation and thus avoiding detection and removal in the spleen.[117,118]

Changes in the adherence properties of the infected erythrocyte are mediated by the expression at the host cell surface of PfEMP1 (see Section 8.2.3), predominantly, although STEVORs and RIFINs may also be involved.[70,119–122] Adhesion has also been associated with the modification of the endogenous erythrocyte membrane transport protein, Band 3. In mature *P. falciparum*-infected erythrocytes, Band 3 modification and/or clustering on the erythrocyte cell surface has been proposed to expose cryptic adherent residues.[122–124] Surface exposure of phosphatidylserine on infected erythrocytes may also mediate cytoadherence to host receptors such as CD36.[124]

8.3.5 Transport and metabolism

The intra-erythrocytic development of *Plasmodium* parasites involves rapid cell growth and replication, requiring high metabolic activity. Therefore, the host erythrocyte must act as a conduit for the provision of nutrients to and removal of waste products from the parasite, as well as providing an appropriate ionic environment for growth. It has long been known that *Plasmodium* parasites alter the permeability of their host erythrocytes,[125] initially inferred from the study of Na^+ homeostasis.[126] In the case of *P. falciparum*-infected erythrocytes (and most likely other plasmodial parasites), it has been suggested that this process is required for nutrient uptake, metabolite removal, and volume and ion regulation, and thus parasite survival.[125,127]

Several functional studies have described novel alterations to the permeability of infected erythrocytes, including reports of upregulated activity of endogenous solute carriers[128–130] and a novel ATP-dependent Ca^{2+} transport pathway.[131,132] There is also evidence, primarily determined by immunofluorescence, that *P. falciparum*-encoded transport proteins are localised to the erythrocyte membrane. These are a copper pump, PfCuP-ATPase,[133] a V-type H^+ pump[134] and a putative K^+ channel, Pfkch1/PfK1.[135] However, the predominant mechanism responsible for the permeability changes observed in the host plasma membrane of *Plasmodium*-infected erythrocytes is due to the induction of a single or closely-related set of channel-like pathways, as the parasite matures.[9,136] Termed the new permeation pathways (NPP), they are induced at approximately 12–15 h post-invasion in the case of *P. falciparum*-infected erythrocytes.[137] The NPP are permeable to a wide range of low molecular weight solutes, including inorganic and organic cations and anions, sugars, amino acids and nucleosides, but demonstrate a significant preference for anions over electroneutral solutes over cations.[136] Furthermore, they are inhibited by a variety of compounds, many of which are non-selective anion transport inhibitors such as 5-nitro-2-(3-phenylpropylamino)-benzoic acid (NPPB) and furosemide.[138,139]

Over the past decade the transport properties of the NPP have been extensively characterised using electrophysiological techniques,[140–145] yet a number of unresolved questions remain.[146] These include whether one or more pathways underlies the NPP, and the molecular nature of the pathway(s) (and, thus, whether the single or multiple underlying transport mechanisms are endogenous or parasite-derived). There is evidence that NPP-like activity can be activated in uninfected erythrocytes by oxidation[138,139] and by phosphorylation[142] in support of a host-derived transport mechanism. There is also evidence for the involvement of parasite-derived proteins that are required for induction and/or formation of the NPP.[147–150] Recently, parasite mutants have been developed with altered NPP activity (specifically in a novel channel type termed the plasmodial surface anion channel, PSAC) and it will be interesting to see in the future what parasite proteins are mutated and how they relate to NPP formation.[149]

In terms of nutrients, the plasmodial parasite uses various uptake mechanisms to scavenge from the host, including endocytosis of the host cytosol (predominantly haemoglobin, which is digested to provide amino acids[13]) and specific transmembrane transport pathways.[125,151,152] Several essential substrates are required in the extracellular medium of *P. falciparum*-infected erythrocytes for growth, including glucose, purines, the amino acid isoleucine (the only amino acid not present in haemoglobin) and the vitamin pantothenic acid.[153,154] To reach the parasite, these nutrients must first gain access to the erythrocyte cytosol. While erythrocytes are considered to be relatively inert metabolically, the endogenous plasma membrane transport pathways in erythrocytes are still able to supply the majority of the parasite's nutrient requirements. In the case of human erythrocytes, this includes glucose *via* the facilitative glucose transporter, GLUT 1, and, to a lesser degree, isoleucine.[155,156] However, there is no endogenous transporter in human erythrocytes for pantothenic acid, which is transported *via* the NPP.[157]

Asexual blood stage parasites are reliant on glycolysis to derive their ATP requirements, with the TCA cycle playing a minimal role.[158] This results in the metabolism of glucose into lactate, which is removed from the parasite to the host cytosol, predominantly by a parasite monocarboxylate transporter.[159] Lactate can cross the erythrocyte plasma membrane *via* its own monocarboxylate transporter (MCT1), the anion exchanger, Band 3, and by simple diffusion. However, given the large quantities produced by the parasite relative to the erythrocyte, it has been calculated that these pathways are unable to cope without an additional efflux route such as the NPP.[160,161] Furthermore, the digestion of haemoglobin results in the production of large quantities of amino acids, many of which are not utilised by the parasite and need to be removed.[13] If left to accumulate (in the absence of the NPP), these waste products would become toxic and exert a large osmotic effect that would result in cell swelling and lysis before the parasite has developed fully.[11] The question of volume regulation in *Plasmodium*-infected erythrocytes is an interesting one, given the metabolic processes that occur, ionic changes (see below) and parasite growth, and the NPP seem integral to this process.

In fact the volume of *P. falciparum*-infected erythrocytes throughout parasite development changes little compared with uninfected erythrocytes. This raises questions over how the parasite obtains the space it requires to grow physically, and why infected erythrocytes are more fragile osmotically than uninfected erythrocytes.[162–164]

Human erythrocytes maintain an inward Na^+ and an outward K^+ concentration gradient, as a result of the activity of Na^+/K^+ pumps. The NPP transport both Na^+ and K^+ at physiologically significant rates and their induction has two consequences. Firstly, there is a host cytosol dehydrating effect, as Na^+ influx *via* the NPP is half the rate of K^+ efflux.[137] This adds further support for the NPP playing a volume regulatory role (see above). Secondly, the host cytosolic concentrations of Na^+ and K^+ in mature *P. falciparum*-infected erythrocytes change to those measured in the plasma.[165] This results in the generation of large Na^+ and K^+ gradients across the parasite plasma membrane, which is known to contain Na^+-dependent (secondary active) transporters, such as a Na^+-dependent inorganic phosphate cotransporter.[166]

8.3.6 *Parasite egress*

Parasite egress is an explosive process occurring over a matter of seconds, resulting in the release of merozoites for further cycles of infection. Although egress is far from understood, several events occur within the parasite and parasitophorous vacuole,[167–169] including essential kinase activity of a parasite-derived plant-like calcium-dependent protein kinase, PfCDK5. Additional processes occur in the host cell that, in part, underlie egress. First, shrinkage of the erythrocyte cytoplasm in parallel with swelling of the parasitophorous vacuole has been observed in *P. falciparum*-infected erythrocytes.[170,171] The importance of this redistribution of water has been demonstrated by the fact that parasites have difficulty egressing from dehydrated erythrocytes (both dehydrated normal erythrocytes and those from sickle cell patients). Second, the erythrocyte plasma membrane loses tension,[170] which is suggested to involve cytoskeleton digestion by cysteine proteases, predominantly.[172–174] In addition to possible roles of plasmodial cysteine proteases,[175] it has been shown that a host protease, calpain 1, is recruited by the parasite and is essential for egress.[172] Its depletion from lysed and resealed erythrocytes does not stop *P. falciparum* invasion or development, but traps fully developed merozoites within the host. The exact role of calpain 1 is unknown, but calpains aid remodelling of the cytoskeleton and plasma membrane during mammalian cell migration,[176] and calpain 1 can degrade the erythrocyte cytoskeleton *in vitro* and *in vivo* during *Plasmodium* infection.[172] Furthermore, calpain 1 is calcium regulated, which suggests host free calcium cytosolic levels are altered by the parasite during egress. Third, a few seconds before cell rupture the erythrocyte plasma membrane becomes porous and the host cytoplasm is released.[170] The proteins responsible for this poration are thought to be parasite-derived perforin-like proteins.[177]

8.4 Erythrocyte Genetic Polymorphisms Conferring Resistance to Malaria

Malaria has been described 'as the strongest known force for evolutionary selection in the recent history of the human genome'.[178] This is possibly unsurprising given the billions of clinical infections that have occurred year on year and the number of countries to which malaria is, and has been, endemic. Mutations in over 30 human genes are known to associate with innate resistance to malaria and this includes a large number that are involved in the formation of erythrocytes, as discussed below.

8.4.1 *Haemoglobinopathies*

Haemoglobin is the molecule responsible for transporting oxygen and carbon dioxide around the body. It is formed in adults by a tetramer of 2 α-subunits and 2 β-subunits, transcribed from two *HBA* (1 and 2) and one *HBB* gene, respectively. Haemoglobinopathies fall into two groups: abnormal haemoglobins and thalassaemias, with the latter referring to altered levels of globin synthesis. They have long been associated with innate malaria resistance, given their distribution in malaria endemic areas, and they led Haldane to first propose a link between them and protection against malaria.[179]

8.4.1.1 *Haemoglobin C (HbC)*

HbC results from a point mutation in the β-globin gene, which leads to the substitution of a glutamate for lysine at position 6. Predominantly found in West Africa, the heterozygous state (HbAC) is asymptomatic, while the homozygous state (HbCC) can produce a mild anaemia. Generally, it is thought that HbC protects against severe malaria, with HbCC giving far more protection (>90%) than HbAC (~30%),[180–183] but providing little or no protection against asymptomatic or uncomplicated malaria.[184,185] While a definitive answer awaits, several mechanisms for protection have been put forward. These include (i) the reduced ability of *P. falciparum* to multiply in HbCC erythrocytes, as demonstrated *in vitro* (e.g.);[186] (ii) the reduced expression of PfEMP1 in infected erythrocytes containing HbC (*in vitro*), which would lead to reduced sequestration and rosetting (see Section 8.2.4.) and their consequences;[187] and (iii) an increased risk of homozygous individuals being infected by *P. malariae*, which may modulate subsequent *falciparum* malaria by a vaccine-like effect.[184] An interesting recent observation is that while HbC (and possibly HbS, see below) can protect against malaria, its presence increases the levels of gametocytes in the blood and subsequent transmission to the mosquito vector, suggesting that haemoglobinopathies can benefit both host and parasite.[188]

8.4.1.2 *Haemoglobin E (HbE)*

HbE is the most prevalent haemoglobinopathy in Southeast Asia, reaching its present level in less than 5 millennia.[189] As with HbC, HbAE is asymptomatic, while HbEE causes a mild anaemia. It is caused by a nonsynonymous point mutation in the β-globin gene, resulting in a substitution of a glutamate for a lysine at position 26. In Thailand, HbAE is associated with reduced disease severity.[190] A subsequent report demonstrated that HbE trait (but interestingly not HbEE) reduces *P. falciparum* invasion *in vitro* by an undefined mechanism, resulting in lower parasite density and thus protection against severe malaria.[191] At present, further work is required to understand fully the role of HbE in innate malaria resistance.

8.4.1.3 *Haemoglobin S (HbS)*

The cause of HbS is very similar to HbC, with mutation leading to a valine rather than a lysine at position 6 of the β-globin molecule[192,193] While the heterozygous state is asymptomatic, the homozygous state causes sickle cell disease. HbS polymerises upon deoxygenation, a process that leads to distortion of erythrocyte shape (and the observation of sickle shaped cells). Sickled erythrocytes cause vascular occlusion, with associated sequelae including ischaemia, organ dysfunction, pain and, ultimately, death (before adulthood without appropriate treatment). The disease is also characterised by a prevailing anaemia, in part due to the reduced lifespan of HbS erythrocytes. Even so, HbS is one of the most prevalent haemoglobinopathies in malaria-endemic areas and the first to be studied in relation to innate resistance.[194] Unlike HbAC and HbAE, HbAS is protective against all symptomatic forms of malaria, with the degree of protection increasing with disease severity.[185,195–197] However, the mechanism(s) underlying the striking protection provided by HbS are not fully understood. Possible mechanisms include reduced parasite growth,[198] increased phagocytosis[199] (also see Section 8.3.3.1), reduced expression of PfEMP1[200] (also see section on HbC above) and development of acquired immunity.[197,201]

8.4.1.4 *Thalassaemias*

Thalassaemias are widely distributed around the world, including in malaria-endemic regions and are caused by numerous mutations. α- and β-thalassaemia individuals have altered levels of synthesis of their respective globins (with α^+ or β^+ and α^0 or β^0 denoting reduced levels and absence of globins, respectively). Only β-thalassaemia in the homozygous state produces a severe disease, while α^0-thalassaemia is lethal in the homozygous state. Of the two, α-thalassaemia has received the greater attention and it is generally accepted that α-thalassaemia confers protection against severe malaria and severe anaemia in particular.[202–205] Interestingly, it was reported that cases of mild malaria are increased in young children with α-thalassaemia compared with control groups from the Pacific.[206] It was suggested that early contraction of vivax malaria may protect these individuals from later cases of severe *falciparum* malaria. More recently, it has been observed that α-thalassaemia erythrocytes have reduced surface expression of the complement receptor 1 (CR1; see Section 8.3.2.3[207]). This receptor is required for rosetting, which itself is linked to disease severity (see Section 8.3.2.2).

8.4.2 *Erythrocyte plasma membrane proteins*

8.4.2.1 *Duffy antigen receptor for chemokines (DARC)*

The Duffy blood group antigen was identified as an innate resistance factor for erythrocyte infection by *P. knowlesi* and *P. vivax* in the mid-1970s.[36,208] The Duffy antigen is encoded by the *FY* gene, which was cloned by Chaudhuri et al.,[209] and is a 336 amino acid acidic glycoprotein. Subsequently, it was identified as a receptor for chemokines,[210] hence the acronym DARC. In sub-Saharan Africa, the majority of the population carry a point mutation in the *FY* gene promoter region, which impairs promoter activity in erythroid cells by disrupting a binding site for the GATA-1 erythroid transcription factor, and this results in a Duffy negative phenotype.[211] This blood group, which has reached fixation in most of West and Central Africa, accounts for the absence of *P. vivax* infections in the region.[212] A second independent mutation that could result in a Duffy negative phenotype was identified in Papua New Guinea (PNG), an area in which *P. vivax* is of clinical importance.[213] However, only heterozygotes are observed in the population and this is associated with 50% less antigen expression, increased resistance to *P. vivax* and less severe disease,[214] which is consistent with the importance of the antigen for infection. The mechanism by which DARC confers resistance to *P. vivax* (and *P. knowlesi*) is by stopping parasite invasion in its absence (see Section 8.2.1). While it is clear that DARC is by far the most important invasion receptor for *P. vivax* infection, there are a growing number of reports of *P. vivax* infection in Duffy negative individuals,[51–53] suggesting the parasite is able to use alternative, as yet unidentified, invasion receptors on the erythrocyte surface.

8.4.2.2 *Anion exchanger 1 (AE1 or Band 3)*

Band 3 is the most abundant protein in the erythrocyte plasma membrane and acts as both an anion transporter and a structural link between the plasma membrane and the underlying cytoskeleton (see Chapters 2 and 4). Southeast Asian ovalocytosis (SAO) has long been associated with malaria resistance due to its high prevalence in Melanesian populations in areas where malaria is endemic[215] and is caused by a Band 3 mutation. While the homozygous state is not observed and presumed lethal, the heterozygous state is asymptomatic. The mutation is a 27 base pair deletion in the Band 3 gene (*slc4a1*). This results in the loss of nine amino acids[400–408] at the boundary between the cytoplasmic N-terminal region and the first of its multiple transmembrane domains, and is often linked to a common point mutation that results in a lysine to glutamate substitution at position 56.[216–218] While still trafficked to the plasma membrane and able to interact with normal Band 3, the mutated Band 3 is unable to transport anions[219] and its presence leads to rigid, oval shaped erythrocytes (reviewed in).[220] Studies in PNG determined that the condition leads to protection against severe malaria only, specifically cerebral malaria[221,222] and it was hypothesised that infected SAO erythrocytes are less able to adhere to the microvasculature in the brain. Cortes and colleagues have suggested an alternative mechanism of action. First, they demonstrated *in vitro* that fresh SAO erythrocytes were selectively resistant to invasion by a range of parasite isolates, suggesting that these erythrocytes could be resistant to invasion by parasites that cause cerebral malaria.[223] Second, they demonstrated that infected SAO erythrocytes bind to CD36, a receptor not found in the microvasculature of the brain, under flow conditions at a higher level than control infected erythrocytes, suggesting protection is conferred by an altered distribution of sequestration.[224] While the later hypothesis is highly plausible, a full understanding of the mechanism of action will require further study.

Band 3 also carries the majority of the ABO blood group antigen, formed by post-translational modification by glycosyltransferases. The responsible glycosyltransferase[225] transfers *N*-acetyl D-galactosamine in the case of group A, and D-galactose in the case of group B, to glycans on glycoproteins and glycolipids and is inactive in the case of group O. Two recent studies have demonstrated an association between the ABO blood group and malaria resistance. The first reported a protective effect of the O blood group against severe *falciparum* malaria in West Africa,[226] while the second reported that non-O alleles are risk factors for severe *falciparum* malaria in sub-Saharan Africa.[227] Given the association between rosetting (see Section 8.2.4) and disease severity,[108] the reduced ability of erythrocytes carrying the O blood group to rosette (as shown *in vitro*)[228,229] is suggested to underlie the protective effect.

8.4.2.3 *Others*

Complement receptor 1 (CR1), which contains the Knops blood group antigens, is a glycoprotein found in the erythrocyte plasma membrane that is required for the removal of immune complexes coated with activated complement components and for the control of complement-activating enzymes (reviewed by).[230] CR1 is a receptor involved in rosetting[110] and it has recently been identified as a sialic-acid-independent *P. falciparum* invasion receptor[231] (see Section 8.2.1). Mutations in its gene are associated with protection against severe malaria (as with the ABO group) and are found at high frequency in malaria endemic regions (e.g. PNG).[207] However, results from the field have been contradictory[232] and further studies are required to understand the role of CR1 in innate malaria resistance.

Glycophorins are a group of negatively charged glycoproteins, which are expressed in the erythrocyte plasma membrane at high levels. Glycophorins A, B and C have all been demonstrated *in vitro* to act as receptors for *P. falciparum* invasion[28,233,234] (see Section 8.2.1). The absence of glycophorin C (Gerbich negative, which is caused by the deletion of exon 3 in the gene), for example, is a common phenotype in PNG.[235] Therefore, it has been hypothesised that mutation in the genes of these proteins might link with malaria resistance, although definitive proof remains elusive.

8.4.3 *Erythrocyte enzymes*

8.4.3.1 *Glucose-6-phosphate-dehydrogenase (G6PD)*

G6PD deficiency came to prominence when it was discovered as the cause of the haemolytic anaemia that developed in some individuals treated with the antimalarial drug primaquine. The X-linked enzymopathy is highly prevalent in malaria-endemic regions[236] and is protective against severe malaria.[237,238] The protective mechanism is reported to involve phagocytosis of immature *P. falciparum*-infected erythrocytes, which is caused by increased oxidative membrane damage.[199,239]

8.4.3.2 *Pyruvate kinase deficiency*

Pyruvate kinase is an enzyme in the glycolytic pathway, which the erythrocyte uses to generate its ATP requirements. Pyruvate kinase deficiency is a common enzymopathy that causes haemolytic anaemia (see Chapter 7), although whether its worldwide distribution is the result of malaria pressure is open to debate. However, using pyruvate kinase-deficient mice infected with *P. chabaudi*, it has been demonstrated that the deficiency protects against infection *in vivo*.[240] More recent *in vitro* studies, using erythrocytes from pyruvate kinase-deficient individuals infected with *P. falciparum*, have shown that the enzymopathy confers resistance to infection in the homozygous state. Interestingly, it also increases macrophage clearance of immature parasitised erythrocytes in both the homozygous and heterozygous states.[241,242]

8.5 Concluding Remarks

Malaria is a devastating disease. Presently, it is entirely treatable if diagnosed in time but the parasites responsible are continually evolving resistance mechanisms to antimalarial drugs. Thus, there is a need to develop new therapeutic strategies. Here we have described some of the numerous and complex interactions that occur between the host erythrocyte and plasmodial parasite during blood stage infection. It is clear that the parasite is highly reliant on its host for survival and that malaria has played a major role in shaping the polymorphisms found in human erythrocytes. Excitingly, it has recently become feasible to perform *in vitro* erythrocyte genetic alterations[243] that will open new avenues to enable a deeper understanding of the interactions that occur. Indeed, host-targeted therapeutics are attractive, as such approaches would limit the development of drug-resistance and keep us one step ahead of this deadly foe.

Bibliography

1. Snow RW, Guerra CA, Noor AM, Myint HY, Hay SI. The global distribution of clinical episodes of *Plasmodium falciparum* malaria. *Nature*. 2005;434:214–217.
2. Guerra CA, Howes RE, Patil AP, Gething PW, Van Boeckel TP, Temperley WH *et al*. The international limits and population at risk of *Plasmodium vivax* transmission in 2009. *PLoS Negl Trop Dis*. 2010; 4:e774.
3. Singh B, Kim Sung L, Matusop A, Radhakrishnan A, Shamsul SS, Cox-Singh J *et al*. A large focus of naturally acquired *Plasmodium knowlesi* infections in human beings. *Lancet*. 2004;363:1017–1024.
4. Cox-Singh J, Hiu J, Lucas SB, Divis PC, Zulkarnaen M, Chandran P *et al*. Severe malaria — a case of fatal *Plasmodium knowlesi* infection with post-mortem findings: a case report. *Malar J*. 2010;9:10.
5. Genton B, D'Acremont V, Rare L, Baea K, Reeder JC, Alpers MP *et al*. *Plasmodium vivax* and mixed infections are associated with severe malaria in children: a prospective cohort study from Papua New Guinea. *PLoS Med*. 2008;5:e127.
6. Tjitra E, Anstey NM, Sugiarto P, Warikar N, Kenangalem E, Karyana M *et al*. Multidrug-resistant *Plasmodium vivax* associated with severe and fatal malaria: a prospective study in Papua, Indonesia. *PLoS Med*. 2008;5:e128.
7. Trampuz A, Jereb M, Muzlovic I, Prabhu RM. Clinical review: severe malaria. *Crit Care*. 2003;7:315–323.
8. Cesbron-Delauw MF, Gendrin C, Travier L, Ruffiot P, Mercier C. *Apicomplexa* in mammalian cells: trafficking to the parasitophorous vacuole. *Traffic*. 2008;9:657–664.
9. Ginsburg H, Krugliak M, Eidelman O, Cabantchik ZI. New permeability pathways induced in membranes of *Plasmodium falciparum* infected erythrocytes. *Mol Biochem Parasitol*. 1983;8:177–190.
10. Maier AG, Cooke BM, Cowman AF, Tilley L. Malaria parasite proteins that remodel the host erythrocyte. *Nat Rev Microbiol*. 2009;7:341–354.
11. Lew VL, Tiffert T, Ginsburg H. Excess hemoglobin digestion and the osmotic stability of *Plasmodium falciparum*-infected red blood cells. *Blood*. 2003;101:4189–4194.

12. Pantaleo A, Ferru E, Carta F, Mannu F, Giribaldi G, Vono R et al. Analysis of changes in tyrosine and serine phosphorylation of red cell membrane proteins induced by *P. falciparum* growth. *Proteomics.* 2010;10:3469–3479.
13. Zarchin S, Krugliak M, Ginsburg H. Digestion of the host erythrocyte by malaria parasites is the primary target for quinoline-containing antimalarials. *Biochem Pharmacol.* 1986;35:2435–2442.
14. Bannister LH, Hopkins JM, Fowler RE, Krishna S, Mitchell GH. Ultrastructure of rhoptry development in *Plasmodium falciparum* erythrocytic schizonts. *Parasitology.* 2000;121:273–287.
15. Gilson PR, Crabb BS. Morphology and kinetics of the three distinct phases of red blood cell invasion by *Plasmodium falciparum* merozoites. *Int J Parasitol.* 2009;39:91–96.
16. Nagamune K, Moreno SN, Chini EN, Sibley LD. Calcium regulation and signaling in apicomplexan parasites. *Subcell Biochem.* 2008;47:70–81.
17. Jones ML, Kitson EL, Rayner JC. *Plasmodium falciparum* erythrocyte invasion: a conserved myosin associated complex. *Mol Biochem Parasitol.* 2006;147:74–84.
18. Baum J, Richard D, Healer J, Rug M, Krnajski Z, Gilberger TW et al. A conserved molecular motor drives cell invasion and gliding motility across malaria life cycle stages and other apicomplexan parasites. *J Biol Chem.* 2006;281:5197–5208.
19. Gerold P, Schofield L, Blackman MJ, Holder AA, Schwarz RT. Structural analysis of the glycosyl-phosphatidylinositol membrane anchor of the merozoite surface proteins-1 and -2 of *Plasmodium falciparum*. *Mol Biochem Parasitol.* 1996;75:131–143.
20. Goel VK, Li X, Chen H, Liu SC, Chishti AH, Oh SS. Band 3 is a host receptor binding merozoite surface protein 1 during the *Plasmodium falciparum* invasion of erythrocytes. *Proc Natl Acad Sci USA.* 2003;100:5164–5169.
21. Sanders PR, Gilson PR, Cantin GT, Greenbaum DC, Nebl T, Carucci DJ et al. Distinct protein classes including novel merozoite surface antigens in Raft-like membranes of *Plasmodium falciparum*. *J Biol Chem.* 2005;280:40169–40176.
22. Kadekoppala M, Holder AA. Merozoite surface proteins of the malaria parasite: the MSP1 complex and the MSP7 family. *Int J Parasitol.* 2010;40:1155–1161.

23. Crawford J, Tonkin ML, Grujic O, Boulanger MJ. Structural characterization of apical membrane antigen 1 (AMA1) from *Toxoplasma gondii*. *J Biol Chem*. 2010;285:15644–15652.
24. Mital J, Meissner M, Soldati D, Ward GE. Conditional expression of *Toxoplasma gondii* apical membrane antigen-1 (TgAMA1) demonstrates that TgAMA1 plays a critical role in host cell invasion. *Mol Biol Cell*. 2005;16:4341–4349.
25. Iyer J, Gruner AC, Renia L, Snounou G, Preiser PR. Invasion of host cells by malaria parasites: a tale of two protein families. *Mol Microbiol*. 2007;65:231–249.
26. Sim BK, Chitnis CE, Wasniowska K, Hadley TJ, Miller LH. Receptor and ligand domains for invasion of erythrocytes by *Plasmodium falciparum*. *Science*. 1994;264:1941–1944.
27. Lobo CA, Rodriguez M, Reid M, Lustigman S. Glycophorin C is the receptor for the *Plasmodium falciparum* erythrocyte binding ligand PfEBP-2 (baebl). *Blood*. 2003;101:4628–4631.
28. Mayer DC, Cofie J, Jiang L, Hartl DL, Tracy E, Kabat J, *et al*. Glycophorin B is the erythrocyte receptor of *Plasmodium falciparum* erythrocyte-binding ligand, EBL-1. *Proc Natl Acad Sci USA*. 2009;106:5348–5352.
29. Tham WH, Wilson DW, Lopaticki S, Schmidt CQ, Tetteh-Quarcoo PB, Barlow PN, *et al*. Complement receptor 1 is the host erythrocyte receptor for *Plasmodium falciparum* PfRh4 invasion ligand. *Proc Natl Acad Sci USA*. 2010;107:17327–17332.
30. Baum J, Chen L, Healer J, Lopaticki S, Boyle M, Triglia T *et al*. Reticulocyte-binding protein homologue 5 — an essential adhesin involved in invasion of human erythrocytes by *Plasmodium falciparum*. *Int J Parasitol*. 2009;39:371–380.
31. Gao X, Yeo KP, Aw SS, Kuss C, Iyer JK, Genesan S *et al*. Antibodies targeting the PfRH1 binding domain inhibit invasion of *Plasmodium falciparum* merozoites. *PLoS Pathog*. 2008;4:e1000104.
32. Gaur D, Singh S, Singh S, Jiang L, Diouf A, Miller LH. Recombinant *Plasmodium falciparum* reticulocyte homology protein 4 binds to erythrocytes and blocks invasion. *Proc Natl Acad Sci USA*. 2007;104:17789–17794.
33. Hayton K, Gaur D, Liu A, Takahashi J, Henschen B, Singh S *et al*. Erythrocyte binding protein PfRH5 polymorphisms determine species-specific

pathways of *Plasmodium falciparum* invasion. *Cell Host Microbe.* 2008;4: 40–51.

34. Rodriguez M, Lustigman S, Montero E, Oksov Y, Lobo CA. PfRH5: a novel reticulocyte-binding family homolog of *Plasmodium falciparum* that binds to the erythrocyte, and an investigation of its receptor. *PLoS One.* 2008;3:e3300.

35. Kato K, Mayer DC, Singh S, Reid M, Miller LH. Domain III of *Plasmodium falciparum* apical membrane antigen 1 binds to the erythrocyte membrane protein Kx. *Proc Natl Acad Sci USA.* 2005;102:5552–5557.

36. Miller LH, Mason SJ, Clyde DF, McGinniss MH. The resistance factor to *Plasmodium vivax* in blacks. The Duffy-blood-group genotype, *FyFy. N Engl J Med.* 1976;295:302–304.

37. Galinski MR, Medina CC, Ingravallo P, Barnwell JW. A reticulocyte-binding protein complex of *Plasmodium vivax* merozoites. *Cell.* 1992; 69:1213–1226.

38. Dolan SA, Miller LH, Wellems TE. Evidence for a switching mechanism in the invasion of erythrocytes by *Plasmodium falciparum. J Clin Invest.* 1990;86:618–624.

39. Dolan SA, Proctor JL, Alling DW, Okubo Y, Wellems TE, Miller LH. Glycophorin B as an EBA-175 independent *Plasmodium falciparum* receptor of human erythrocytes. *Mol Biochem Parasitol.* 1994;64:55–63.

40. Mitchell GH, Hadley TJ, McGinniss MH, Klotz FW, Miller LH. Invasion of erythrocytes by *Plasmodium falciparum* malaria parasites: evidence for receptor heterogeneity and two receptors. *Blood.* 1986;67:1519–1521.

41. Bei AK, Membi CD, Rayner JC, Mubi M, Ngasala B, Sultan AA *et al.* Variant merozoite protein expression is associated with erythrocyte invasion phenotypes in *Plasmodium falciparum* isolates from Tanzania. *Mol Biochem Parasitol.* 2007;153:66–71.

42. Duraisingh MT, Maier AG, Triglia T, Cowman AF. Erythrocyte-binding antigen 175 mediates invasion in *Plasmodium falciparum* utilizing sialic acid-dependent and -independent pathways. *Proc Natl Acad Sci USA.* 2003;100:4796–4801.

43. Duraisingh MT, Triglia T, Ralph SA, Rayner JC, Barnwell JW, McFadden GI *et al.* Phenotypic variation of *Plasmodium falciparum* merozoite proteins directs receptor targeting for invasion of human erythrocytes. *EMBO J.* 2003;22:1047–1057.

44. Jennings CV, Ahouidi AD, Zilversmit M, Bei AK, Rayner J, Sarr O et al. Molecular analysis of erythrocyte invasion in *Plasmodium falciparum* isolates from Senegal. *Infect Immun.* 2007;75:3531–3538.
45. Lobo CA, de Frazao K, Rodriguez M, Reid M, Zalis M, Lustigman S. Invasion profiles of Brazilian field isolates of *Plasmodium falciparum*: phenotypic and genotypic analyses. *Infect Immun.* 2004;72:5886–5891.
46. Lobo CA, Rodriguez M, Hou G, Perkins M, Oskov Y, Lustigman S. Characterization of PfRhop148, a novel rhoptry protein of *Plasmodium falciparum*. *Mol Biochem Parasitol.* 2003;128:59–65.
47. Lobo CA, Rodriguez M, Struchiner CJ, Zalis MG, Lustigman S. Associations between defined polymorphic variants in the PfRH ligand family and the invasion pathways used by *P. falciparum* field isolates from Brazil. *Mol Biochem Parasitol.* 2006;149:246–251.
48. Nery S, Deans AM, Mosobo M, Marsh K, Rowe JA, Conway DJ. Expression of *Plasmodium falciparum* genes involved in erythrocyte invasion varies among isolates cultured directly from patients. *Mol Biochem Parasitol.* 2006;149:208–215.
49. Triglia T, Duraisingh MT, Good RT, Cowman AF. Reticulocyte-binding protein homologue 1 is required for sialic acid-dependent invasion into human erythrocytes by *Plasmodium falciparum*. *Mol Microbiol.* 2005;55: 162–174.
50. Persson KE, McCallum FJ, Reiling L, Lister NA, Stubbs J, Cowman AF et al. Variation in use of erythrocyte invasion pathways by *Plasmodium falciparum* mediates evasion of human inhibitory antibodies. *J Clin Invest.* 2008;118:342–351.
51. Cavasini CE, Mattos LC, Couto AA, Bonini-Domingos CR, Valencia SH, Neiras WC et al. *Plasmodium vivax* infection among Duffy antigen-negative individuals from the Brazilian Amazon region: an exception? *Trans R Soc Trop Med Hyg.* 2007;101:1042–1044.
52. Menard D, Barnadas C, Bouchier C, Henry-Halldin C, Gray LR, Ratsimbasoa A et al. *Plasmodium vivax* clinical malaria is commonly observed in Duffy-negative Malagasy people. *Proc Natl Acad Sci USA.* 2010;107:5967–5971.
53. Ryan JR, Stoute JA, Amon J, Dunton RF, Mtalib R, Koros J et al. Evidence for transmission of *Plasmodium vivax* among a duffy antigen negative population in Western Kenya. *Am J Trop Med Hyg.* 2006;75:575–581.

54. Elmendorf HG, Haldar K. *Plasmodium falciparum* exports the Golgi marker sphingomyelin synthase into a tubovesicular network in the cytoplasm of mature erythrocytes. *J Cell Biol*. 1994;124:449–462.
55. Langreth SG, Jensen JB, Reese RT, Trager W. Fine structure of human malaria *in vitro*. *J Protozool*. 1978;25:443–452.
56. Wickert H, Krohne G. The complex morphology of Maurer's clefts: from discovery to three-dimensional reconstructions. *Trends Parasitol*. 2007;23:502–509.
57. Hanssen E, Hawthorne P, Dixon MW, Trenholme KR, McMillan PJ, Spielmann T *et al*. Targeted mutagenesis of the ring-exported protein-1 of *Plasmodium falciparum* disrupts the architecture of Maurer's cleft organelles. *Mol Microbiol*. 2008;69:938–953.
58. Kulzer S, Rug M, Brinkmann K, Cannon P, Cowman A, Lingelbach K *et al*. Parasite-encoded Hsp40 proteins define novel mobile structures in the cytosol of the *P. falciparum*-infected erythrocyte. *Cell Microbiol*. 2010;12: 1398–1420.
59. Hiller NL, Bhattacharjee S, van Ooij C, Liolios K, Harrison T, Lopez-Estrano C *et al*. A host-targeting signal in virulence proteins reveals a secretome in malarial infection. *Science*. 2004;306:1934–1937.
60. Marti M, Good RT, Rug M, Knuepfer E, Cowman AF. Targeting malaria virulence and remodeling proteins to the host erythrocyte. *Science*. 2004; 306:1930–1933.
61. Boddey JA, Moritz RL, Simpson RJ, Cowman AF. Role of the *Plasmodium* export element in trafficking parasite proteins to the infected erythrocyte. *Traffic*. 2009;10:285–299.
62. Chang HH, Falick AM, Carlton PM, Sedat JW, DeRisi JL, Marletta MA. N-terminal processing of proteins exported by malaria parasites. *Mol Biochem Parasitol*. 2008;160:107–115.
63. Sargeant TJ, Marti M, Caler E, Carlton JM, Simpson K, Speed TP *et al*. Lineage-specific expansion of proteins exported to erythrocytes in malaria parasites. *Genome Biol*. 2006;7:R12.
64. de Koning-Ward TF, Gilson PR, Boddey JA, Rug M, Smith BJ, Papenfuss AT *et al*. A newly discovered protein export machine in malaria parasites. *Nature*. 2009;459:945–949.
65. Crabb BS, de Koning-Ward TF, Gilson PR. Protein export in *Plasmodium* parasites: from the endoplasmic reticulum to the vacuolar export machine. *Int J Parasitol*. 2010;40:509–513.

66. Haase S, de Koning-Ward TF. New insights into protein export in malaria parasites. *Cell Microbiol.* 2010;12:580–587.
67. Smith JD, Chitnis CE, Craig AG, Roberts DJ, Hudson-Taylor DE, Peterson DS *et al.* Switches in expression of *Plasmodium falciparum var* genes correlate with changes in antigenic and cytoadherent phenotypes of infected erythrocytes. *Cell.* 1995;82:101–110.
68. Su XZ, Heatwole VM, Wertheimer SP, Guinet F, Herrfeldt JA, Peterson DS *et al.* The large diverse gene family *var* encodes proteins involved in cytoadherence and antigenic variation of *Plasmodium falciparum*-infected erythrocytes. *Cell.* 1995;82:89–100.
69. Scherf A, Lopez-Rubio JJ, Riviere L. Antigenic variation in *Plasmodium falciparum. Annu Rev Microbiol.* 2008;62:445–470.
70. Kyes SA, Rowe JA, Kriek N, Newbold CI. Rifins: a second family of clonally variant proteins expressed on the surface of red cells infected with *Plasmodium falciparum. Proc Natl Acad Sci USA.* 1999;96:9333–9338.
71. Lavazec C, Sanyal S, Templeton TJ. Hypervariability within the Rifin, Stevor and *Pf*mc-2TM superfamilies in *Plasmodium falciparum. Nucleic Acids Res.* 2006;34:6696–6707.
72. Horrocks P, Pinches RA, Chakravorty SJ, Papakrivos J, Christodoulou Z, Kyes SA *et al.* PfEMP1 expression is reduced on the surface of knobless *Plasmodium falciparum* infected erythrocytes. *J Cell Sci.* 2005;118:2507–2518.
73. Crabb BS, Cooke BM, Reeder JC, Waller RF, Caruana SR, Davern KM *et al.* Targeted gene disruption shows that knobs enable malaria-infected red cells to cytoadhere under physiological shear stress. *Cell.* 1997;89:287–296.
74. Taylor DW, Parra M, Chapman GB, Stearns ME, Rener J, Aikawa M *et al.* Localization of *Plasmodium falciparum* histidine-rich protein 1 in the erythrocyte skeleton under knobs. *Mol Biochem Parasitol.* 1987;25:165–174.
75. Pei X, An X, Guo X, Tarnawski M, Coppel R, Mohandas N. Structural and functional studies of interaction between *Plasmodium falciparum* knob-associated histidine-rich protein (KAHRP) and erythrocyte spectrin. *J Biol Chem.* 2005;280:31166–31171.
76. Hora R, Bridges DJ, Craig A, Sharma A. Erythrocytic casein kinase II regulates cytoadherence of *Plasmodium falciparum*-infected red blood cells. *J Biol Chem.* 2009;284:6260–6269.

77. Da Silva E, Foley M, Dluzewski AR, Murray LJ, Anders RF, Tilley L. The *Plasmodium falciparum* protein RESA interacts with the erythrocyte cytoskeleton and modifies erythrocyte thermal stability. *Mol Biochem Parasitol*. 1994;66:59–69.
78. Pei X, Guo X, Coppel R, Bhattacharjee S, Haldar K, Gratzer W et al. The ring-infected erythrocyte surface antigen (RESA) of *Plasmodium falciparum* stabilizes spectrin tetramers and suppresses further invasion. *Blood*. 2007;110:1036–1042.
79. Mills JP, Diez-Silva M, Quinn DJ, Dao M, Lang MJ, Tan KS et al. Effect of plasmodial RESA protein on deformability of human red blood cells harboring *Plasmodium falciparum*. *Proc Natl Acad Sci USA*. 2007;104:9213–9217.
80. Coppel RL, Lustigman S, Murray L, Anders RF. MESA is a *Plasmodium falciparum* phosphoprotein associated with the erythrocyte membrane skeleton. *Mol Biochem Parasitol*. 1988;31:223–231.
81. Bennett BJ, Mohandas N, Coppel RL. Defining the minimal domain of the *Plasmodium falciparum* protein MESA involved in the interaction with the red cell membrane skeletal protein 4.1. *J Biol Chem*. 1997;272:15299–15306.
82. Black CG, Proellocks NI, Kats LM, Cooke BM, Mohandas N, Coppel RL. *In vivo* studies support the role of trafficking and cytoskeletal-binding motifs in the interaction of MESA with the membrane skeleton of *Plasmodium falciparum*-infected red blood cells. *Mol Biochem Parasitol*. 2008;160:143–147.
83. Waller KL, Nunomura W, An X, Cooke BM, Mohandas N, Coppel RL. Mature parasite-infected erythrocyte surface antigen (MESA) of *Plasmodium falciparum* binds to the 30-kDa domain of protein 4.1 in malaria-infected red blood cells. *Blood*. 2003;102:1911–1914.
84. Pasloske BL, Baruch DI, van Schravendijk MR, Handunnetti SM, Aikawa M, Fujioka H et al. Cloning and characterization of a *Plasmodium falciparum* gene encoding a novel high-molecular weight host membrane-associated protein, PfEMP3. *Mol Biochem Parasitol*. 1993;59:59–72.
85. Pei X, Guo X, Coppel R, Mohandas N, An X. *Plasmodium falciparum* erythrocyte membrane protein 3 (PfEMP3) destabilizes erythrocyte membrane skeleton. *J Biol Chem*. 2007;282:26754–26758.
86. Waller KL, Stubberfield LM, Dubljevic V, Nunomura W, An X, Mason AJ et al. Interactions of *Plasmodium falciparum* erythrocyte membrane protein 3 with the red blood cell membrane skeleton. *Biochim Biophys Acta*. 2007;1768:2145–2156.

87. Maier AG, Rug M, O'Neill MT, Brown M, Chakravorty S, Szestak T et al. Exported proteins required for virulence and rigidity of *Plasmodium falciparum*-infected human erythrocytes. *Cell.* 2008;134:48–61.
88. Maier AG, Baum J, Smith B, Conway DJ, Cowman AF. Polymorphisms in erythrocyte binding antigens 140 and 181 affect function and binding but not receptor specificity in *Plasmodium falciparum. Infect Immun.* 2009;77:1689–1699.
89. Parker PD, Tilley L, Klonis N. *Plasmodium falciparum* induces reorganization of host membrane proteins during intraerythrocytic growth. *Blood.* 2004;103:2404–2406.
90. Giribaldi G, Ulliers D, Mannu F, Arese P, Turrini F. Growth of *Plasmodium falciparum* induces stage-dependent haemichrome formation, oxidative aggregation of band 3, membrane deposition of complement and antibodies, and phagocytosis of parasitized erythrocytes. *Br J Haematol.* 2001;113:492–499.
91. Maguire PA, Sherman IW. Phospholipid composition, cholesterol content and cholesterol exchange in *Plasmodium falciparum*-infected red cells. *Mol Biochem Parasitol.* 1990;38:105–112.
92. Simoes AP, Roelofsen B, Op den Kamp JA. Lipid compartmentalization in erythrocytes parasitized by *Plasmodium* spp. *Parasitol Today.* 1992;8:18–21.
93. Moll GN, Vial HJ, Bevers EM, Ancelin ML, Roelofsen B, Comfurius P et al. Phospholipid asymmetry in the plasma membrane of malaria infected erythrocytes. *Biochem Cell Biol.* 1990;68:579–585.
94. Taverne J, Van Schie R, Playfair J, Reutelingsperger C. Malaria: phosphatidylserine expression is not increased on the surface of parasitized erythrocytes. *Parasitol Today.* 1995;11:298–299.
95. Brand VB, Sandu CD, Duranton C, Tanneur V, Lang KS, Huber SM et al. Dependence of *Plasmodium falciparum in vitro* growth on the cation permeability of the human host erythrocyte. *Cell Physiol Biochem.* 2003;13:347–356.
96. Joshi P, Dutta GP, Gupta CM. An intracellular simian malarial parasite (*Plasmodium knowlesi*) induces stage-dependent alterations in membrane phospholipid organization of its host erythrocyte. *Biochem J.* 1987;246:103–108.

97. Maguire PA, Prudhomme J, Sherman IW. Alterations in erythrocyte membrane phospholipid organization due to the intracellular growth of the human malaria parasite, *Plasmodium falciparum*. *Parasitology*. 1991; 102:179–186.
98. Schwartz RS, Olson JA, Raventos-Suarez C, Yee M, Heath RH, Lubin B *et al.* Altered plasma membrane phospholipid organization in *Plasmodium falciparum*-infected human erythrocytes. *Blood*. 1987;69:401–407.
99. Pattanapanyasat K, Sratongno P, Chimma P, Chitjamnongchai S, Polsrila K, Chotivanich K. Febrile temperature but not proinflammatory cytokines promotes phosphatidylserine expression on *Plasmodium falciparum* malaria-infected red blood cells during parasite maturation. *Cytometry A*. 2010;77:515–523.
100. Wu Y, Nelson MM, Quaile A, Xia D, Wastling JM, Craig A. Identification of phosphorylated proteins in erythrocytes infected by the human malaria parasite *Plasmodium falciparum*. *Malar J*. 2009;8:105.
101. Chishti AH, Maalouf GJ, Marfatia S, Palek J, Wang W, Fisher D *et al.* Phosphorylation of protein 4.1 in *Plasmodium falciparum*-infected human red blood cells. *Blood*. 1994;83:3339–3345.
102. Nunes MC, Okada M, Scheidig-Benatar C, Cooke BM, Scherf A. *Plasmodium falciparum* FIKK kinase members target distinct components of the erythrocyte membrane. *PLoS One*. 2010;5:e11747.
103. Ward P, Equinet L, Packer J, Doerig C. Protein kinases of the human malaria parasite *Plasmodium falciparum*: the kinome of a divergent eukaryote. *BMC Genomics*. 2004;5:79.
104. Blisnick T, Vincensini L, Fall G, Braun-Breton C. Protein phosphatase 1, a *Plasmodium falciparum* essential enzyme, is exported to the host cell and implicated in the release of infectious merozoites. *Cell Microbiol*. 2006;8:591–601.
105. Muller IB, Knockel J, Eschbach ML, Bergmann B, Walter RD, Wrenger C. Secretion of an acid phosphatase provides a possible mechanism to acquire host nutrients by *Plasmodium falciparum*. *Cell Microbiol*. 2010;12: 677–691.
106. Carvalho BO, Lopes SC, Nogueira PA, Orlandi PP, Bargieri DY, Blanco YC *et al.* On the cytoadhesion of *Plasmodium vivax*-infected erythrocytes. *J Infect Dis*. 2010;202:638–647.

107. Chakravorty SJ, Craig A. The role of ICAM-1 in *Plasmodium falciparum* cytoadherence. *Eur J Cell Biol*. 2005;84:15–27.
108. Rowe A, Obeiro J, Newbold CI, Marsh K. *Plasmodium falciparum* rosetting is associated with malaria severity in Kenya. *Infect Immun*. 1995;63: 2323–2326.
109. Rowe JA, Claessens A, Corrigan RA, Arman M. Adhesion of *Plasmodium falciparum*-infected erythrocytes to human cells: molecular mechanisms and therapeutic implications. *Expert Rev Mol Med*. 2009;11:e16.
110. Rowe JA, Moulds JM, Newbold CI, Miller LH. *P. falciparum* rosetting mediated by a parasite-variant erythrocyte membrane protein and complement-receptor 1. *Nature*. 1997;388:292–295.
111. Udeinya IJ, Schmidt JA, Aikawa M, Miller LH, Green I. Falciparum malaria-infected erythrocytes specifically bind to cultured human endothelial cells. *Science*. 1981;213:555–557.
112. Udomsangpetch R, Wahlin B, Carlson J, Berzins K, Torii M, Aikawa M et al. *Plasmodium falciparum*-infected erythrocytes form spontaneous erythrocyte rosettes. *J Exp Med*. 1989;169:1835–1840.
113. Pain A, Ferguson DJ, Kai O, Urban BC, Lowe B, Marsh K et al. Platelet-mediated clumping of *Plasmodium falciparum*-infected erythrocytes is a common adhesive phenotype and is associated with severe malaria. *Proc Natl Acad Sci USA*. 2001;98:1805–1810.
114. Berendt AR, Tumer GD, Newbold CI. Cerebral malaria: the sequestration hypothesis. *Parasitol Today*. 1994;10:412–414.
115. Doumbo OK, Thera MA, Kone AK, Raza A, Tempest LJ, Lyke KE et al. High levels of *Plasmodium falciparum* rosetting in all clinical forms of severe malaria in African children. *Am J Trop Med Hyg*. 2009;81: 987–993.
116. MacPherson GG, Warrell MJ, White NJ, Looareesuwan S, Warrell DA. Human cerebral malaria. A quantitative ultrastructural analysis of parasitized erythrocyte sequestration. *Am J Pathol*. 1985;119:385–401.
117. Barnwell JW, Howard RJ, Coon HG, Miller LH. Splenic requirement for antigenic variation and expression of the variant antigen on the erythrocyte membrane in cloned *Plasmodium knowlesi* malaria. *Infect Immun*. 1983;40: 985–994.
118. David PH, Hommel M, Miller LH, Udeinya IJ, Oligino LD. Parasite sequestration in *Plasmodium falciparum* malaria: spleen and antibody

modulation of cytoadherence of infected erythrocytes. *Proc Natl Acad Sci USA*. 1983;80:5075–5079.
119. Blythe JE, Yam XY, Kuss C, Bozdech Z, Holder AA, Marsh K et al. *Plasmodium falciparum* STEVOR proteins are highly expressed in patient isolates and located in the surface membranes of infected red blood cells and the apical tips of merozoites. *Infect Immun*. 2008;76: 3329–2236.
120. Kraemer SM, Smith JD. A family affair: *var* genes, PfEMP1 binding, and malaria disease. *Curr Opin Microbiol*. 2006;9:374–380.
121. Oleinikov AV, Amos E, Frye IT, Rossnagle E, Mutabingwa TK, Fried M et al. High throughput functional assays of the variant antigen PfEMP1 reveal a single domain in the 3D7 *Plasmodium falciparum* genome that binds ICAM1 with high affinity and is targeted by naturally acquired neutralizing antibodies. *PLoS Pathog*. 2009;5:e1000386.
122. Sherman IW, Crandall IE, Guthrie N, Land KM. The sticky secrets of sequestration. *Parasitol Today*. 1995;11:378–384.
123. Land KM, Crandall IE, Sherman IW. Malaria cytoadherence: binding sites for an anti-adhesive antibody on *Plasmodium falciparum*-infected erythrocytes. *Ann Trop Med Parasitol*. 1995;89:685–686.
124. Sherman IW, Eda S, Winograd E. Cytoadherence and sequestration in *Plasmodium falciparum*: defining the ties that bind. *Microbes Infect*. 2003;5:897–909.
125. Kirk K. Membrane transport in the malaria-infected erythrocyte. *Physiol Rev*. 2001;81:495–537.
126. Overman RR. Reversible cellular permeability alterations in disease; *in vivo* studies on sodium, potassium and chloride concentrations in erythrocytes of the malarious monkey. *Am J Physiol*. 1948;152: 113–121.
127. Staines HM, Powell T, Thomas SL, Ellory JC. *Plasmodium falciparum*-induced channels. *Int J Parasitol*. 2004;34:665–673.
128. Ancelin ML, Parant M, Thuet MJ, Philippot JR, Vial HJ. Increased permeability to choline in simian erythrocytes after *Plasmodium knowlesi* infection. *Biochem J*. 1991;273:701–709.
129. Ginsburg H, Krugliak M. Uptake of L-tryptophan by erythrocytes infected with malaria parasites (*Plasmodium falciparum*). *Biochim Biophys Acta*. 1983;729:97–103.

130. Staines HM, Kirk K. Increased choline transport in erythrocytes from mice infected with the malaria parasite *Plasmodium vinckei vinckei*. *Biochem J*. 1998;334:525–530.
131. Desai SA, McCleskey EW, Schlesinger PH, Krogstad DJ. A novel pathway for Ca^{++} entry into *Plasmodium falciparum*-infected blood cells. *Am J Trop Med Hyg*. 1996;54:464–470.
132 Staines HM, Chang W, Ellory JC, Tiffert T, Kirk K, Lew VL. Passive Ca^{2+} transport and Ca^{2+}-dependent K^+ transport in *Plasmodium falciparum*-infected red cells. *J Membr Biol*. 1999;172:13–24.
133. Rasoloson D, Shi L, Chong CR, Kafsack BF, Sullivan DJ. Copper pathways in *Plasmodium falciparum* infected erythrocytes indicate an efflux role for the copper P-ATPase. *Biochem J*. 2004;381:803–811.
134. Marchesini N, Vieira M, Luo S, Moreno SN, Docampo R. A malaria parasite-encoded vacuolar H^+-ATPase is targeted to the host erythrocyte. *J Biol Chem*. 2005;280:36841–36847.
135. Waller KL, McBride SM, Kim K, McDonald TV. Characterization of two putative potassium channels in *Plasmodium falciparum*. *Malar J*. 2008; 7:19.
136. Kirk K, Horner HA, Elford BC, Ellory JC, Newbold CI. Transport of diverse substrates into malaria-infected erythrocytes via a pathway showing functional characteristics of a chloride channel. *J Biol Chem*. 1994;269:3339–3347.
137. Staines HM, Ellory JC, Kirk K. Perturbation of the pump-leak balance for Na^+ and K^+ in malaria-infected erythrocytes. *Am J Physiol Cell Physiol*. 2001;280:C1576–1587.
138. Kirk K, Horner HA. In search of a selective inhibitor of the induced transport of small solutes in *Plasmodium falciparum*-infected erythrocytes: effects of arylaminobenzoates. *Biochem J*. 1995;311:761–768.
139. Staines HM, Dee BC, O'Brien M, Lang HJ, Englert H, Horner HA *et al*. Furosemide analogues as potent inhibitors of the new permeability pathways of *Plasmodium falciparum*-infected human erythrocytes. *Mol Biochem Parasitol*. 2004;133:315–318.
140. Bouyer G, Egee S, Thomas SL. Three types of spontaneously active anionic channels in malaria-infected human red blood cells. *Blood Cells Mol Dis*. 2006;36:248–254.

141. Desai SA, Bezrukov SM, Zimmerberg J. A voltage-dependent channel involved in nutrient uptake by red blood cells infected with the malaria parasite. *Nature.* 2000;406:1001–1005.
142. Egee S, Lapaix F, Decherf G, Staines HM, Ellory JC, Doerig C et al. A stretch-activated anion channel is up-regulated by the malaria parasite *Plasmodium falciparum. J Physiol.* 2002;542:795–801.
143. Huber SM, Uhlemann AC, Gamper NL, Duranton C, Kremsner PG, Lang F. *Plasmodium falciparum* activates endogenous Cl⁻ channels of human erythrocytes by membrane oxidation. *EMBO J.* 2002;21:22–30.
144. Staines HM, Powell T, Ellory JC, Egee S, Lapaix F, Decherf G et al. Modulation of whole-cell currents in *Plasmodium falciparum*-infected human red blood cells by holding potential and serum. *J Physiol.* 2003;552:177–183.
145. Verloo P, Kocken CH, Van der Wel A, Tilly BC, Hogema BM, Sinaasappel M et al. *Plasmodium falciparum*-activated chloride channels are defective in erythrocytes from cystic fibrosis patients. *J Biol Chem.* 2004;279:10316–10322.
146. Staines HM, Alkhalil A, Allen RJ, De Jonge HR, Derbyshire E, Egee S et al. Electrophysiological studies of malaria parasite-infected erythrocytes: current status. *Int J Parasitol.* 2007;37:475–482.
147. Alkhalil A, Cohn JV, Wagner MA, Cabrera JS, Rajapandi T, Desai SA. *Plasmodium falciparum* likely encodes the principal anion channel on infected human erythrocytes. *Blood.* 2004;104:4279–4286.
148. Baumeister S, Winterberg M, Duranton C, Huber SM, Lang F, Kirk K et al. Evidence for the involvement of *Plasmodium falciparum* proteins in the formation of new permeability pathways in the erythrocyte membrane. *Mol Microbiol.* 2006;60:493–504.
149. Lisk G, Pain M, Gluzman IY, Kambhampati S, Furuya T, Su XZ et al. Changes in the plasmodial surface anion channel reduce leupeptin uptake and can confer drug resistance in *Plasmodium falciparum*-infected erythrocytes. *Antimicrob Agents Chemother.* 2008;52:2346–2354.
150. Merckx A, Nivez MP, Bouyer G, Alano P, Langsley G, Deitsch K et al. *Plasmodium falciparum* regulatory subunit of cAMP-dependent PKA and anion channel conductance. *PLoS Pathog.* 2008;4:e19.
151. Martin RE, Ginsburg H, Kirk K. Membrane transport proteins of the malaria parasite. *Mol Microbiol.* 2009;74:519–528.

152. Martin RE, Henry RI, Abbey JL, Clements JD, Kirk K. The 'permeome' of the malaria parasite: an overview of the membrane transport proteins of *Plasmodium falciparum*. *Genome Biol*. 2005;6:R26.
153. Divo AA, Geary TG, Davis NL, Jensen JB. Nutritional requirements of *Plasmodium falciparum* in culture. I. Exogenously supplied dialyzable components necessary for continuous growth. *J Protozool*. 1985;32:59–64.
154. Liu J, Istvan ES, Gluzman IY, Gross J, Goldberg DE. *Plasmodium falciparum* ensures its amino acid supply with multiple acquisition pathways and redundant proteolytic enzyme systems. *Proc Natl Acad Sci USA*. 2006;103:8840–8845.
155. Kirk K, Horner HA, Kirk J. Glucose uptake in *Plasmodium falciparum*-infected erythrocytes is an equilibrative not an active process. *Mol Biochem Parasitol*. 1996;82:195–205.
156. Martin RE, Kirk K. Transport of the essential nutrient isoleucine in human erythrocytes infected with the malaria parasite *Plasmodium falciparum*. *Blood*. 2007;109:2217–2224.
157. Saliba KJ, Horner HA, Kirk K. Transport and metabolism of the essential vitamin pantothenic acid in human erythrocytes infected with the malaria parasite *Plasmodium falciparum*. *J Biol Chem*. 1998;273:10190–10195.
158. Olszewski KL, Mather MW, Morrisey JM, Garcia BA, Vaidya AB, Rabinowitz JD et al. Branched tricarboxylic acid metabolism in *Plasmodium falciparum*. *Nature*. 2010;466:774–778.
159. Elliott JL, Saliba KJ, Kirk K. Transport of lactate and pyruvate in the intraerythrocytic malaria parasite, *Plasmodium falciparum*. *Biochem J*. 2001;355:733–739.
160. Cranmer SL, Conant AR, Gutteridge WE, Halestrap AP. Characterization of the enhanced transport of L- and D-lactate into human red blood cells infected with *Plasmodium falciparum* suggests the presence of a novel saturable lactate proton cotransporter. *J Biol Chem*. 1995;270:15045–15052.
161. Kanaani J, Ginsburg H. Transport of lactate in *Plasmodium falciparum*-infected human erythrocytes. *J Cell Physiol*. 1991;149:469–476.
162. Allen RJ, Kirk K. Cell volume control in the *Plasmodium*-infected erythrocyte. *Trends Parasitol*. 2004;20:7–10.
163. Esposito A, Choimet JB, Skepper JN, Mauritz JM, Lew VL, Kaminski CF et al. Quantitative imaging of human red blood cells infected with *Plasmodium falciparum*. *Biophys J*. 2010;99:953–960.

164. Lew VL, Tiffert T, Ginsburg H. Response to: Allen and Kirk: cell volume control in the *Plasmodium*-infected erythrocyte. *Trends Parasitol.* 2004;20:10–11.
165. Lee P, Ye Z, Van Dyke K, Kirk RG. X-ray microanalysis of *Plasmodium falciparum* and infected red blood cells: effects of qinghaosu and chloroquine on potassium, sodium, and phosphorus composition. *Am J Trop Med Hyg.* 1988;39:157–165.
166. Saliba KJ, Martin RE, Broer A, Henry RI, McCarthy CS, Downie MJ et al. Sodium-dependent uptake of inorganic phosphate by the intracellular malaria parasite. *Nature.* 2006;443:582–585.
167. Bowyer PW, Simon GM, Cravatt BF, Bogyo M. Global profiling of proteolysis during rupture of *P. falciparum* from the host erythrocyte. *Mol Cell Proteomics.* 2011;10:M110. 001636.
168. Dvorin JD, Martyn DC, Patel SD, Grimley JS, Collins CR, Hopp CS et al. A plant-like kinase in *Plasmodium falciparum* regulates parasite egress from erythrocytes. *Science.* 2010;328:910–912.
169. Yeoh S, O'Donnell RA, Koussis K, Dluzewski AR, Ansell KH, Osborne SA et al. Subcellular discharge of a serine protease mediates release of invasive malaria parasites from host erythrocytes. *Cell.* 2007;131:1072–1083.
170. Glushakova S, Humphrey G, Leikina E, Balaban A, Miller J, Zimmerberg J. New stages in the program of malaria parasite egress imaged in normal and sickle erythrocytes. *Curr Biol.* 2010;20:1117–1121.
171. Glushakova S, Yin D, Li T, Zimmerberg J. Membrane transformation during malaria parasite release from human red blood cells. *Curr Biol.* 2005;15:1645–1650.
172. Chandramohanadas R, Davis PH, Beiting DP, Harbut MB, Darling C, Velmourougane G et al. Apicomplexan parasites co-opt host calpains to facilitate their escape from infected cells. *Science.* 2009;324:794–797.
173. Gelhaus C, Vicik R, Schirmeister T, Leippe M. Blocking effect of a biotinylated protease inhibitor on the egress of *Plasmodium falciparum* merozoites from infected red blood cells. *Biol Chem.* 2005;386:499–502.
174. Glushakova S, Mazar J, Hohmann-Marriott MF, Hama E, Zimmerberg J. Irreversible effect of cysteine protease inhibitors on the release of malaria parasites from infected erythrocytes. *Cell Microbiol.* 2009;11:95–105.

175. Hanspal M, Dua M, Takakuwa Y, Chishti AH, Mizuno A. *Plasmodium falciparum* cysteine protease falcipain-2 cleaves erythrocyte membrane skeletal proteins at late stages of parasite development. *Blood.* 2002;100:1048–1054.
176. Huttenlocher A, Palecek SP, Lu Q, Zhang W, Mellgren RL, Lauffenburger DA *et al.* Regulation of cell migration by the calcium-dependent protease calpain. *J Biol Chem.* 1997;272:32719–32722.
177. Kafsack BF, Carruthers VB. Apicomplexan perforin-like proteins. *Commun Integr Biol.* 2010;3:18–23.
178. Kwiatkowski DP. How malaria has affected the human genome and what human genetics can teach us about malaria. *Am J Hum Genet.* 2005;77: 171–192.
179. Haldane JBS. The rate of mutation of human genes. *Hereditas.* 1949; 35:267–273.
180. Agarwal A, Guindo A, Cissoko Y, Taylor JG, Coulibaly D, Kone A *et al.* Hemoglobin C associated with protection from severe malaria in the Dogon of Mali, a West African population with a low prevalence of hemoglobin S. *Blood.* 2000;96:2358–2363.
181. May J, Evans JA, Timmann C, Ehmen C, Busch W, Thye T *et al.* Hemoglobin variants and disease manifestations in severe falciparum malaria. *Jama.* 2007;297:2220–2226.
182. Mockenhaupt FP, Ehrhardt S, Cramer JP, Otchwemah RN, Anemana SD, Goltz K *et al.* Hemoglobin C and resistance to severe malaria in Ghanaian children. *J Infect Dis.* 2004;190:1006–1009.
183. Modiano D, Luoni G, Sirima BS, Simpore J, Verra F, Konate A *et al.* Haemoglobin C protects against clinical *Plasmodium falciparum* malaria. *Nature.* 2001;414:305–308.
184. Danquah I, Ziniel P, Eggelte TA, Ehrhardt S, Mockenhaupt FP. Influence of haemoglobins S and C on predominantly asymptomatic *Plasmodium* infections in northern Ghana. *Trans R Soc Trop Med Hyg.* 2010;104: 713–719.
185. Kreuels B, Kreuzberg C, Kobbe R, Ayim-Akonor M, Apiah-Thompson P, Thompson B *et al.* Differing effects of HbS and HbC traits on uncomplicated falciparum malaria, anemia, and child growth. *Blood.* 2010;115: 4551–4558.
186. Fairhurst RM, Fujioka H, Hayton K, Collins KF, Wellems TE. Aberrant development of *Plasmodium falciparum* in hemoglobin CC red cells:

implications for the malaria protective effect of the homozygous state. *Blood*. 2003;101:3309–3315.
187. Fairhurst RM, Baruch DI, Brittain NJ, Ostera GR, Wallach JS, Hoang HL et al. Abnormal display of PfEMP-1 on erythrocytes carrying haemoglobin C may protect against malaria. *Nature*. 2005;435:1117–1121.
188. Gouagna LC, Bancone G, Yao F, Yameogo B, Dabire KR, Costantini C et al. Genetic variation in human HBB is associated with *Plasmodium falciparum* transmission. *Nat Genet*. 2010;42:328–331.
189. Ohashi J, Naka I, Patarapotikul J, Hananantachai H, Brittenham G, Looareesuwan S et al. Extended linkage disequilibrium surrounding the hemoglobin E variant due to malarial selection. *Am J Hum Genet*. 2004;74:1198–1208.
190. Hutagalung R, Wilairatana P, Looareesuwan S, Brittenham GM, Aikawa M, Gordeuk VR. Influence of hemoglobin E trait on the severity of Falciparum malaria. *J Infect Dis*. 1999;179:283–286.
191. Chotivanich K, Udomsangpetch R, Pattanapanyasat K, Chierakul W, Simpson J, Looareesuwan S et al. Hemoglobin E: a balanced polymorphism protective against high parasitemias and thus severe *P. falciparum* malaria. *Blood*. 2002;100:1172–1176.
192. Ingram VM. Gene mutations in human haemoglobin: the chemical difference between normal and sickle cell haemoglobin. *Nature*. 1957;180:326–328.
193. Pauling L, Itano HA, Singer SJ, Wells IC. Sickle cell anemia, a molecular disease. *Science*. 1949;110:543–548.
194. Allison AC. The distribution of the sickle-cell trait in East Africa and elsewhere, and its apparent relationship to the incidence of subtertian malaria. *Trans R Soc Trop Med Hyg*. 1954;48:312–318.
195. Ackerman H, Usen S, Jallow M, Sisay-Joof F, Pinder M, Kwiatkowski DP. A comparison of case-control and family-based association methods: the example of sickle-cell and malaria. *Ann Hum Genet*. 2005;69:559–565.
196. Aidoo M, Terlouw DJ, Kolczak MS, McElroy PD, ter Kuile FO, Kariuki S et al. Protective effects of the sickle cell gene against malaria morbidity and mortality. *Lancet*. 2002;359:1311–1312.
197. Williams TN, Mwangi TW, Wambua S, Alexander ND, Kortok M, Snow RW et al. Sickle cell trait and the risk of *Plasmodium falciparum* malaria and other childhood diseases. *J Infect Dis*. 2005;192:178–186.

198. Pasvol G, Weatherall DJ, Wilson RJ. Cellular mechanism for the protective effect of haemoglobin S against *P. falciparum* malaria. *Nature.* 1978;274: 701–703.
199. Ayi K, Turrini F, Piga A, Arese P. Enhanced phagocytosis of ring-parasitized mutant erythrocytes: a common mechanism that may explain protection against falciparum malaria in sickle trait and beta-thalassemia trait. *Blood.* 2004;104:3364–3371.
200. Cholera R, Brittain NJ, Gillrie MR, Lopera-Mesa TM, Diakite SA, Arie T *et al.* Impaired cytoadherence of *Plasmodium falciparum*-infected erythrocytes containing sickle hemoglobin. *Proc Natl Acad Sci USA.* 2008;105: 991–996.
201. Cabrera G, Cot M, Migot-Nabias F, Kremsner PG, Deloron P, Luty AJ. The sickle cell trait is associated with enhanced immunoglobulin G antibody responses to *Plasmodium falciparum* variant surface antigens. *J Infect Dis.* 2005;191:1631–1638.
202. Allen SJ, O'Donnell A, Alexander ND, Alpers MP, Peto TE, Clegg JB *et al.* α^+-Thalassemia protects children against disease caused by other infections as well as malaria. *Proc Natl Acad Sci USA.* 1997;94:14736–14741.
203. Mockenhaupt FP, Ehrhardt S, Gellert S, Otchwemah RN, Dietz E, Anemana SD *et al.* α^+-thalassemia protects African children from severe malaria. *Blood.* 2004;104:2003–2006.
204. Wambua S, Mwangi TW, Kortok M, Uyoga SM, Macharia AW, Mwacharo JK *et al.* The effect of α^+-thalassaemia on the incidence of malaria and other diseases in children living on the coast of Kenya. *PLoS Med.* 2006;3:e158.
205. Williams TN, Wambua S, Uyoga S, Macharia A, Mwacharo JK, Newton CR *et al.* Both heterozygous and homozygous α^+ thalassemias protect against severe and fatal *Plasmodium falciparum* malaria on the coast of Kenya. *Blood.* 2005;106:368–371.
206. Williams TN, Maitland K, Bennett S, Ganczakowski M, Peto TE, Newbold CI *et al.* High incidence of malaria in α-thalassaemic children. *Nature.* 1996;383: 522–525.
207. Cockburn IA, Mackinnon MJ, O'Donnell A, Allen SJ, Moulds JM, Baisor M *et al.* A human complement receptor 1 polymorphism that reduces *Plasmodium falciparum* rosetting confers protection against severe malaria. *Proc Natl Acad Sci USA.* 2004;101:272–277.

208. Miller LH, Mason SJ, Dvorak JA, McGinniss MH, Rothman IK. Erythrocyte receptors for (*Plasmodium knowlesi*) malaria: duffy blood group determinants. *Science*. 1975;189:561–563.
209. Chaudhuri A, Polyakova J, Zbrzezna V, Williams K, Gulati S, Pogo AO. Cloning of glycoprotein D cDNA, which encodes the major subunit of the Duffy blood group system and the receptor for the *Plasmodium vivax* malaria parasite. *Proc Natl Acad Sci USA*. 1993;90:10793–10797.
210. Chaudhuri A, Zbrzezna V, Polyakova J, Pogo AO, Hesselgesser J, Horuk R. Expression of the Duffy antigen in K562 cells. Evidence that it is the human erythrocyte chemokine receptor. *J Biol Chem*. 1994;269: 7835–7838.
211. Tournamille C, Colin Y, Cartron JP, Le Van Kim C. Disruption of a GATA motif in the Duffy gene promoter abolishes erythroid gene expression in Duffy-negative individuals. *Nat Genet*. 1995;10:224–228.
212. Culleton RL, Mita T, Ndounga M, Unger H, Cravo PV, Paganotti GM et al. Failure to detect *Plasmodium vivax* in West and Central Africa by PCR species typing. *Malar J*. 2008;7:174.
213. Zimmerman PA, Woolley I, Masinde GL, Miller SM, McNamara DT, Hazlett F et al. Emergence of $FY*A^{null}$ in a *Plasmodium vivax*-endemic region of Papua New Guinea. *Proc Natl Acad Sci USA*. 1999;96: 13973–13977.
214. Kasehagen LJ, Mueller I, Kiniboro B, Bockarie MJ, Reeder JC, Kazura JW et al. Reduced *Plasmodium vivax* erythrocyte infection in PNG Duffy-negative heterozygotes. *PLoS One*. 2007;2:e336.
215. Amato D, Booth PB. Hereditary ovalocytosis in Melanesians. *P N G Med J*. 1977;20:26–32.
216. Jarolim P, Palek J, Amato D, Hassan K, Sapak P, Nurse GT et al. Deletion in erythrocyte band 3 gene in malaria-resistant Southeast Asian ovalocytosis. *Proc Natl Acad Sci USA*. 1991;88:11022–11026.
217. Mohandas N, Winardi R, Knowles D, Leung A, Parra M, George E et al. Molecular basis for membrane rigidity of hereditary ovalocytosis. A novel mechanism involving the cytoplasmic domain of band 3. *J Clin Invest*. 1992;89:686–692.
218. Tanner MJ, Bruce L, Martin PG, Rearden DM, Jones GL. Melanesian hereditary ovalocytes have a deletion in red cell band 3. *Blood*. 1991;78:2785–2786.

219. Schofield AE, Reardon DM, Tanner MJ. Defective anion transport activity of the abnormal band 3 in hereditary ovalocytic red blood cells. *Nature*. 1992;355:836–838.
220. Bruce LJ, Tanner MJ. Erythroid band 3 variants and disease. *Baillieres Best Pract Res Clin Haematol*. 1999;12:637–654.
221. Allen SJ, O'Donnell A, Alexander ND, Mgone CS, Peto TE, Clegg JB et al. Prevention of cerebral malaria in children in Papua New Guinea by southeast Asian ovalocytosis band 3. *Am J Trop Med Hyg*. 1999;60:1056–1060.
222. Genton B, al-Yaman F, Mgone CS, Alexander N, Paniu MM, Alpers MP et al. Ovalocytosis and cerebral malaria. *Nature*. 1995;378:564–565.
223. Cortes A, Benet A, Cooke BM, Barnwell JW, Reeder JC. Ability of *Plasmodium falciparum* to invade Southeast Asian ovalocytes varies between parasite lines. *Blood*. 2004;104:2961–2966.
224. Cortes A, Mellombo M, Mgone CS, Beck HP, Reeder JC, Cooke BM. Adhesion of *Plasmodium falciparum*-infected red blood cells to CD36 under flow is enhanced by the cerebral malaria-protective trait South-East Asian ovalocytosis. *Mol Biochem Parasitol*. 2005;142:252–257.
225. Yamamoto F, Clausen H, White T, Marken J, Hakomori S. Molecular genetic basis of the histo-blood group ABO system. *Nature*. 1990;345: 229–233.
226. Rowe JA, Handel IG, Thera MA, Deans AM, Lyke KE, Kone A et al. Blood group O protects against severe *Plasmodium falciparum* malaria through the mechanism of reduced rosetting. *Proc Natl Acad Sci USA*. 2007;104: 17471–17476.
227. Fry AE, Griffiths MJ, Auburn S, Diakite M, Forton JT, Green A et al. Common variation in the ABO glycosyltransferase is associated with susceptibility to severe *Plasmodium falciparum* malaria. *Hum Mol Genet*. 2008;17:567–576.
228. Carlson J, Wahlgren M. *Plasmodium falciparum* erythrocyte rosetting is mediated by promiscuous lectin-like interactions. *J Exp Med*. 1992;176:1311–1317.
229. Udomsangpetch R, Todd J, Carlson J, Greenwood BM. The effects of hemoglobin genotype and ABO blood group on the formation of rosettes by *Plasmodium falciparum*-infected red blood cells. *Am J Trop Med Hyg*. 1993;48:149–153.
230. Khera R, Das N. Complement receptor 1: disease associations and therapeutic implications. *Mol Immunol*. 2009;46:761–772.

231. Spadafora C, Awandare GA, Kopydlowski KM, Czege J, Moch JK, Finberg RW et al. Complement receptor 1 is a sialic acid-independent erythrocyte receptor of Plasmodium falciparum. *PLoS Pathog.* 2010;6: e1000968.

232. Rowe JA, Opi DH, Williams TN. Blood groups and malaria: fresh insights into pathogenesis and identification of targets for intervention. *Curr Opin Hematol.* 2009;16:480–487.

233. Maier AG, Duraisingh MT, Reeder JC, Patel SS, Kazura JW, Zimmerman PA et al. Plasmodium falciparum erythrocyte invasion through glycophorin C and selection for Gerbich negativity in human populations. *Nat Med.* 2003;9: 87–92.

234. Mayer DC, Jiang L, Achur RN, Kakizaki I, Gowda DC, Miller LH. The glycophorin C N-linked glycan is a critical component of the ligand for the Plasmodium falciparum erythrocyte receptor BAEBL. *Proc Natl Acad Sci USA.* 2006;103:2358–2362.

235. Patel SS, King CL, Mgone CS, Kazura JW, Zimmerman PA. Glycophorin C (Gerbich antigen blood group) and band 3 polymorphisms in two malaria holoendemic regions of Papua New Guinea. *Am J Hematol.* 2004;75:1–5.

236. Nkhoma ET, Poole C, Vannappagari V, Hall SA, Beutler E. The global prevalence of glucose-6-phosphate dehydrogenase deficiency: a systematic review and meta-analysis. *Blood Cells Mol Dis.* 2009;42:267–278.

237. Guindo A, Fairhurst RM, Doumbo OK, Wellems TE, Diallo DA. X-linked G6PD deficiency protects hemizygous males but not heterozygous females against severe malaria. *PLoS Med.* 2007;4:e66.

238. Ruwende C, Khoo SC, Snow RW, Yates SN, Kwiatkowski D, Gupta S et al. Natural selection of hemi- and heterozygotes for G6PD deficiency in Africa by resistance to severe malaria. *Nature.* 1995;376:246–249.

239. Cappadoro M, Giribaldi G, O'Brien E, Turrini F, Mannu F, Ulliers D et al. Early phagocytosis of glucose-6-phosphate dehydrogenase (G6PD)-deficient erythrocytes parasitized by Plasmodium falciparum may explain malaria protection in G6PD deficiency. *Blood.* 1998;92: 2527–2534.

240. Min-Oo G, Fortin A, Tam MF, Nantel A, Stevenson MM, Gros P. Pyruvate kinase deficiency in mice protects against malaria. *Nat Genet.* 2003;35: 357–362.

241. Ayi K, Min-Oo G, Serghides L, Crockett M, Kirby-Allen M, Quirt I et al. Pyruvate kinase deficiency and malaria. *N Engl J Med*. 2008;358: 1805–1810.
242. Durand PM, Coetzer TL. Pyruvate kinase deficiency protects against malaria in humans. *Haematologica*. 2008;93:939–940.
243. Bei AK, Brugnara C, Duraisingh MT. *In vitro* genetic analysis of an erythrocyte determinant of malaria infection. *J Infect Dis*. 2010;202: 1722–1727.

Index

2,3-bisphosphoglycerate (2,3-BPG) 5, 17, 18, 34, 38
4,4′-diisothiocyano-2,2′-stillbene-disulfonic acid (DIDS) 60, 109, 110, 112, 117, 125, 127, 130–132, 134, 135, 142–145, 149
    DIDS-p 80
    DIDS-sensitive 134
    H$_2$DIDS 110, 127, 130, 135
5-(N-ethyl-N-isopropyl)-amiloride (EIPA) 331
5-nitro-2-(3-phenylpropylamino)-benzoic acid (NPPB) 125, 144, 331, 398
15-lipooxygenase (15-LOX) 245

A23187 (Ca ionophore) 70, 71, 88, 89, 115, 121, 132
acanthocytes 34
acetazolamide 40, 133
acid sensitive outwardly rectifying (ASOR) anion channel 125, 144
actin 13, 24, 66, 129, 135, 243, 368, 394, 395
    cytoskeleton 243
acute mountain sickness (AMS) 39
adducin 66, 129
adenosine 152, 331
adrenoceptor 94

agglutinogens 8
allostery 15
α-globin 235, 250–252
α-hemoglobin-stabilizing protein (AHSP) 251, 252
α-hemolysin 125
α-thalassemia 34
aluminium 329
amantadine 329
amiloride 84, 91, 95, 96, 98, 143, 144
amiodarone 329
amitriptyline 331
AMP-activated protein kinase deficiency 329
Amphiuma RBCs 68, 97
amphotericin B 329
Amt/MEP 141
amyloid 329
anandamide 329
anemia 2, 14, 22, 28, 29–35, 100, 125, 230, 240, 247, 249, 253, 254, 256, 258, 259, 261, 262, 328, 333, 353, 357, 359, 244
    anemia of chronic disease (ACD) 35, 36, 374, 378
    anemia of inflammation 374, 378
    microcytic anemia 34, 35, 373, 376

Index

anion exchanger (AE1)   13, 20, 23, 33, 58, 60, 61, 66–69, 73, 97, 125–135, 138, 140, 143, 145–149, 241, 310, 311, 314, 316–318, 327, 328, 329, 368, 380, 381, 392, 396, 397, 399, 404, 405
   anti-band 3 antibodies   311
   band 3-ankyrin   133
   band 3-directed IgG   313
anisocytosis   31, 33
ankyrin   13, 33, 66, 67, 129, 135, 147–149, 241, 314, 368, 380, 394, 396
annexin A7   329
anti-A IgG   329
antigen
   apical membrane antigen 1 (AMA-1)   391, 392
   Colton antigen   137
   Diego blood group antigen   135
   Duffy antigen receptor for chemokines (DARC)   404
   Kidd antigen   13
   L antigen   70, 82
   L_1 antigen   82, 118
   L_p antigen   82–84, 118
   Landsteiner–Wiener (LW) antigen   133, 138, 139, 147
   M antigen   70, 81, 82
   Rh antigen   13, 58, 133, 136, 138
   RhD antigen   138, 139
   ring-infected erythrocyte surface antigen (RESA)   395
   senescent cell antigen (SCA)   310, 311, 316, 318
   Ter119 antigen   242, 263
   Waldner blood group antigen   135
Apicomplexa   387, 390
aplastic anemia   7, 30
   Diamond–Blackfan   30
   Fanconi   30
aquaporin   13, 66, 68, 73, 136–138, 148, 246
Arrestin   94
aryl hydrocarbon receptor nuclear translocator (ARNT)   354, 355
   ARNT/HIF   356
Atg7   244
aurothiomalate   329
auto-immune hemolytic anaemia (AHA)   370, 378, 379
autoantibodies   301, 310, 311
autophagy   243–245, 251
azathioprine   329

band 3 *see* anion exchanger, band 3
basal cell adhesion molecule (BCAM)   13
basophilic erythroblast   24, 241, 253, 254
Bay-Y5884   330
BCL11A   237, 238
BCL-2   244
Bcl-xl   253, 358
β adrenoceptors   151
β-chains   15, 19
β-globin   5, 15, 34, 35, 38, 234, 236, 237, 241, 250, 251, 259–262, 402, 403
β-thalassemia   35, 237, 250–252, 261
   reticulocytes   252

β1 integrin   233, 246
bilayer   59, 64
bile pigment   303
bilirubin   27, 34, 370, 373, 378, 382
biotinylation   303
bismuth   330
blebbing   327, 332
blood
    blood film   370, 371, 384
    blood groups   137
    blood transfusion   7–9, 377
    blood volume   9, 28, 353
    bloodletting   1, 6, 40
Bohr effect   4, 17, 19
bone marrow   1, 3, 23, 24, 26, 27, 30, 34–38, 151, 231, 233–235, 239–241, 243, 246, 247, 254, 256, 263, 302, 309, 351, 352, 358, 359, 368, 369, 371, 373, 375–377, 379, 384
    BFU-E   255
    erythroblastic islands   247, 248
    hypoplasia   30
    megaloblasts   31
bone morphogenetic protein 4 (Bmp4)   255, 256
burst-forming units-erythroid (BFU-E)   24, 240, 254, 255
burst-promoting activity (BPA)   254

c-Kit (CD117)   249, 254, 258
C-reactive protein (CRP)   374, 378, 382
cadmium   330
caffeine   331

calcium
    Ca ATPase   84, 86, 87, 127
    Ca pumps   89
    Ca-calmodulin   72, 86, 89, 92, 122, 149
    Ca^{2+}-ATPase   13, 73, 318
    Ca^{2+} activity   332
    Ca^{2+} entry   328, 329, 331, 332
calmodulin   86, 87, 89, 92, 94, 116, 121, 122, 138
calpain   332, 400, 401
calyculin A   95, 97, 105, 107, 112, 115
carbamate   21
carbon dioxide   1, 58, 401
    CO_2-dissociation curve   21
carbonic anhydrase   12, 20, 40, 58, 94, 127, 128, 133, 135, 142, 147, 367, 368
carboxyhemoglobin (CO-Hb)   15, 38
casein kinase II   395, 396
caspase   26, 243, 249, 251, 333
catecholamines   26, 151, 331
cation chloride cotransporters (CCC)   69
ceramide   246, 328–332
cGMP   23, 116, 137
    cGMP-dependent protein kinase type I   329
chaperone   246, 251
charybdotoxin (CTX)   122
chlorine
    Cl/HCO_3 66, 127, 131
    Cl/HCO_3 anion exchange   126, 127
    Cl/HCO_3 exchange   61, 73, 95, 129, 132
    Cl^- removal   332, 334, 335

chlorpromazine 330
chlotrimazole (CTZ) 122, 123, 126
chromuim
   ^{51}Cr 9, 302, 303, 315
   ^{51}CrO$_4$ 302
chronic kidney disease (CKD) 30, 32
chronic mountain sickness (CMS) 6, 39
Chuvash polycythemia 39
ciglitazone 330
Cisplatin 330
clearance 125, 126, 244, 316, 327, 334, 360, 407
clotrimazole 61
clusters of differentiation (CD)
   CD34$^+$ 23, 263, 264
   CD44 242
   CD47 13, 24, 67, 139, 140, 147, 148, 242, 312, 316, 331, 360
   CD95 (Fas) 333
   CD117 *see* c-Kit
cobalamin 382
colony forming units (CFU)
   CFU-E 240–242, 253–255
   CFU-E-proerythroblast 242
   CFU-GEMM 23
common lymphoid progenitor (CLP) 239
common myeloid progenitor (CMP) 39, 40, 239
complement C3 327
complement receptor 1 (CR1) 392, 404, 406
Coombs' test *see* direct anti-globin test
cooperativity 15

copper 96, 103, 330
cordycepin 330
cotransport 73, 110
cotransporter KCC1 61, 109
crystalline 368
curcumin 330
cyanosis 38
cyclosporine 330
cytoadhesion 394, 397
cytoskeleton 63, 118, 119, 129, 147, 148, 151, 317, 318, 332, 380, 394–396, 400, 401, 405

DAG test 370
δ-aminolevulinic acid (δ-ALA) 14, 24
dematin *see* proteins, protein 4.9
deoxygenation 14, 71, 105, 123, 144, 403
   Hb 23
di-isopropyl phosphofluoridate (DF^{32}P) 303
diphosphoglycerates 69
direct anti-globulin test (DAT test) 373, 378, 382
Donnan distribution 68
Duffy antigen receptor *see* antigen, Duffy antigen receptor for chemokines

echinocyte 11, 34, 64, 317
electroneutral cation cotransporters 69
elliptocytosis 67, 134
Embden–Meyerhoff hexose-monophosphate 59
embryonic Hb 19
endocytic vesicles 243

endoplasmic reticulum 12, 122, 243, 367, 393
endothelin 331
   endothelin B receptor 329
   endothelin-1 (ET-1) 126
enucleation 231, 233, 235, 241–243, 245, 248, 263, 264
ephrin-2 249
eryptosis 26, 62, 65, 89, 125, 126, 327–329, 331–335, 361, 396
erythroblast 2, 242, 243, 247, 248, 250, 358
   enucleation 242, 262
   erythroblast macrophage protein (Emp) 247, 248
   erythroblast-niche interactions 249
   orthochromatic erythroblast 24, 241, 242, 249
   polychromatic erythroblast 24, 241
erythroblastic islands 24, 233, 246–249, 256, 360
erythrocyte 104
   binding antigen-175 (EBA-175) 392
   lifespan 10, 301–303, 305, 307, 308, 313
   mean corpuscular volume (MCV) 10, 119, 304, 306, 307
   numbers 229, 306
erythrocytograms 305
erythrocytosis 30, 36–41, 360
erythrophagocytosis 360
erythropoiesis 4, 6, 23–25, 26, 27, 30, 32, 33, 129, 229–236, 238, 240, 242, 244, 246, 249, 251–259, 262–265, 303, 309, 358, 360, 368, 369, 373, 377, 380–383

erythropoiesis-stimulating agents (ESA) 22, 32
erthropoietic niche 146, 253–255, 264
erythropoietin (Epo) 6, 7, 23, 25, 26, 28, 30, 32, 35–41, 234, 240, 246, 249, 251, 253–256, 263, 264, 331, 335, 351–355, 357–361, 381, 382
   Epo excess 30, 329
   Epo-receptor (EPO-R) 26, 37, 151, 253, 254–256, 358
ESCRT protein complex 246
etoposide 89
Evans blue 9
exosome 242, 245, 246, 265, 309

factor inhibiting HIF (FIH) 354, 356
Fas ligand 249
Fas/caspase 8/caspase 3 333
Fas/Fas ligand 256
ferritin 27, 31, 34, 370, 374, 376, 378, 382
ferroportin 27, 374, 383
fibroblast-like type-1 interstitial cells 351
fibronectin 250, 254
flippase 64, 88, 312, 318
Flt3 ligand 240, 264
flufenamic acid 331
fluid mosaic model 59, 60, 65
FOG-1 259, 260
folic acid 23, 30, 31, 382
furosemide 61, 101, 104, 398

γ-chains 19
GAPDH 66, 148

Index

Gardos channel (Ca^{2+}-activated K$^+$ channel) 13, 73, 120, 121, 123, 125, 126, 132, 152 *see also* intermediate conductance K channel; KCa3.1; KCNN4; small conductance 4
Gardos effect 58, 121, 125, 135, 146
GATA 357
    GATA-1 26, 28, 237, 241, 249, 251, 254, 255, 257–260, 357–359, 404
ghost 60, 74, 79, 118, 139, 142, 146
glucocorticoid 151
    glucocorticoid receptor (GR) 256, 257
glucose depletion 328
glucose-6-phosphate dehydrogenase (G6PD or G6PDH) 34, 144, 370, 376, 380
    G6PD deficiency 33, 328, 406
Glut 1 13, 138, 399
glutathione (GSH) 12, 59, 84, 110–112, 115, 125, 380
glycation 305, 330
glycocalyx 13
glycophorin 13, 24, 62, 66, 67, 134, 241–243, 315, 318, 330, 392, 395, 396, 406
gold chloride 330
Golgi cisternae 243
Gower-1 19, 235
Gower-2 19, 235
granulocyte-macrophage progenitors (GMP) 239

Haldane effect 5, 21, 127
Ham's acid haemolysis test 373, 379, 383
haptoglobin 33, 371, 378
Hck 116
Heinz bodies 33, 261, 310, 372
hemangioblasts 231, 260
hematocrit (Hct) 7, 9, 10, 21, 22, 32, 36–38, 40, 247, 352, 353, 360
hematopoietic niche 246, 254
hematopoietic stem cell (HSC) 2, 23, 35, 150, 229, 232, 238, 244, 254–256, 260, 264
    long term (LT)-HSC 238
heme 4, 11, 12, 15, 27, 235, 250, 251, 261
    heme regulated inhibitor of translation (HRI) 250
    protoporphyrin 14
hemiglobin (Met-Hb) 15
hemin 302, 330
hemocytometer 3
hemocytometry 304, 306
hemoglobin (Hb) 1, 3–5, 9–12, 14, 15, 20, 21, 24, 25, 27–29, 31, 33–38, 40, 60, 62, 67–69, 71, 72, 74, 94, 118, 119, 123, 128, 132, 139, 148, 235–238, 261, 314, 333, 334, 353, 367–369, 371, 372, 375, 378, 382, 383
    embryonic Hb 19
    Hb electrophoresis 34, 35, 38
    Hb Gower-2 19
    Hb oxygenation 133
    Hb Portland-1 19, 235
    Hb Portland-2 19, 235
    Hb-O$_2$ binding curve 3, 18
    Hb4 10, 14
    HbA 15, 17, 19, 126, 235, 236, 252
    HbA0 15, 305–308, 314
    HbA1 15, 308, 310, 314

HbA1a   305, 314
HbA1b   305, 314
HbA1c   15, 305–308, 314, 306
HbA1e1   314
HbA1e2   314
HbA2   35, 314
HbAC   402, 403
HbAE   402, 403
HbAS   403
HbC   382, 402, 403
HbE   15, 402
HbF   35, 235–238, 262, 314, 334
HBG1   235
HBG2   235
HbS   34, 61, 71, 123, 124, 126, 144, 357, 403
hemoglobinopathy   379, 402
hemoglobinuria   8
hemolysin   330
hemolytic anemia   33, 244, 252, 261, 328
hemolytic disorder   32, 140, 144, 234
   hemolytic disorder of the newborn (HDN)   138
hemolytic uremic syndrome   328, 332
Henderson–Hasselbalch equation   5
Henry's law   14
hepatic nuclear factor 4 (HNF-4) *see* nuclear factors, nuclear factor-4
hepcidin   27, 35, 255, 374, 378, 383
hereditary   67
   hereditary elliptocytosis (HE)   33
   hereditary motor and sensory neuropathy with agnesis of corpus callosum (HMSN-ACC)   120
   hereditary persistence of fetal hemoglobin (HPFH)   237, 261
   hereditary spherocytosis (HS)   33, 70, 126, 133, 372, 380
Hill coefficient   4, 17, 18, 117
histone   242
homocysteine   125, 151
Howell–Jolly bodies   310
Hsc70 molecular chaperone   246
Hüfner, Carl Gustav von   4
Hüfner's number   15, 29
hyperthermia   328
hypochromic microcytic anemia   31
hypoxemia   38, 351
hypoxia responsive elements (HRE)   355, 356, 357
hypoxia-inducible transcription factor (HIF)   25, 26, 38, 354–358
   HIF-1α 354, 355
   HIF-1β 354
   HIF-2α 38, 39, 354
   HIF-3α 354
   HIF-α 39
   HIF-stabilizers   28

immunoglobin A (IgA)   330
inhibitory PAS protein (iPAS)   354
insulin   96, 98, 100, 101, 150
insulin-like growth factor 1 (IGF-1)   26, 40, 254, 263
integrin-associated protein (IAP)   67, 139, 140, 147, 148
intercellular adhesion molecule (ICAM)   67, 139, 140, 147, 148
   ICAM-4   13, 24, 247
   ICAM-4S   247
interferon (IFN)-γ 240, 253, 255

interleukin
    IL-3   23, 240, 253, 264
    IL-6   240, 255
    IL-11   240, 253
intermediate conductance K channel (IK)   28, 61, 120, 121 *see also* Gardos channel; KCa3.1; KCNN4; small conductance 4
intrinsic factor   4, 32, 377, 383, 384
iron   3, 4, 6, 11, 12, 24, 27, 30, 31, 34, 35, 235, 242, 245, 250, 251, 255, 302, 303, 368, 369, 370, 373, 374, 376, 377, 381–385
    ^{59}Fe   302, 304–306, 308
    deficiency   27, 30, 308, 328, 334, 370, 371, 373, 374, 376, 381, 382
    erythropoiesis   23
    iron-deficiency anemia   31
    overload   34, 35, 374, 380, 382
    total iron-binding capacity (TIBC)   31
isoagglutinogens   8
isoleucine   399

Jacobs–Parpart–Stewart cycle,   58
Janus kinases 2 (JAK2)   26, 30, 37, 38, 253, 358–360, 377
    JAK2V617F   37

K-Cl cotransport (KCC)   61, 63, 64, 69, 71, 73, 81, 82, 84, 97, 107, 109–112, 114, 117–119, 149, 261
    K$^+$-Cl$^-$ cotransporter   13
    KCC1   149

KCa3.1   58, 72, 73, 120, 121, 361 *see also* Gardos channel; intermediate conductance K channel; KCNN4; small conductance 4
KCNN4   120, 121 *see also* Gardos channel; intermediate conductance K channel; KCa3.1; small conductance 4
Klotho deficiency   329
knob associated histidine-rich protein (KAHRP)   394, 395
knob formation   394
Krebs citric acid cycle   59
Krüppel-like factor (KLF)   260, 262
    KLF1   237, 239, 257, 261, 262

lanthanum chloride (LaCl3)   89
LC3 *see* proteins, protein light chain 3
lead   125, 330
leptin   150
leptocytes   308
leukotriene   330
lipid bilayer   13, 33, 58, 59, 117, 129, 137, 142, 146, 317, 318, 380
lipopeptides   330
listeriolysin   330
lithium   77, 90, 330
loop diuretics   101, 104, 109, 110
low-ionic strength stimulated cation fluxes (LISCF)   144, 145
LT-HSC *see* hematopoietic stem cell, long term
Lu *see* basal cell adhesion molecule
Lutheran blood group   13

lymphoid primed multipotent progenitor (LMPP) 239
lysophosphatidic acid (LPA) 316

macrocytosis 369, 383
malaria 34, 134, 328, 334, 376, 381, 383, 387, 388, 390, 395, 397, 401–403, 405–407
mDia1 243
mean cell volume (MCV) 369, 370, 382, 383 *see also* erythrocyte, mean corpuscular volume
mean corpuscular Hb concentration (MCHC) 10, 11, 33, 304, 307
mean osmotic fragility (MOF) 63, 64, 72, 81, 121
megakaryocyte-erythroid progenitors (MEP) 239
megalocytes 31
menadione 330
mercury 68, 74, 330
merozoite 388–393, 400
Met-Hb *see* hemoglobin
methyldopa 330
methylglyoxal 330
microcytic anemia *see* anemia, microcytic anemia
microcytic cells 369, 370
microRNA 263
mitochondria 11, 12, 19, 243, 244, 327, 367
mitogen-activated protein kinase (MAPK) 358, 359
monensin 330
Monge's disease 39
monocarboxylate transporter (MCT1) 399
mononuclear phagocyte system 309, 310

mountain sickness 39
multipotential progenitors (MPPs) 239, 254, 260
multivesicular endosome 245
MYB gene 237–239, 242
myelodysplasia 377
myelodysplastic syndrome 328
myelofibrosis 38, 370, 377

N-ethylmaleimide (NEM) 96, 107, 109–111, 114, 115, 117 145
N-methyl-D-aspartate (NMDA) 125, 151
Na-K-2Cl 61, 73
Na-K-Cl cotransporter (NKCC) 13, 62, 98–105, 107, 109, 110, 113–116, 119, 149
Na/H exchangers 61
Na/HX 90–92, 95–98, 100, 116, 119, 150
Na/K 89
Na/K-TPase 13, 24, 73–77, 80, 83, 85–87
Na/Li exchanger 90, 99
Na^+-Cl-cotransporter 13
Na^+-H^+-exchanger 1 24
NaCl cotransporter (NCC) 90, 101, 102
NBQX/CNQX 331
neocytolysis 328, 335, 361
new permeation pathways (NPP) 398–400
NF-E4 gene 238
NHE proteins 73, 90–92, 94–96, 98, 119, 144
    NHE1 98
niche extracellular matrix proteins 250

nicotinamide adenine dinnucleotide phosphate (NADPH)   12, 33, 59, 380
niflumic acid   125, 144, 331
nitrendipine   123
nitroprusside (NO)   33, 331, 353
Nix   244
NKCC1   73, 102, 103, 105, 106, 108, 109, 116, 118
non-selective ion channels (NSIC)   143, 144
normoblast   24, 368
nuclear factors   238, 257, 258, 355, 356
   nuclear factor Bach I   250
   nuclear factor-4 (HNF-4)   355
nucleated red blood cells (NRBC)   372

okadaic acid   96, 107, 112
orthochromatic erythroblast *see* erythroblast, orthochromatic erythroblast
osmotic pressure   14, 67
osmotic resistance   14, 308, 309
osmotic shock   332, 334
oxidative stress   33, 106, 332–334, 396
   OSR1   106, 107, 116
oxygenation   5, 14–16, 71, 133, 334, 352, 353, 355
   oxygenation-deoxygenation   118, 132

P2X   151
P2Y purinergic   152
p95   241
p105   241
paclitaxel   330

palytoxin (PTX)   85
pancytopenia   30, 377, 384
pantothenic acid   399
para-nitrophenyl phosphate (pNPP)   88
parasite egress   395, 396, 400
parasite invasion   390, 392, 404
parasitophorous vacuole membrane (PVM)   388, 389, 393, 394
paroxysmal nocturnal haemoglobinuria (PNH)   328, 370, 373, 375, 378, 379, 383
partial oxygen pressure (pO_2)   4, 6, 14, 15, 17, 18, 25, 39, 40, 112, 353, 354
PDK1   329
peptidoglycan   330
PER-ARNT-SIM (PAS) family   354
peripheral blood stem cells (PBSC)   24
Pernicious anaemia   4, 32, 377, 384
PEXEL/HT motif   393
phagocytosis   147, 249, 301, 310, 312, 315, 316, 361, 403, 407
phosphate depletion   328
phosphatidylinositol 3 kinase (PI3K)   358, 359
phosphatidylserine   126, 243, 248, 312, 313, 315, 317, 327, 333, 334, 360, 361, 396, 397
phospholamban   87
phospholipase A2   316
phospholipid bilayer   57, 60, 65, 146
phytic acid   330
plasma membrane Ca (PMCA) pump   73, 74, 86–89, 124, 126, 150

plasmodial surface anion channel (PSAC)   398
Plasmodium falciparum   334, 387, 388, 389, 391–402, 406, 407
   FIK kinases   396
   P. falciparum erythrocyte membrane protein 3 (PfEMP3)   395
   P. falciparum infected   400
   P. falciparum RBL   392
   P. falciparum reticulocyte homology protein 4 (PfRh4)   392
Plasmodium knowlesi   388, 404
Plasmodium malariae   388, 392, 402
Plasmodium ovale   388, 392
Plasmodium translocon of exported proteins (PTEX)   394
Plasmodium vivax   388, 392, 393, 404
   P. vivax RBL   392
platelet-activating factor (PAF)   330, 332
   PAF receptor deficiency   329
poikilocytosis   31, 34
polychromatic   2
   polychromatic erythroblast *see* erythroblast, polychromatic erythroblast
polycythemia   30, 230
   polycythemia vera (PV)   37, 38
polyglobulia   6
polyribosomes   24
Post–Albers
   Post–Albers canonical cycle   59, 77, 89

Post–Albers canonical modes   85
Post–Albers sequential canonical model   74
Post–Albers transport cycle   78
postrenal transplant erythrocytosis (PTE)   40
proerythroblast   24, 104, 241, 242, 254, 256, 259
prolyl hydroxylase domain (PHD)   25, 354, 356, 357
   PHD-2   38, 39
prostaglandin E2 (PGE_2)   125, 330, 334
proteins
   GABARAP   244
   protein 4.1   13, 33, 66, 133, 149
   protein 4.2   13, 33, 380
   protein 4.2 (palladin)   66
   protein 4.9 (dematin)   66
   protein export element/host targeting (PEXEL/HT) motif   393
   protein inhibitors of activated STAT (PIAS)   360
   protein kinase C (PKC)   92, 94, 96, 116, 122, 253
   protein light chain 3 (LC3)   244
   protein phosphatase 1 (PP1)   104–106, 114–116
   protein phosphatase 2A (PP2A)   115
   protein skeleton   13
   protein tyrosine phosphatases (PTP)   360

protoporphyrin 14, 24, 303
pure red cell aplasia 30
pyridoxine 375, 384
pyrrole 14
pyruvate kinase (PK) 34, 71, 133, 149, 380, 407

Rac GTPases 243
radiocontrast agents 330
receptor 152
renal insufficiency 328
resveratrol 331
reticulocyte 10, 23, 24–27, 31, 33, 34, 83, 97, 100, 104, 118, 119, 121, 151, 241–246, 252, 254, 301, 302, 304, 308–310, 368, 370, 371, 378, 382, 384
    reticulocyte binding-like (RBL) proteins 391–393
    reticulocyte exosomes 246
retinoic acid 330
Rh-associated glycoproteins (RhAG) 8, 13, 33, 67, 73, 128, 133, 139–143, 147, 148, 381
ribosome 24, 243, 244, 245
rosetting 397, 402, 406

sarafotoxin 6c 331
SCL 257, 260
    SCL/TAL1 260
    SCL12A 108
scramblase 64, 313, 318
selenium 331
sentrin/SUMO-specific protease (SENP) 357
sepsis 9, 328, 332
shear stress 22, 23
SHP-1 26, 37

sickle cell 30, 71, 100, 329, 370–372, 375, 384, 400
    sickle cell anaemia (SCA) 119, 237, 263, 312, 360, 372, 375, 376, 379, 384
    sickle cell disease 34, 328, 379, 403
signal transducer and activator of transcription protein 5 (STAT5) 253, 358–360
silver ions 331
SLC genes
    SLC4A1 91, 126, 129, 133, 381, 405
    SLC9A 90, 96
    SLC12A 102
    SLC12A7 108
Smad5 255, 256
small conductance 4 (sK4) 120, 121 *see also* Gardos channel; intermediate conductance K channel; KCa3.1; KCNN4
South East Asian ovalocytosis (SAO) 129, 131, 134, 135, 376, 405
spectrin 13, 33, 66, 119, 129, 134, 135, 241, 314, 368, 380, 394–396
    spectrin cytoskeleton 147
    spectrin-actin 22
    spectrin/actin cytoskeleton 13
    spectrins 65–67, 149
spherocytes 11, 14, 64, 370–372
spherocytosis 67
sphingomyelinase 332
splenectomy 34, 308–310, 375, 381, 384
    splenectomy-associated 309
splenomegaly 33–35
staurosporine 89, 115, 145, 331

Ste20-related proline-alanine-rich kinases (SPAK) 104–108, 114, 116
stem cell factor (SCF) 234, 240, 249, 253–255, 258, 263, 264
stercobilin 27, 302–304, 308
Stoke–Einstein equation 62
stomatin 66, 368
stomatocytes 139, 371
sub-phospholipid bilayer 65
SUMO (small ubiquitin-like modifier) 356
    SUMOylated HIF-1 357
    SUMOylation 356, 357
suppressors of cytokine signalling (SOCS) 26, 360
    SOCS3 37

T-structure 17
TAL1 *see* SCL, SCL/TAL1
TAUT deficiency 329
TCL5 260
thalassemia 30, 34, 250, 328, 329, 334
thrombopoietin 240
thrombospondin-1-receptor 331
thrombospondin-related adhesive protein (TRAP) family 392
thrombus formation 22, 125
thymol 331
thymoquinone 331
tin 331
total iron binding capacity (TIBC) 384
TRAM-34 122
transcription factor PU.1 241, 259, 260
transferrin 27, 31, 302, 306, 384, 385

transferrin receptor (CD71) 27, 240, 242, 243, 245, 250, 263, 309
transforming growth factor (TGF)-β 253, 255, 240
transgenic sickle (SAD) mice 120, 126
tropomodulin 66
tropomyosin 24, 66
TRPC6 329, 332
tumor necrosis factor (TNF)-α 35, 36, 240, 253, 255

ubiquitin 252, 265, 354, 356
    ubiquitination 246, 257
Ulk1 244
urea 13, 59, 104, 110, 138, 331
    urea transporter B (UTB) 138

vacuoles 309, 310
valinomycin 70, 132, 331
van't Hoff–Morse equation 63
vanadate 74, 88, 123, 331
vascular cell adhesion protein 1 (VCAM-1) 247
vascular endothelial growth factor (VEGF) 352
vimentin 241
vitamin B_{12} 4, 23, 30–32, 377, 382, 385
volume 104
von Hippel–Lindau syndrome 40
    von Hippel–Lindau tumor suppressor (VHL) 356
    von Hippel–Lindau tumor suppressor protein (pVHL) 25, 38, 39, 41, 354, 357
    VHL/HIF pathway 352

Index

Wilms' tumor   40
Wilson's disease   328, 332
WNK kinases   107, 116
    WNK1   108, 114
    WNK3   116
    WNK4   105, 106, 116

xanthohumol   331

xenobiotic   110, 124, 328, 332

yolk sac   23, 231–233, 235, 261

zidovudine   32, 331
zinc   72, 331